Springer Monographs *in* Mathematics

T0238974

Amnon Jakimovski, Ambikeshwar Sharma and
József Szabados

Walsh Equiconvergence of Complex Interpolating Polynomials

 Springer

Amnon Jakimovski
Tel-Aviv University
Tel-Aviv, Israel

Ambikeshwar Sharma
The University of Alberta
Edmonton, Canada

József Szabados
Hungarian Academy of Sciences
Budapest, Hungary

A C.I.P. Catalogue record for this book is available from the Library of Congress.

ISBN   978-90-481-7060-9
ISBN   978-1-4020-4175-4 (e-Book)

Published by Springer,
P.O. Box 17, 3300 AA Dordrecht, The Netherlands.

www.springer.com

Printed on acid-free paper

DEDICATION

And one might therefore say
of me that in this book, I
have only made up a bunch
of other people's flowers and
that of my own I have only
provided the string that ties
them together.

(Book III, Chapter XVI
of Physiognomy)
Signeur de Montaigne

CONTENTS

PREFACE

This monograph is centered around a simple and beautiful observation of J.L. Walsh, in 1932, that if a function is analytic in a disc of radius ρ $(\rho > 1)$ but not in $|z| \leq \rho$, then the difference between the Lagrange interpolant to it in the nth roots of unity and the partial sums of degree $n - 1$ of the Taylor series about the origin, tends to zero in a larger disc of radius ρ^2, although both operators converge to $f(z)$ only for $|z| < \rho$. This result was stated by Walsh in 1932 in a short paper [304] and proved in [87]. A precise formulation of this interesting result appears in 1935 in the first edition of his book *Interpolation and Approximation by Rational Functions in the Complex Domain* [88, p. 153]).

One of the reasons why this result of Walsh was not noticed until 1980 seems to be that it is sharp in the sense that if $z = \rho^2$, then there exists a function $f(z)$ analytic in $|z| < \rho$, for which the difference, between its Lagrange interpolant on the nth roots of unity and the partial sum of degree $n - 1$ of its Taylor series about the origin, does not tend to zero for $z = \rho^2$. The function which provides this phenomenon is $\frac{1}{\rho - z}$. In 1980 a paper authored by A.S. Cavaretta, A. Sharma and R.S. Varga [27] gives an extension of the above result in many new directions.

The object of this monograph is to collect the various results stemming from this theorem of Walsh which have appeared in the literature, and to give as well some new results. The first work which gave publicity to this subject was a paper by R.S. Varga [82] which appeared in 1982 and later a survey paper by A. Sharma [72] in 1986. T.J. Rivlin, E.B. Saff and R.S. Varga (all students of Walsh) made significant contributions to extend this result. New directions were due to V. Totik [85], K. Ivanov and A. Sharma [43], J. Szabados [80], Lou Yuanren [202], M.P. Stojanova [76], A. Jakimovski and A. Sharma [48] and others.

T.J. Rivlin in his brief comment on the above result in the selected papers of Walsh, says that "...by the mid nineties the interest in this theorem had almost disappeared. The result was probably about 200 published papers".

This comment encouraged us to write this monograph and to present a unified presentation of the significant results and extensions of this theorem along with a complete bibliography. (How T.J. Rivlin arrived at the figure of about 200 published papers is not clear to us.)

This book is easily accessible to students who have had a course in complex variables and have gone, for example, through the book *Theory of Approximation* by P.J. Davis, or the book *Approximation of Functions* by G.G. Lorentz. Our book is divided into 12 chapters. Chapter 1 begins with elementary results on Lagrange interpolation to functions defined on $|z| \leq \rho$ and gives a proof of the Theorem of Walsh which is the object of the present study. Chapter 2 deals with an extension of Walsh's theorem to Hermite interpolation. Chapter 3 is concerned with an extension of Walsh's theorem to rational functions with given poles outside the circle $|z| < \rho$. Chapters 4 and 5 deal with sharpness and converse results respectively. Chapter 6 is concerned with Padé approximation and Walsh equiconvergence for meromorphic functions with a finite number of given poles. Chapter 7 deals with quantitative results in the overconvergence of meromorphic functions of Chapter 6. In Chapter 8, we turn to the study of equiconvergence of Lagrange and Hermite interpolation for functions analytic in an ellipse. In Chapter 9 we extend the Walsh equiconvergence by application of methods of regular summability, which was initiated by R. Brück [16] and continued by A. Jakimovski and A. Sharma [46]. Chapter 10 deals with Faber expansions of analytic functions and extensions of Walsh equiconvergence results for differences of approximation operators on Fejér and Faber nodes. Chapter 11 is concerned with corresponding results for equiconvergence on lemniscates.

We can never thank Prof. R.S. Varga enough for his kindness and constant encouragement, advice and suggestions over several years. He has been kind enough to go through the manuscript with constructive corrections and amendments.

We are grateful to Prof. A.S. Cavaretta for his kindness and help by reading part of this book with care and to Prof. M.G. de Bruin for his critical and constructive help in Chapters 6 and 7. Without their help we could not complete these chapters in their present forms.

A. Sharma is particularly grateful to his family for their encouragement and patience with him during the preparation of this monograph. His son Raja and his wife Sarla went "the extra mile" beyond their filial duties in caring for him, and ungrudgingly endured his eccentricities. He records his gratefulness to the Good Samaritan Society (Mount Pleasant Choice Center) for his care and nursing during his illness while the work was in preparation.

We want to thank specially Professor Zeev Ditzian (University of Alberta) whose constant help and encouragement during our cooperative work in Edmonton created a friendly and fruitful atmosphere.

The authors acknowledge with thanks the support from NSERC grants over the past few years for our continued collaboration to this work at Edmonton. We deeply appreciate with thanks the scrupulous care of Vivian Spak in typing the manuscript.

Edmonton, July 2003

A. Jakimovski, A. Sharma, and J. Szabados

Ambikeshwar Sharma died on December 22, 2003, after a long illness. It is our honor and duty to finish the work on this book he initiated so enthusiastically.

Edmonton, July 2004

A. Jakimovski and J. Szabados

CHAPTER 1

LAGRANGE INTERPOLATION
AND WALSH EQUICONVERGENCE

1.1 Introduction

Let $f(z)$ be a function analytic in an open domain D and continuous on the boundary of this domain. Further let n be a positive integer, and $z_1, \ldots, z_n$ pairwise different points from D. We shall denote the (unique) Lagrange polynomial interpolant, of degree at most $n-1$, of $f(z)$ in these n zeros by $L_{n-1}(f; z)$. With the notation $\omega_n(z) := \prod_{k=1}^{n}(z - z_k)$, this polynomial can be represented in the form

$$L_{n-1}(f; z) = \frac{1}{2\pi i} \int_C \frac{\omega_n(t) - \omega_n(z)}{\omega_n(t)} \frac{f(t)}{t - z} dt,$$

where C may be any rectifiable Jordan curve in D containing the points $z_1, \ldots, z_n$ and z in the interior of the domain bounded by C. Indeed, this is a polynomial of degree at most $n - 1$, and by Cauchy's theorem

$$L_{n-1}(f, z_k) = \frac{1}{2\pi i} \int_C \frac{f(t)}{t - z_k} dt = f(z_k), \qquad k = 1, \ldots, n.$$

The uniqueness of this interpolant follows from the fundamental theorem of algebra: if there existed two different interpolating polynomials, then their difference, a polynomial of degree at most $n - 1$ not identically zero, would vanish at n points, which is impossible. Most often in this book, we will be concerned with the special case when the nodes of interpolation are the nth roots of unity, i.e., when $\omega_n(z) = z^n - 1$. In 1884, Méray gave a very instructive example of a function whose Lagrange interpolant in the nth roots of unity does not converge to it anywhere except at the point 1. Thus if $f(z) = 1/z$ then $L_{n-1}(f; z) = z^{n-1}$ is the polynomial of degree $n - 1$ which interpolates $f(z)$ in the zeros of $z^n - 1$. For $|z| > 1$, $\lim_{n\to\infty} z^{n-1}$ does not exist and for $|z| < 1$, $\lim_{n\to\infty} z^{n-1} = 0$, while for $|z| = 1$, $z \neq 1$ it diverges so that $L_{n-1}(f; z)$ converges to $f(z) = z^{-1}$ only at the point 1. The same applies to the case when $f(z) = z^{-k}$, $k > 0$. Even for analytic functions in the closed unit circle $|z| \leq 1$, the condition

$$\lim_{n\to\infty} \prod_{k=1}^{n} |z - z_k|^{\frac{1}{n}} = |z| \quad \text{for} \quad |z| > 1 \tag{1.0}$$

1

must be satisfied for the nodes of interpolation z_k, $|z_k| = 1$, $k = 1, \ldots, n$, in order to have uniform convergence of the corresponding Lagrange interpolants in $|z| \leq 1$. (For the roots of unity, this is obviously satisfied.) For functions which are not analytic, we have the following theorem.

THEOREM 1. *Let $f(z)$ be defined and continuous (or R-integrable, i.e., Riemann integrable) on the circumference of the unit circle $\Gamma := \{z : |z| = 1\}$. If $L_{n-1}(f; z)$ is the Lagrange interpolant to $f(z)$ in the zeros of $z^n - 1$, then*

$$\lim_{n \to \infty} L_{n-1}(f; z) = \frac{1}{2\pi i} \int_\Gamma \frac{f(t)}{t - z} \, dt, \quad |z| < 1, \tag{1.1}$$

uniformly for $|z| \leq \delta < 1$.

PROOF. Denoting $w_n = \exp 2\pi i / n$, the Lagrange interpolant has the following representation:

$$L_{n-1}(f; z) = \sum_{k=1}^{n} f(w_n^k) \cdot \frac{w_n^k(z^n - 1)}{(z - w_n^k)n} . \tag{1.2}$$

Namely, this is indeed a polynomial of degree at most $n - 1$, since each w_n^k is a root of the polynomial $z^n - 1$. Moreover,

$$\lim_{z \to w_n^j} \frac{z^n - 1}{z - w_n^k} = \begin{cases} 0 & \text{if } j \neq k, \\ \frac{n}{w_n^k} & \text{if } j = k, \end{cases}$$

i.e., $L_{n-1}(f; w_n^j) = f(w_n^j)$, $j = 1, \ldots, n$ as stated. From the definition of the Riemann integral, we have

$$F(z) := \frac{1}{2\pi i} \int_\Gamma \frac{f(t)}{t - z} dt = \lim_{n \to \infty} \frac{1}{2\pi i} \sum_{k=1}^{n} \frac{f(w_n^k)(w_n^{k+1} - w_n^k)}{w_n^k - z}, \quad |z| < 1 \tag{1.3}$$

and

$$\lim_{n \to \infty} [F(z) - L_n(f; z)] = \lim_{n \to \infty} \left[\frac{1}{2\pi i} + \frac{z^n - 1}{n(w_n - 1)} \right] \sum_{k=1}^{n} \frac{w_n^k(w_n - 1)f(w_n^k)}{w_n^k - z} .$$

Since $\lim_{n \to \infty} n(w_n - 1) = 2\pi i$, we see that for $|z| < 1$, we have (1.1). The uniform convergence for $|z| \leq \delta < 1$ is also clear from the last formula. $\square$

If $f(z)$ is analytic for $|z| < 1$ and continuous for $|z| = 1$, then $f(z) = F(z)$. In order to extend this result to other operators, we shall need the following

LEMMA 1. *Let $f(z)$ be R-integrable on Γ and let $L_{n-1}(f;z)$ be the Lagrange interpolant to f on the zeros of $z^n - 1$. Then, for any fixed nonnegative integer p*

$$\lim_{n \to \infty} L_{n-1}^{(p)}(f;z) = \frac{p!}{2\pi i} \int_{\Gamma} \frac{f(t)}{(t-z)^{p+1}}\, dt, \quad |z| < 1 \qquad (1.4)$$

the convergence being uniform for $|z| \le \delta < 1$.

PROOF. Since

$$L_{n-1}(f;z) = \frac{1}{n}\sum_{k=0}^{n-1} f(w_n^k)\sum_{j=0}^{n-1} w_n^{-kj}z^j.$$

Differentiating the above p times with respect to z, we get

$$L_{n-1}^{(p)}(f;z) = \frac{p!}{n}\sum_{k=0}^{n-1}\frac{f(w_n^k)w_n^k}{(w_n^k - z)^{p+1}}$$
$$- \frac{z^n}{n}\sum_{k=0}^{p}\binom{p}{k}(n)_k z^{-k}(p-k)!\sum_{\ell=0}^{n-2}\frac{f(w_n^\ell)w_n^\ell}{(w_n^\ell - z)^{p-k+1}} \qquad (1.5)$$

where $(n)_k = n(n-1)\ldots(n-k+1)$. We notice that

$$\lim_{n \to \infty}\frac{1}{n}\sum_{k=0}^{n-1}\frac{f(w_n^k)w_n^k}{(w_n^k - z)^{p-k+1}} = \frac{1}{2\pi i}\int_{\Gamma}\frac{f(t)}{(t-z)^{p-k+1}}\, dt,$$

and that for any $k > 0$, $|z|^n n^k \to 0$ uniformly for $|z| \le \delta < 1$ as $n \to \infty$. (1.4) now follows from (1.5). □

If $f^{(j)}(z)$ exists along Γ for $j = 0, 1, \ldots, r-1$, we denote by $h_{rn-1}(f;z)$ the polynomial of degree $rn - 1$ which satisfies the conditions:

$$h_{rn-1}^{(j)}(f;w_n^k) = f^{(j)}(w_n^k), \quad k = 1,\ldots,n; \quad j = 0,1,\ldots,r-1. \qquad (1.6)$$

Then we have

THEOREM 2. *Let $f^{(r-1)}(z)$ exist and be R-integrable along Γ. If $h_{rn-1}(f;z)$ is the Hermite interpolant to f satisfying (1.6), then*

$$\lim_{n \to \infty} h_{rn-1}(f;z) = \frac{1}{2\pi i}\int_{\Gamma}\frac{f(t)}{t-z}\, dt, \quad |z| < 1 \qquad (1.7)$$

and uniformly for $|z| \le \delta < 1$.

PROOF. For $r = 1$, the theorem is the same as Theorem 1; so it is enough to consider the case when $r > 1$. Set

$$h_{rn-1}(f;z) = L_{n-1}(f;z) + \sum_{j=1}^{r-1}(1 - z^n)^j P_{n,j}(f;z) \qquad (1.8)$$

where each $P_{n,j}(f;z)$ is a polynomial of degree $\leq n-1$. Thus it is enough to prove that

$$\lim_{n\to\infty} P_{n,j}(f;z) = 0, \quad j = 1,\ldots,r-1, \quad |z| < 1.$$

However we shall prove the stronger result that

$$\lim_{n\to\infty} P_{n,j}^{(\ell)}(f;z) = 0, \quad j = 1,2,\ldots,r-1, \quad \ell = 0,1,\ldots, \quad |z| < 1. \qquad (1.9)$$

We use induction on j. First let $j = 1$. Differentiating (1.8) at $z = w_n^k$, we obtain

$$w_n h'_{rn-1}(f;w_n^k) = w_n f'(w_n^k) = w_n L'_{n-1}(f;w_n^k) - n P_{n,1}(f;w_n^k)$$
$$(k = 0,1,\ldots,n-1)$$

whence we have

$$P_{n,1}(f;z) = \frac{1}{n}\left[z L'_{n-1}(f;z) - L_{n-1}(zf';z)\right]. \qquad (1.10)$$

Differentiating this ℓ times gives

$$P_{n,1}^{(\ell)}(f;z) = \frac{1}{n}[z L_{n-1}^{(\ell+1)}(f;z) + \ell L_{n-1}^{(\ell)}(f;z) - L_{n-1}^{(\ell)}(zf';z)] \qquad (1.11)$$

so that by Lemma 1, we see that (1.9) holds for $j = 1$. Now suppose that (1.9) has been proved for j, $1 \leq j \leq r-2$. From (1.8), we deduce

$$w_n^{(j+1)k} h_{rn-1}^{(j+1)}(f;w_n^k) = w_n^{(j+1)k} L_{n-1}^{(j+1)}(f;w_n^k) + (-1)^{j+1}(j+1)! n^{j+1} \times$$
$$\times P_{n,j+1}(f;w_n^k)$$
$$+ w_n^{(j+1)k} \sum_{\ell=1}^{j} \sum_{s=\ell}^{j+1} \binom{j+1}{s} \left(\frac{d^s}{dz^s}(1-z^n)^\ell\right)_{z=w_n^k} P_{n,\ell}^{(j+1-s)}(f;w_n^k)$$
$$(k = 0,1,\ldots,n-1).$$

Because of (1.6), we obtain

$$(-1)^{j+1}(j+1)! P_{n,j+1}(f;z) = \frac{1}{n^{j+1}} L_{n-1}(z^{j+1} f^{(j+1)};z) - \left(\frac{z}{n}\right)^{j+1} L_{n-1}^{(j+1)}(f;z)$$
$$- \sum_{\ell=1}^{j} \sum_{s=j}^{r-1} \binom{j+1}{s} \frac{z^{j+1-s}}{n^{j+1}} P_{n,\ell}^{(j+1-s)}(f;z) \sum_{t=1}^{\ell} \binom{\ell}{t}(-1)^t (nt)_s.$$

Differentiating ℓ times, using the induction hypothesis and Lemma 1, we see that (1.9) holds for $j+1$ and the proof is complete. $\qquad\square$

If we set

$$h_{rn-1}(f;z) = \sum_{k=0}^{rn-1} \gamma_k z^k,$$

then we can define the average of the partial sums of $h_{rn-1}(f;z)$ and set

$$A_{rn-1}(f;z) = \frac{1}{rn} \sum_{j=0}^{rn-1} \sum_{k=0}^{j} \gamma_k z^k.$$

In a similar fashion, one can establish

THEOREM 3. *Let $f^{(r-1)}$ exist along Γ and be R-integrable on Γ. Let $A_{rn-1}(f;z)$ be the average of the partial sums of the Hermite interpolant of $f(z)$ satisfying (1.6). Then*

$$\lim_{n\to\infty} A_{rn-1}(f;z) = \frac{1}{2\pi i}\int_\Gamma \frac{f(t)}{t-z}\,dt,$$

for $|z| < 1$ and uniformly for $|z| \le \delta < 1$.

PROOF. By a change in the order of summation we see that

$$A_{rn-1}(f;z) = zL'_{n-1}(f;z) + z\sum_{k=0}^{r-1}(rn-k)\gamma_k z^k = h_{rn-1}(f;z) - \frac{1}{nr}zh'_{rn-1}(f;z).$$

Now we see from (1.8) that

$$zh'_{rn-1}(f;z) = zL'_{n-1}(f;z) + z\sum_{k=0}^{r-1}(1-z^n)^j P'_{n,j}(f;z) - nz^n\sum_{j=1}^{r-1}j(1-z^n)^{j-1}\times$$
$$\times P_{n,j}(f;z).$$

Thus, using Lemma 1 and

$$\lim_{n\to\infty}[h_{rn-1}(f;z) - L_{n-1}(f;z)] = 0, \qquad |z| < 1$$

(which follows from (1.8)-(1.9)), we see that

$$\lim_{n\to\infty}\frac{z}{n}h'_{rn-1}(f;z) = 0.$$

This, combined with Theorem 1 proves Theorem 3.

1.2. Least-Square Minimization

For $m \ge n$, let $Q_{n-1}(f;z)$ denote the unique polynomial of degree $\le n-1$ which minimizes

$$\sum_{k=0}^{m-1}|f(w_m^k) - p(w_m^k)|^2, \quad w_m^m = 1, \tag{2.1}$$

over all $p(z) \in \pi_{n-1}$. If $m = n$, $Q_{n-1}(f;z)$ is the Lagrange interpolant to f at the nth roots of unity. If $m > n$, then it is easy to see that $Q_{n-1}(f;z)$ is obtained by truncating $L_{m-1}(f;z)$. More precisely if

$$L_{m-1}(f;z) = \sum_{k=0}^{m-1}c_k z^k, \quad m > n \quad \text{then} \quad Q_{n-1}(f;z) = \sum_{\nu=0}^{n-1}c_\nu z^\nu,$$

where

$$c_\nu = \frac{1}{m} \sum_{k=0}^{m-1} f(w_m^k) w_m^{-\nu k}, \quad \nu = 0, 1, \ldots, n-1. \tag{2.2}$$

To see this, we first observe that from

$$L_{m-1}(f; z) = \frac{1}{m} \sum_{n=0}^{m-1} \frac{f(w_m^k)(z^m - 1)w^k}{(z - w_m^k)}$$

it follows that the coefficient of z^ν in $L_{m-1}(f; z)$ is given by (2.2). If we want to minimize (2.1) and set $p(z) = \sum_{\nu=0}^{n-1} p_\nu z^\nu$, then in order to minimize

$$\sum_{k=0}^{m-1} \left| f(w_m^k) - \sum_{\nu=0}^{n-1} p_\nu w_m^{k\nu} \right|^2$$

we need the orthogonality conditions

$$\sum_{k=0}^{m-1} \left(f(w_m^k) - \sum_{\nu=0}^{n-1} p_\nu w_m^{\nu k} \right) w_m^{-\mu k} = 0, \quad \mu = 0, 1, \ldots, n-1.$$

Simplifying, we see that

$$\sum_{k=0}^{m-1} f(w_m^k) w_m^{-\mu k} = \sum_{\nu=0}^{m-1} p_\nu \sum_{k=0}^{m-1} w_m^{\nu k - \mu k} = m p_\mu$$

which proves that $p_\mu = c_\mu$ in (2.2) and proves the assertion. From (2.2) we can see that

$$Q_{n-1}(f; z) = \frac{1}{m(w_m - 1)} \sum_{k=0}^{m-1} \frac{f(w_m^k)(w_m^{k+1} - w_m^k)}{w_m^k - z}$$
$$+ \frac{z^n}{m} \sum_{k=0}^{m-1} \frac{f(w_m^k) w_m^{-(n-1)k}}{w_m^k - z} = S_1 + S_2.$$

We notice that as $n \to \infty$

$$|S_2| = O(|z|^n) = o(1) \quad \text{uniformly for} \quad |z| \leq \delta < 1.$$

Since $m > n$, and $\lim_{n \to \infty} m(w_m - 1) = 2\pi i$ we have proved

THEOREM 4. *If $f(z)$ is R-integrable on Γ and if $Q_{n-1}(f; z)$ is the unique polynomial which minimizes (2.1), then*

$$\lim_{n \to \infty} Q_{n-1}(f; z) = \frac{1}{2\pi i} \int_\Gamma \frac{f(t)}{t - z} \, dt, \quad |z| < 1 \tag{2.3}$$

uniformly for $|z| \leq \delta < 1$.

The above theorems have a corresponding analogue for Laurent development.

THEOREM 5. *Let $f(z)$ be R-integrable on the unit circle Γ and let $Q_{n,n}(z)$ be the polynomial in z and $\frac{1}{z}$ of degree n in each, which interpolates $f(z)$ in the zeros of $z^{2n+1} - 1$. If $Q_{n,n}(z) = q_n(z) + r_n(z^{-1})$, where*

$$q_n(z) = a_0 + a_1 z + \cdots + a_n z^n, \quad r_n(z^{-1}) = a_{-1} z^{-1} + a_{-2} z^{-2} + \cdots + a_{-n} z^{-n},$$

then

$$\begin{cases} \lim\limits_{n \to \infty} q_n(z) = \frac{1}{2\pi i} \int_\Gamma \frac{f(t)}{t-z}\, dt, & |z| < 1 \\[4mm] \lim\limits_{n \to \infty} r_n(z^{-1}) = \frac{1}{2\pi i} \int_\Gamma \frac{f(t)}{t-z}\, dt, & |z| > 1. \end{cases} \tag{2.4}$$

The convergence is uniform in $\frac{1}{\delta} \leq |z| \leq \delta < 1$.

If $f(z)$ is analytic in an annulus $\rho^{-1} < |z| < \rho$, $\rho > 1$ then the equations (2.4) are valid respectively for $|z| < \rho$ and for $|z| > \frac{1}{\rho}$ and uniformly for $|z| \leq R < \rho$ and $|z| \geq \frac{1}{R} > \frac{1}{\rho}$. Moreover

$$q_n(z) + r_n(z^{-1}) \to f(z) \quad \text{for} \quad \frac{1}{\rho} < |z| < \rho,$$

and uniformly in

$$\frac{1}{R} \leq |z| \leq R < \rho.$$

1.3. Functions Analytic in $\Gamma_\rho = \{z : |z| = \rho\}$

We shall now consider functions which are analytic in the disc of radius ρ ($\rho > 1$) but not in Γ_ρ. We shall denote this class of functions by A_ρ. It is known that if $f(z) \in A_\rho$ and if

$$f(z) = \sum_{k=0}^{\infty} a_k z^k$$

is the power-series expansion of $f(z)$, then the right side converges in $|z| < \rho$ and

$$\varlimsup_{n \to \infty} |a_n|^{1/n} = \frac{1}{\rho}.$$

If we set $p_{n-1}(f; z) = \sum\limits_{k=0}^{n-1} a_k z^k$, the Taylor expansion of f then $p_{n-1}(f; z)$ converges to $f(z)$ for $|z| < \rho$, if $f(z) \in A_\rho$. Similarly $L_{n-1}(f; z)$ (the Lagrange interpolant to f on the zeros of $z^n - 1$) also converges to $f(z)$ only for $|z| < \rho$. However the difference of $L_{n-1}(f; z)$ and $p_{n-1}(f; z)$ converges to 0 for $|z| < \rho^2$. This beautiful observation is formulated as

THEOREM 6. *Let* $f(z) \in A_\rho$ ($\rho > 1$) *and let* $L_{n-1}(f;z)$ *be the Lagrange interpolant to* f *on the zeros of* $z^n - 1$. *Then the sequence* $L_{n-1}(f;z)$ *converges geometrically to* $f(z)$ *in any closed subdomain of* $|z| < \rho$. *Moreover if* $p_{n-1}(f;z)$ *is the Taylor section of* $f(z)$ *of degree* $n - 1$, *then*

$$\lim_{n \to \infty} [L_{n-1}(f;z) - p_{n-1}(f;z)] = 0, \tag{3.1}$$

geometrically for any closed subdoamin of $|z| < \rho^2$.

PROOF. Since $f(z) = \frac{1}{2\pi i} \int_{\Gamma_R} \frac{f(t)}{t-z} dt$ where $R < \rho$, and since

$$L_{n-1}(f;z) = \frac{1}{2\pi i} \int_{\Gamma_R} \frac{f(t)(t^n - z^n)}{(t^n - 1)(t - z)} dt,$$

we obtain

$$f(z) - L_{n-1}(f;z) = \frac{1}{2\pi i} \int_{\Gamma_R} \frac{(z^n - 1)f(t)}{(t^n - 1)(t - z)} dt, \quad |z| < R.$$

We see from the above that

$$\overline{\lim}_{n \to \infty} |f(z) - L_{n-1}(f;z)|^{1/n} \leq \frac{|z|}{R},$$

which proves the geometric convergence for closed subdomains of $|z| < \rho$ (since $R < \rho$ was arbitrary). Similarly, we have

$$L_{n-1}(f;z) - p_{n-1}(f;z) = \frac{1}{2\pi i} \int_{\Gamma_R} \frac{(t^n - z^n)f(t)}{t^n(t^n - 1)(t - z)} dt. \tag{3.2}$$

Hence

$$\overline{\lim}_{n \to \infty} |L_{n-1}(f;z) - p_{n-1}(f;z)|^{1/n} \leq \frac{\max\{R, |z|\}}{R^2}, \quad R < \rho.$$

The result follows from this immediately. □

The quantity ρ^2 is the best possible, in the sense that for any point z on $|z| = \rho^2$, there is a function $f(z) \in A_\rho$ for which (3.1) does not hold. The function $f(z) = \frac{1}{z-\rho}$ is a natural example since in this case

$$L_{n-1}(f;z) - p_{n-1}(f;z) = \frac{\rho^n - z^n}{\rho^n(\rho^n - 1)(z - \rho)}$$

when $z = \rho^2$, and we see that this difference becomes $1/(\rho - \rho^2)$. Many extensions of Theorem 6 have recently been given. We begin with a straightforward extension. Let us set

$$p_{n-1,j}(f;z) := \sum_{k=0}^{n-1} a_{k+jn} z^k, \quad j = 0, 1, 2, \ldots \tag{3.3}$$

where the function $f(z) \in A_\rho$ has the Taylor-series expansion $\sum_0^\infty a_k z^k$. We shall prove below the following

THEOREM 7. *If $f(z) \in A_\rho$ and if $\ell \geq 1$ is any given integer, then*

$$\lim_{n \to \infty} \max_{|z| \leq \mu} \left| L_{n-1}(f; z) - \sum_{j=0}^{\ell-1} p_{n-1,j}(f; z) \right|^{1/n} \leq \frac{\mu}{\rho^{\ell+1}}, \quad \mu < \rho^{\ell+1} \qquad (3.4)$$

i.e. the convergence is uniform and geometric for all $|z| \leq \mu < \rho^{\ell+1}$. Moreover the region $|z| < \rho^{\ell+1}$ is best possible in the sense that for any point z_0 with $|z_0| = \rho^{\ell+1}$, there exists a function $f_0(z) \in A_0$ for which (3.4) does not hold for $z = z_0$.

Thus if we take $z_0 = \rho$, and $f_0(z) = (\rho - z)^{-1}$, then

$$p_{n-1,j}(f_0, z) = \frac{\rho^n - z^n}{(\rho - z)\rho^{(j+1)n}}$$

and

$$\sum_{j=0}^{\ell-1} p_{n-1,j}(f_0, z) = \frac{(\rho^n - z^n)(\rho^{\ell n} - 1)}{(\rho - z)\rho^{\ell n}(\rho^n - 1)}.$$

It is easy to see that

$$\lim_{n \to \infty} \min_{|z|=\rho^{\ell+1}} \left| L_{n-1}(f_0; z) - \sum_{j=0}^{\ell-1} p_{n-1,j}(f_0; z) \right|^{1/n} \geq \frac{1}{\rho^{\ell+1} + \rho} > 0.$$

PROOF. As in the proof of Theorem 6, we can express the difference on the left in (3.4) as a contour integral

$$\frac{1}{2\pi i} \int_{\Gamma_R} \frac{f(t)(t^n - z^n)}{(t - z)(t^n - 1)t^{\ell n}} \, dt. \qquad (3.5)$$

For $|t| = R$ and for all $|z| \leq \mu < R < \rho^{\ell+1}$ $(\mu \geq \rho)$, we have

$$\left| \frac{t^n - z^n}{t - z} \right| \leq \frac{\mu^n + R^n}{R - \mu},$$

so that the above integral is bounded above in modulus by

$$\frac{MR(\mu^n + R^n)}{(R - \mu)(R^n - 1)R^{\ell n}}$$

where $M := \max_{z \in \Gamma_R} |f(z)|$. Taking n^{th} roots we see that

$$\overline{\lim}_{n \to \infty} \left\{ \max_{|z| \leq \mu} \left| L_{n-1}(f; z) - \sum_{j=0}^{\ell-1} p_{n-1,j}(f; z) \right| \right\}^{1/n} \leq \frac{\mu}{R^{\ell+1}},$$

which proves the desired uniform and geometric convergence of (2.3).   □

On letting $\ell \to \infty$ in (3.4), we see that

$$L_{n-1}(f; z) = \sum_{j=0}^{\infty} p_{n-1,j}(f; z)$$

which shows that if $L_{n-1}(f; z) = \sum_{\nu=0}^{n-1} c_\nu z^\nu$, then, it can be verified that

$$c_\nu = \sum_{\lambda=0}^{\infty} a_{\nu+\lambda n}.$$

In Theorems 6 and 7, we compared two processes of interpolation each of which separately converges to $f(z)$ only for $|z| < \rho$, while their difference converges to zero in a larger region. In view of this, the above phenomenon is often termed as "overconvergence" or "equiconvergence." It is natural to ask whether the Taylor polynomial $p_{n-1}(f; z)$ can be replaced by the polynomial $\widehat{p}_{n-1}(f; z)$ which is the polynomial of best uniform approximation to $f(z)$ in $|z| \le 1$. If $f_0(z) = (\rho - z)^{-1}$, then

$$\widehat{p}_{n-1}(f_0; z) = \frac{\rho^{n-1} - z^{n-1}}{(\rho - z)\rho^{n-1}} + \frac{z^{n-1}}{(\rho^2 - 1)\rho^{n-2}}$$

for all $n \ge 2$. Then

$$L_{n-1}(f_0; z) - \widehat{p}_{n-1}(f_0; z) = \frac{\rho^{n-1} - z^{n-1}}{(\rho - z)(\rho^n - 1)\rho^{n-1}} - \frac{z^{n-1}(\rho^{n-2} - 1)}{(\rho^n - 1)(\rho^2 - 1)\rho^{n-2}}$$

and

$$\widehat{p}_{n-1}(f_0; z) - p_{n-1}(f_0; z) = \frac{1}{\rho(\rho^2 - 1)} \left(\frac{z}{\rho}\right)^{n-1}$$

which converges to zero only for $|z| < \rho$. If $f(z) \in A_\rho$ and is also continuous in $D_\rho := \{z : |z| \le \rho\}$, it is natural to ask if this stronger hypothesis on the function would make the equiconvergence region larger. The answer to this question is given by

THEOREM 8. *Let $f(z) \in A_\rho \cap C(D_\rho)$. Then for each positive integer ℓ, we have*

$$\lim_{n \to \infty} \left\{ L_{n-1}(f; z) - \sum_{j=0}^{\ell-1} p_{n-1,j}(f; z) \right\} = 0, \quad |z| \le \rho^{\ell+1},$$

the convergence being uniform for all $|z| \le \rho^{\ell+1}$ and geometric for all $|z| \le r < \rho^{\ell+1}$.

PROOF. For any $f(z) \in A_\rho \cap C(D_\rho)$, let $s_{n-1}(f;z)$ be the polynomial of best approximation to f from π_{n-1} on the circle $D_\rho = \{z : |z| \leq \rho\}$. Then

$$E_{n-1}(f) := \inf_{q \in \pi_{n-1}} \|f - q\|_{D_n} = \|f - s_{n-1}\|_{D_\rho}$$

and it is known that $\lim_{n \to \infty} E_{n-1}(f) = 0$. From the linearity of the Lagrange and Taylor polynomials, we have

$$L_{n-1}(f;z) - \sum_{j=0}^{\ell-1} p_{n-1,j}(f;z) = L_{n-1}(f - s_{n-1};z) - \sum_{j=0}^{\ell-1} p_{n-1,j}(f - s_{n-1};z),$$

so that from (3.4), we obtain for $R < \rho$

$$L_{n-1}(f;z) - \sum_{j=0}^{\ell-1} p_{n-1,j}(f;z) = \frac{1}{2\pi i} \int_{\Gamma_R} \frac{(f(t) - s_{n-1}(f;t)) \cdot (t^n - z^n)}{(t-z)(t^n - 1)t^{\ell n}} \, dt.$$

This shows that

$$\max_{|z| \leq \rho^{\ell+1}} \left| L_{n-1}(f;z) - \sum_{j=0}^{\ell-1} p_{n-1,j}(f;z) \right| \leq \frac{E_{n-1}(f)(\rho^{n(\ell+1)} + R^n)R}{(\rho^{\ell+1} - R)(R^n - 1)R^{\ell n}}.$$

Since the left side is independent of R, we get on letting R tend to ρ,

$$\max_{|z| \leq \rho^{\ell+1}} \left| L_{n-1}(f;z) - \sum_{j=0}^{\ell-1} p_{n-1,j}(f;z) \right| \leq \frac{E_{n-1}(f)(1 + \rho^{-n\ell})}{\rho^\ell(1 - \rho^{-\ell})(1 - \rho^{-n})}.$$

But the right side tends to zero as $n \to \infty$, which proves the result. Uniform and geometric convergence for $|z| \leq r < \rho^{\ell+1}$ follows as in Theorem 7.

1.4. An extension of Walsh's Theorem

We claim that the sum

$$\sum_{j=0}^{\ell-1} p_{n-1,j}(f;z)$$

in (3.3) (Theorem 7) is the Lagrange interpolant in the nth roots of unity of the polynomial $p_{\ell n-1}(f;z) = \sum_{k=0}^{\ell n-1} a_k z^k$. This is easily seen since

$$p_{\ell n-1}(f;z) = \sum_{\lambda=0}^{\ell-1} \sum_{k=0}^{n-1} a_{k+\lambda n} z^{k+\lambda n} = \sum_{\lambda=0}^{\ell-1} \sum_{k=0}^{n-1} a_{k+\lambda n}(z^{\lambda n} - 1)z^k$$

$$+ \sum_{\lambda=0}^{\ell-1} \sum_{k=0}^{n-1} a_{k+\lambda n} z^k$$

so that

$$L_{n-1}(p_{\ell n-1}; z) = \sum_{\lambda=0}^{\ell-1} \sum_{k=0}^{n-1} a_{k+\lambda n} z^k = \sum_{\lambda=0}^{\ell-1} p_{n-1,\lambda}(f; z).$$

With this simple observation, one can write the formula (3.3) in the equivalent form

$$\lim_{n \to \infty} \left[L_{n-1}(f; z) - L_{n-1}\big(p_{\ell n-1}(f; z); z\big) \right] = 0 \quad \text{for} \quad |z| < \rho^{\ell+1}. \tag{4.1}$$

If we denote by $L_{n-1}(f; \alpha, z)$ the Lagrange interpolant in the zeros of $z^n - \alpha^n$, when $\alpha \neq 0$, and the Hermite interpolant of order n at 0 when $\alpha = 0$, then (4.1) is also equivalent to

$$\lim_{n \to \infty} \left[L_{n-1}(f; 1, z) - L_{n-1}\big(L_{\ell n-1}(f; 0, z); 1, z\big) \right] = 0, \quad |z| < \rho^{\ell+1}.$$

This train of ideas amply justifies the following

THEOREM 9. If $m = rn + q$, $s \leq \frac{q}{n} < 1$ and $\frac{q}{n} = s + O(\frac{1}{n})$ then for each $f(z) \subset A_\rho$ and for each $\alpha, \beta \in D_\rho$ $(\alpha \neq \beta)$, we have

$$\lim_{n \to \infty} \Delta_{n,m}^{\alpha,\beta}(f; z) := \lim_{n \to \infty} \left[L_{n-1}(f, \alpha, z) - L_{n-1}\big(L_{m-1}(f, \beta, z), \alpha, z\big) \right] = 0 \tag{4.2}$$

for $|z| < \sigma$, where

$$\sigma := \rho / \max \left(\left(\frac{|\alpha|}{\rho} \right)^r, \left(\frac{|\beta|}{\rho} \right)^{r+s} \right). \tag{4.3}$$

More precisely, for any μ with $\rho < \mu < \infty$, we have

$$\lim_{n \to \infty} \left\{ \max_{z \in D_\mu} |\Delta_{n,m}^{\alpha,\beta}(f; z)| \right\}^{1/n} \leq \frac{\mu}{\sigma}.$$

Moreover if α, β, m satisfy neither $\alpha = \beta = 0$ nor $\alpha^r = \beta^r$ when $m = rn$, then (4.3) is best possible in the sense that for any z_0 with $|z_0| = \sigma$, there is a function $f_0 \in A_\rho$ such that (4.2) fails to hold for f_0 at z_0.

When $\alpha = 1$, $\beta = 0$ and $m = \ell n$, (4.2) yields Theorem 7.

PROOF. Since $\alpha, \beta \in D_\rho$, we may write

$$L_{m-1}(f, \beta, z) = \frac{1}{2\pi i} \int_{\Gamma_R} \frac{f(t)(t^m - z^m)}{(t - z)(t^m - \beta^m)} \, dt.$$

In order to find a similar representation for $L_{n-1}\big(L_{m-1}(f, \beta, z), \alpha, z\big)$, it is enough to evaluate

$$L_{n-1} \left(\frac{t^m - z^m}{t - z}, \alpha, z \right).$$

Since $m = rn + q$, we have

$$\frac{t^m - z^m}{t - z} = \frac{t^{rn+q} - z^{rn+q}}{t - z}$$

$$= \frac{t^{nr} - z^{rn}}{t - z} t^q + z^{rn} \frac{t^q - z^q}{t - z}$$

$$= t^q \cdot \frac{t^{rn} - z^{rn}}{t^n - z^n} \cdot \frac{t^n - z^n}{t - z} + z^{rn} \cdot \frac{t^q - z^q}{t - z}.$$

From this it is clear that the Lagrange interpolant of $(t^m - z^m)/(t - z)$ in the zeros of $z^n - \alpha^n$ will be given as below:

$$L_{n-1}\left(\frac{t^m - z^m}{t - z}, \alpha, z\right) = t^q \left(\frac{t^{rn} - \alpha^{rn}}{t^n - \alpha^n}\right)\frac{t^n - z^n}{t - z} + \alpha^{rn} \cdot \left(\frac{t^q - z^q}{t - z}\right), \text{ as } q < n.$$

Hence

$$L_{n-1}(f, \alpha, z) = \frac{1}{2\pi i}\int_{\Gamma_R} \frac{f(t)(t^n - z^n)}{(t - z)(t^n - \alpha^n)} \, dt,$$

$$L_{n-1}\left(L_{m-1}(f, \beta, z), \alpha, z\right) = \frac{1}{2\pi i}\int \frac{f(t)}{t^m - \beta^m} L_{n-1}\left(\frac{t^m - z^m}{t - z}, \alpha, z\right) dt.$$

From this we obtain the representation

$$\Delta_{n,m}^{\alpha,\beta}(f; z) = \frac{1}{2\pi i}\int_{\Gamma_R} f(t)K(t, z)dt \tag{4.4}$$

where

$$K(t, z) := \frac{1}{t^{rn+q} - \beta^{rn+q}}\left[t^q \cdot \frac{t^{rn} - \alpha^{rn}}{t^n - \alpha^n} \cdot \frac{t^n - z^n}{t - z} + \alpha^{rn} \frac{t^q - z^q}{t - z}\right]$$

$$- \frac{t^n - z^n}{t^n - \alpha^n} \cdot \frac{1}{t - z}$$

$$= \frac{\beta^{rn+q} - \alpha^{rn} \cdot t^q}{(t^{rn+q} - \beta^{rn+q})(t^n - \alpha^n)} \cdot \frac{t^n - z^n}{t - z}$$

$$+ \frac{\alpha^{rn}}{t^{nr+q} - \beta^{rn+q}} \cdot \frac{t^q - z^q}{t - z}.$$

Since

$$|K(t, z)| \le c \frac{R^n - |z|^n}{R - |z|} \cdot \frac{\max(|\beta|^{rn+q}, |\alpha|^{rn}R^q)}{R^{rn+q}R^n} + c \frac{|\alpha|^{rn}}{R^{rn+q}} \cdot \frac{R^q - |z|^q}{R - |z|} \quad (|z| < R),$$

it follows that (4.2) will be proved if

$$|z|^n \frac{\max(|\beta|^{rn+q}, |\alpha|^{rn}R^q)}{R^{rn+q+n}} < 1 \quad \text{and} \quad \frac{|z|^q|\alpha|^{rn}}{R^{rn+q}} < 1,$$

where $q = sn + O(1)$. In other words taking the nth roots of both sides above, and letting $n \to \infty$, we see that (4.2) is proved if

$$|z| < \rho/\max\left(\left(\frac{|\beta|}{\rho}\right)^{r+s}, \left(\frac{|\alpha|}{\rho}\right)^r\right) =: \sigma_1 \quad \text{and} \quad |z| < \rho/\left(\frac{|\alpha|}{\rho}\right)^{r/s} =: \sigma.$$

Since $\frac{|\alpha|}{\rho} < 1$ and $s < 1$, we have

$$\left(\frac{|\alpha|}{\rho}\right)^{r/s} > \left(\frac{|\alpha|}{\rho}\right)^r$$

so that $\sigma_1 < \sigma$ and this completes the proof. $\square$

COROLLARY. *Let* $m = rn + q$, $s \le \frac{q}{n} < 1$ *and* $\frac{q}{n} = s + O(\frac{1}{n})$. *If* $p_{n,m(n)}(f; z)$ *denotes the polynomial of degree* $n - 1$ *which minimizes*

$$\sum_{k=0}^{m-1} |f(\beta w_m^k) - p(\beta w_m^k)|^2, \quad \beta \in D_\rho,$$

over all polynomials $p(z) \in \pi_{n-1}$, *then*

$$p_{n,m(n)}(f; z) - S_{n-1}(f; z) \to 0 \quad \text{for} \quad |z| < \rho/\left(\frac{|\beta|}{\rho}\right)^{r+s}$$

and the bound for $|z|$ *above is best possible in the same sense as in* Theorem 9.

This corollary follows from Theorem 8 on taking $\alpha = 0$, $\beta = 1$.

1.5. Multivariate Extensions of Walsh's Theorem

In the multivariate case the domain of analyticity of a function $f(z)$, where $z = (z_1, \ldots, z_m) \in \mathbb{C}^m$, can be defined in two different ways. One possibility is to consider the *ball*, i.e., the set defined by $\sum_{j=1}^{m} |z_j|^2 < \rho^2$. The other possibility is to take the *polydisc* $|z_j| < \rho_j$, $j = 1, \ldots, m$. These two definitions lead to entirely different theories, since there is no equivalence (i.e. holomorphic mapping) between the ball and the polydisc. For our purposes, the setup based on a polydisc is more suitable and convenient. We begin with some fundamental definitions. Let

$$1 < \rho_1 \le \rho_2 \le \cdots \le \rho_m, \quad \boldsymbol{\rho} = (\rho_1, \rho_2, \ldots, \rho_m); \quad (5.1)$$

we remark that the ordering in (5.1) can be achieved, without loss of generality, by simply renumbering the components of $\boldsymbol{\rho}$. Then, denote by $A(\boldsymbol{\rho})$ the set of functions analytic in the polydisc

$$D(\boldsymbol{\rho}) := \{z = (z_1, \ldots, z_m) : |z_j| < \rho_j, \ j = 1, \ldots, m\}.$$

where each such function has a singularity on each of the circles $|z_j| = \rho_j$, $j = 1, \ldots, m$. (Here singularity may involve either poles or branchpoints on the circle $|z| = \rho$.) The multivariate Cauchy formula

$$f(z) = \frac{1}{(2\pi i)^m} \int_D \frac{f(t)}{\prod\limits_{j=1}^m (t_j - z_j)}\, dt, \quad z \in D(\rho) \tag{5.2}$$

where the integration is taken over a polydisc D in $D(\rho)$ which contains the point z and, with $dt := dt_1 \ldots dt_m$, is valid for all $f(z) \in A(\rho)$. Let Γ_n^m denote set of all complex polynomials $p(z)$ of m variables which are of degree at most n in each of the variables z_j, $j = 1, \ldots, m$. (This set differs from the usual definition of a polynomial of several variables, having degree at most n, which means that the total degree of each term is at most n, but our definition here serves a more useful purpose later.) The $(n-1)^{\text{th}}$ Taylor section of an $f(z) \in A(\rho)$ is then defined as

$$S_{n-1}(f; z) := \frac{1}{(2\pi i)^m} \int_D f(t) \prod_{j=1}^m \frac{z_j^n - t_j^n}{t_j^n (z_j - t_j)}\, dt \tag{5.3}$$

which, in the sense of the above definition, is an element of Γ_{n-1}^m.

THEOREM 10. *For any $f(z) \in A(\rho)$, the Taylor sections $S_{n-1}(f; z)$ of (5.3) converge to $f(z)$, uniformly and geometrically in each closed subset of $D(\rho)$.*

PROOF. We have from (5.2) and (5.3) that

$$f(z) - S_{n-1}(f; z) = \frac{1}{(2\pi i)^m} \int_D \frac{f(t)}{\prod\limits_{j=1}^m (t_j - z_j)} \left[1 - \prod_{j=1}^m \left(1 - \frac{z_j^n}{t_j^n} \right) \right] dt.$$

Here,

$$\left| 1 - \prod_{j=1}^m \left(1 - \frac{z_j^n}{t_j^n} \right) \right| \leq C \max_{1 \leq j \leq m} \left| \frac{z_j}{t_j} \right|^n \to 0 \quad \text{as} \quad n \to \infty$$

in any closed subset of D, and this proves the theorem. $\qquad \square$

We now turn to the definition of the interpolation operator. The problem of interpolation in the multivariate case is more difficult (in general, existence and uniqueness are not guaranteed), but, with our definition of the set Γ_n^m, the situation simplifies. Consider the polynomial

$$L_{n-1}(f; z) = \frac{1}{(2\pi i)^m} \int_D f(t) \prod_{j=1}^m \frac{z_j^n - t_j^n}{(t_j^n - 1)(z_j - t_j)}\, dt \in \Gamma_{n-1}^m \tag{5.4}$$

where $L_{n_1}(f; z) \in \Gamma_{n-1}^m$, for any $f(z) \in A(\rho)$. As usual, let ω be a primitive n^{th} root of unity. From the representation (5.4) we can see that, at the points $z = (\omega^{k_1}, \ldots, \omega^{k_m})$ where $0 \leq k_j \leq n - 1$, $j = 1, \ldots, m$, are arbitrary integers, the polynomial (5.4) has the same values as $f(z)$. It will follow from the next lemma that this interpolation polynomial $L_{n-1}(f; z)$ is uniquely determined

LEMMA 2. *If $p(z) \in \Gamma_{n-1}^m$ has n^m different roots $z = (z_1, \ldots, z_m)$ such that each z_j takes n different values, then $p(z) \equiv 0$.*

PROOF. We use induction on m. For $m = 1$, the statement follows from the fundamental theorem of algebra. Assume it is true for $m - 1$, and represent $p(z)$ in the form

$$p(z) = \sum_{k=0}^{n-1} z_1^k p_k(z^*) \tag{5.5}$$

where $z^* = (z_2, \ldots, z_p) \in \mathbb{C}^{m-1}$ and $p_k(z^*) \in \Gamma_{n-1}^{m-1}$. Fixing an arbitrary $z^* = (z_2', \ldots, z_m')$ where z_j', $j = 2, \ldots, m - 1$, are coordinates of the roots of $p(z)$, then according to our assumption (5.5) vanishes for n different values of z_1. But then

$$p_k(z^*) = 0, \quad k = 0, \ldots, n - 1.$$

Here, by our assumption, z^* takes n^{m-1} different values, and thus, by the induction hypothesis, the p_k are identically zero for $k = 0, \ldots, n - 1$. This proves the statement for m.   □

Since the interpolation points for the polynomial (5.4) satisfy the condition of Lemma 2, $L_{n-1}(f, z)$ is uniquely determined. The uniform convergence of $L_{n-1}(f, z)$ to $f(z)$ in every closed subset of $D(\rho)$ will follow from Theorem 10 and the following overconvergence theorem:

THEOREM 11. *We have*

$$\overline{\lim_{n \to \infty}} |L_{n-1}(f, z) - S_{n-1}(f, z)|^{1/n} \leq \frac{1}{\rho_1} \prod_{|z_j| > \rho_j} \frac{|z_j|}{\rho_j} \tag{5.6}$$

for all $f(z) \in A(\rho)$ and $z \in \mathbb{C}^m$. (Here, the empty product is defined as unity.)

REMARKS. 1. In particular if $z \in D(\rho)$, then, as the product is unity in (5.6), the right hand side of (5.6) is $1/\rho_1 < 1$ which, coupled with Theorem 10, yields the uniform convergence of $L_{n-1}(f; z)$ to $f(z)$.

2. If

$$\prod_{|z_j| > \rho_j} \frac{|z_j|}{\rho_j} < \rho_1, \tag{5.7}$$

then we have the overconvergence of the difference $L_{n-1} - S_{n-1}$. Condition (5.7) gives an intrinsic relation between the coordinates $z_1, \ldots, z_m$. The larger we choose some $|z_j|$'s, the smaller we have to make the remaining $|z_j|$'s. In order to see more clearly how this works, consider the special case $\rho_1 = \cdots = \rho_m := \rho$. Then note that (5.7) allows us to select either

$$|z_j| < \rho^{1+\frac{1}{m}}, \quad j = 1, \ldots, m,$$

or

$$|z_1| = \cdots = |z_{m-1}| = \rho, \quad |z_m| < \rho^2.$$

In the first case, one has overconvergence in *each* coordinates (but with a smaller radius), while the second case gives *no* overconvergence in $m - 1$ variables, but *optimal* overconvergence in the final coordinate. Of course, other choices are also possible.

PROOF OF THEOREM 11. Equations (5.3) and (5.4) imply

$$\Delta_{n-1}(f; z) := L_{n-1}(f; z) - S_{n-1}(f; z)$$

$$= \frac{1}{(2\pi i)^m} \int_D f(t) \prod_{j=1}^m \frac{z_j^n - t_j^n}{z_j - t_j} \left(\prod_{j=1}^m \frac{1}{t_j^n - 1} - \prod_{j=1}^m \frac{1}{t_j^n} \right) dt. \tag{5.8}$$

Here,

$$\prod_{j=1}^m \frac{1}{t_j^n - 1} - \prod_{j=1}^m \frac{1}{t_j^n} = \prod_{j=1}^m \frac{1}{t_j^n} \left[\prod_{j=1}^m \left(1 + \frac{1}{t_j^n - 1} \right) - 1 \right]$$

$$= O\left(\frac{1}{(\rho_1 - \varepsilon)^n \prod\limits_{j=1}^n (\rho_j - \varepsilon)^n} \right)$$

where $\varepsilon > 0$ is an arbitrary small fixed number. Thus, we obtain from (5.8) that

$$\Delta_{n-1}(t; z) = O\left(\left(\frac{\sum\limits_{j=1}^m \max(|z_j|, \rho_j - \varepsilon)}{(\rho_1 - \varepsilon) \prod\limits_{j=1}^m (\rho_j - \varepsilon)} \right)^n \right),$$

i.e.,

$$\varlimsup_{n \to \infty} |\Delta_{n-1}(f; z)|^{1/n} \le \frac{1}{\rho_1 - \varepsilon} \prod_{|z_j| > \rho_j - \varepsilon} \frac{|z_j|}{\rho_j - \varepsilon}$$

whence, the statement of the theorem follows, since $\varepsilon > 0$ was arbitrary. $\square$

The estimate in (5.6) of Theorem 11 is *sharp* in the following sense. Consider the function

$$f_0(z) = \prod_{j=1}^{m} \frac{1}{z_j - \rho_j} \in A(\boldsymbol{\rho}).$$

Evidently

$$L_{n-1}(f_0; \boldsymbol{z}) = \prod_{j=1}^{m} \frac{z_j^m - \rho_j^n}{(1 - \rho_j^n)(z_j - \rho_j)},$$

and from (5.3)

$$S_{n-1}(f_0; \boldsymbol{z}) = \prod_{j=1}^{m} \frac{z_j^n - \rho_j^n}{\rho_j^n(\rho_j - z_j)}.$$

Thus

$$\Delta_{n-1}(f_0; \boldsymbol{z}) = \prod_{j=1}^{m} \frac{z_j^n - \rho_j^n}{\rho_j - z_j} \left(\prod_{j=1}^{m} \frac{1}{\rho_j^n - 1} - \prod_{j=1}^{m} \frac{1}{\rho_j^n} \right)$$

$$= \prod_{j=1}^{m} \frac{z_j^n - \rho_j^n}{(\rho_j - z_j)\rho_j^n} O\left(\frac{1}{\rho_1^{2n} \rho_2^n \cdots \rho_m^n} \right),$$

whence by (5.1)

$$\varlimsup_{n \to \infty} |\Delta_{n-1}(f; \boldsymbol{z})|^{1/n} = \frac{1}{\rho_1} \prod_{|z_j| > \rho_j} \frac{|z_j|}{\rho_j}.$$

Thus for *some* functions in $A(\boldsymbol{\rho})$, the result of (5.6) is sharp. However, we can ask for the following stronger version of sharpness: is it true that

$$\varlimsup_{n \to \infty} \max_{\substack{|z_j| = r_h \\ j = 1, \ldots, m}} |\Delta_{n-1}(f; \boldsymbol{z})|^{1/n} = \frac{1}{\rho_1} \prod_{r_j > \rho_j} \frac{r_j}{\rho_j}$$

for any $r_j > 0$, $j = 1, \ldots, m$ and $f(z) \in A(\boldsymbol{\rho})$? The answer to this question is *no*, and this is in sharp contrast to the univariate case (cf. Chapter 4). This can be seen from the following example.

EXAMPLE. Let $m = 2$, $1 < \rho_1 < \rho_2$, and consider the function

$$f_1(z) = \sum_{k=0}^{\infty} \left(\frac{z_1}{\rho_1} \right)^{3^k} \sum_{k=0}^{\infty} \left(\frac{z_2}{\rho_2} \right)^{3^k} \in A(\boldsymbol{\rho}).$$

We shall examine the overconvergence case of $|z_1| = r_1 > \rho_1$, $|z_2| = r_2 > \rho_2$.

Formula (5.8) in this case gives

$$\Delta_{n-1}(f_1; z) = \frac{1}{(2\pi i)^2} \int_D f_1(t) \prod_{j=1}^{2} \frac{z_j^n - t_j^n}{(z_j - t_j)t_j^n} \left[\frac{1}{t_1^n} \right.$$

$$+ O\left(\frac{1}{(\rho_1 - \varepsilon)^{2n}} + \frac{1}{(\rho_2 - \varepsilon)^n} \right) \Bigg] dt$$

$$= \frac{1}{(2\pi i)^2} \int_D \prod_{j=1}^{2} \left[\sum_{k=0}^{\infty} \left(\frac{t_j}{\rho_j} \right)^{3^k} \sum_{k=0}^{n-1} \frac{z_j^k}{t_j^{k+1}} \right] \frac{1}{t_1^n} dt$$

$$+ O\left(\left(\frac{r_1 r_2}{(\rho_1 - \varepsilon)^3(\rho_2 - \varepsilon)} + \frac{r_1 r_2}{(\rho_1 - \varepsilon)(\rho_2 - \varepsilon)^2} \right)^n \right)$$

$$= \sum_{n \le 3^k \le 2n-1} \frac{z_1^{3^k - n}}{\rho_1^{3^k}} \cdot \sum_{3^k \le n-1} \left(\frac{z_2}{\rho_2} \right)^{3^k} + o\left(\left(\frac{r_1 r_2}{\rho^2 \rho_2} \right)^n \right),$$

provided $\varepsilon > 0$ is small enough. Now, assume that the integers λ_n, μ_n satisfy

$$3^{\lambda_n} < 2n < 3^{\lambda_n + 1} \quad \text{and} \quad 3^{\mu_n} < n \le 3^{\mu_n + 1}. \tag{5.9}$$

Evidently, in the sum $\sum_{n \le 3^k \le 2n-1}$ above there is at most one term (for $k = \lambda_n$; otherwise it may be empty). Then we can write

$$|\Delta_{n-1}(f_1; z)| = O\left(\frac{r_1^{3^{\lambda_n} - n} r_2^{3^{\mu_n}}}{\rho_1^{3^{\lambda_n}} \rho_2^{3^{\mu_n}}} \right) + o\left(\left(\frac{r_1 r_2}{\rho^2 \rho_2} \right)^n \right). \tag{5.10}$$

By (5.9), $\mu_n \le \lambda_n - 1$, and therefore $3^{\mu_n} \le 3^{\lambda_n - 1} < \frac{2n}{3}$, whence

$$\frac{r_1^{3^{\lambda_n} - n} r_2^{3^{\mu_n}}}{\rho_1^{3^{\lambda_n}} \rho_2^{3^{\mu_n}}} \le \left(\frac{r_1 r_2^{2/3}}{\rho_1^2 \rho_2^{2/3}} \right)^n = o\left(\left(\frac{r_1 r_2}{\rho^2 \rho_2} \right)^n \right).$$

Thus, (5.10) yields

$$|\Delta_{n-1}(f_1; z)| = o\left(\left(\frac{r_1 r_2}{\rho^2 \rho_2} \right)^n \right),$$

i.e., for this function the error estimate in case $r_1 > \rho_1$, $r_2 > \rho_2$ is indeed better than the one provided by Theorem 11.

If we iterate interpolation operators and Taylor series, we can obtain different types of overconvergence results. (For a detailed account on this subject in the univariate case, see Ch. 2.) Here we restrict ourselves to one particular case. Instead of the interpolating polynomial (5.4), let us introduce the operator

$$L_{n-1}(f; \boldsymbol{\alpha}; z) := \frac{1}{(2\pi i)^m} \int_D f(t) \prod_{j=1}^{m} \frac{z_j^n - t_j^n}{(t_j^n - \alpha_j^n)(z_j - t_j)} \, dt \in \Gamma_{n-1}^m, \tag{5.11}$$

where $\boldsymbol{\alpha} = (\alpha_1, \ldots, \alpha_m) \in D(\boldsymbol{\rho})$, $\alpha_j > 0$, $j = 1, \ldots, m$, for any $f(z) \in A(\boldsymbol{\rho})$. (5.11) interpolates f at the points $\boldsymbol{z} = (\alpha_1 \omega^{k_1}, \ldots, \alpha_m \omega^{k_m})$, where $0 \le k_j \le n - 1$, $j = 1, \ldots, m$, are arbitrary integers. This polynomial, just like (5.4), is uniquely determined. Using also the notation (5.3), we now state

THEOREM 12. *If $\lambda > 1$ is a fixed integer, then we have*

$$\limsup_{n \to \infty} |L_{n-1}(f - S_{\lambda n-1}(f); \boldsymbol{\alpha}; \boldsymbol{z})|^{1/n} \le \max_{1 \le j \le m} \left(\frac{\alpha_j}{\rho_j} \right)^{\lambda} \cdot \prod_{j=1}^{m} \frac{|z_j|}{\alpha_j}$$

for any $f \in A(\boldsymbol{\rho})$ and any $\boldsymbol{z} = (z_1, \ldots, z_m)$.

The result shows that we have convergence if

$$\prod_{j=1}^{m} |z_j| < \frac{\prod_{j=1}^{m} \alpha_j}{\max_{1 \le j \le m} \left| \frac{\alpha_j}{\rho_j} \right|^{\lambda}}.$$

In particular, if $0 < \alpha_1 = \cdots = \alpha_m = \alpha < \rho_1 = \cdots = \rho_m = \rho$, then this condition takes the form

$$|z_j| < \frac{\rho^{\lambda/m}}{\alpha^{\lambda/m-1}}, \qquad j = 1, \ldots, m,$$

i.e. we have overconvergence provided $\lambda > m$.

PROOF OF THEOREM 12. (5.2), (5.3) and (5.11) yield

$$L_{n-1}(f - S_{\lambda n-1}(f); \boldsymbol{\alpha}; \boldsymbol{z}) = \frac{1}{(2\pi i)^{2m}} \int_{D_2} \int_{D_1} \frac{f(\boldsymbol{u})}{\prod_{j=1}^{m} (u_j - t_j)} \times$$

$$\times \left[1 - \prod_{j=1}^{m} \left(1 - \frac{t_j^{\lambda n}}{u_j^{\lambda n}} \right) \right] d\boldsymbol{u} \prod_{j=1}^{m} \frac{z_j^m - t_j^m}{(t_j^n - \alpha_j^n)(z_j - t_j)} d\boldsymbol{t}$$

$$= \frac{1}{(2\pi i)^{2m}} \int_{D_1} f(\boldsymbol{u}) \int_{D_2} \prod_{j=1}^{m} \frac{z_j^n - t_j^n}{(t_j^n - \alpha_j^n)(u_j - t_j)(z_j - t_j)} \times$$

$$\times \left[1 - \prod_{j=1}^{m} \left(1 - \frac{t_j^{\lambda n}}{u_j^{\lambda n}} \right) \right] d\boldsymbol{t} d\boldsymbol{u},$$

where

$$D_1 = \{(t_1, \ldots, t_m) : |t_j| = \alpha_j + \varepsilon, \, j = 1, \ldots, m\}$$

and

$$D_2 = \{(u_1, \ldots, u_m) : |u_j| = \rho_j - \varepsilon, \, j = 1, \ldots, m\}$$

with an arbitrarily small $\varepsilon > 0$. Here

$$\left| \prod_{j=1}^{m} \frac{z_j^n - t_j^n}{(t_j^n - \alpha_j^n)(u_j - t_j)(z_j - t_j)} \right| \leq \frac{c}{\varepsilon^m} \prod_{j=1}^{m} \frac{|z_j|^n}{\alpha_j^{n-1}}$$

and

$$\left| 1 - \prod_{j=1}^{m} \left(1 - \frac{t_j^{\lambda n}}{u_j^{\lambda n}} \right) \right| \leq c \max_{1 \leq j \leq m} \left(\frac{\alpha_j + \varepsilon}{\rho_j - \varepsilon} \right)^{\lambda n}.$$

Hence

$$\limsup_{n \to \infty} |L_{n-1}(f - S_{\lambda n-1}(f); \boldsymbol{\alpha}; \boldsymbol{z})|^{1/n} \leq \max_{1 \leq j \leq m} \left| \frac{\alpha_j + \varepsilon}{\rho_j - \varepsilon} \right|^{\lambda} \cdot \prod_{j=1}^{m} \frac{|z_j|}{\alpha_j}.$$

Since $\varepsilon > 0$ was arbitrary, this yields the statement. $\square$

1.6. Historical Remarks

(a) The nodes satisfying (1.0) are called *asymptotically uniformly distributed*. This condition is due to Kalmár [62] (see Gaier [42, p. 64]). Theorems 1 and 5 are due to Walsh [115 , Ch. 7, Theorems 10 and 11], just like the beautiful observation formulated as Theorem 6 (cf. Walsh [115, Ch. 7, Theorem 1]). The sharpness of Theorem 6 was shown again in [115, p. 154]. Theorem 3 can be found in Cavaretta, Dikshit and Sharma [25]. The extension formulated in Theorem 7 is due to Cavaretta, Sharma and Varga [30]. That the Taylor sections in the overconvergence theorem cannot be replaced by the polynomials of best uniform approximation was shown by D. J. Newman (1980) by means of an example at the end of a lecture by A. Sharma where he was present. The determination of best approximating polynomial to $(\rho - z)^{-1}$ is due to Al'per [3] (see also Rivlin [87]), who estimated the order of best approximation of some other, related functions as well. In connection with Theorem 8, it was V. Totik [111] who showed that the additional property of continuity of the function in A_ρ does not essentially improve the error estimate in the overconvergence, thus answering a question of Szabados raised in [106]. The Corollary is Rivlin's generalization [88] of Walsh's theorem. Section 1.5 is a recent result of Sharma and Szabados (cf. [97]). A different approach (using balls instead of the polydisc) was done by Cavaretta, Micchelli and Sharma [26]. With the norm $||\boldsymbol{z}|| = \sqrt{\sum_{j=1}^{m} |z_j|^2}$, for a function analytic in the ball $||\boldsymbol{z}|| < 1$, they introduced the operators

$$S_{n-1}(f; \boldsymbol{z}) := \frac{(m-1)!}{(2\pi i)^m} \int_{||\boldsymbol{t}||=1} f(\boldsymbol{t}) \sum_{\ell=0}^{n-1} \binom{m+\ell-1}{\ell} \left(\sum_{j=1}^{m} z_j t_j \right)^{\ell} d\sigma(\boldsymbol{t}),$$

and for $\boldsymbol{r} = (r_1, \ldots, r_m)$, $0 < r_j < 1$, $j = 1, \ldots, m$,

$$L_{n-1}(f, \boldsymbol{r}, \boldsymbol{z}) := \frac{(m-1)!}{(2\pi i)^m} \int_{||\boldsymbol{t}||=1} f(\boldsymbol{t}) \frac{\sum_{j=1}^m \binom{m+\ell-1}{\ell} \left(\sum_{j=1}^m z_j t_j \right)^\ell}{\prod_{j=1}^m (1 - r_j^n t_j^n)} \, d\sigma(\boldsymbol{z}),$$

(6.1)

where $d\sigma(\boldsymbol{z})$ is the Lebesgue measure on $||\boldsymbol{z}|| = 1$. These operators are the analogues of the Taylor series and Lagrange interpolation, respectively. (In fact, for $m = 1$, (6.1) is indeed a Lagrange interpolation polynomial, but for $m > 1$ no such interpretation exists.) It is proved in [26] that

$$\lim_{n \to \infty} [L_{n-1}(f, \boldsymbol{r}, \boldsymbol{z}) - S_{n-1}(f, \boldsymbol{z})] = 0$$

for

$$||\boldsymbol{z}|| < \frac{1}{\max_{1 \le j \le m} r_j},$$

i.e., we have overconvergence.

(b) Concerning overconvergence, Lagrange interpolation on nodes other than the roots of unity was first considered by Brück. Let

$$\omega_n^\alpha(z) := \frac{(z+\alpha)^{n+1} - (\alpha z + 1)^{n+1}}{1 - \alpha^{n+1}} := \prod_{k=0}^n (z - z_{kn}^\alpha), \qquad 0 < \alpha < 1,$$

be a monic polynomial of degree $n+1$, where

$$z_{kn}^\alpha = \frac{w_{kn} - \alpha}{1 - \alpha w_{kn}}, \qquad w_{kn} := exp \frac{2\pi i k}{n+1}, \qquad k = 0, \ldots, n.$$

From this representation, it is easy to see that for $|z| > 1$,

$$\lim_{n \to \infty} \omega_n^\alpha(z)^{\frac{1}{n+1}} = z + \alpha$$

which implies that the interpolation nodes z_{kn}^α are not asymptotically uniformly distributed (see Gaier [42, p. 64]). Now if $f(z)$ is analytic in $K(R, \alpha) := \{z \in \mathbf{C} : |z + \alpha| < R + \alpha\}$, $L_n^\alpha(f; z)$ denotes its Lagrange interpolation to f on the zeros of $\omega_n^\alpha(z)$ and $S_n^\alpha(f; z) := \sum_{k=0}^n a_k z^k$ is the Taylor expansion of $f(z)$ about $-\alpha$, then

$$\lim_{n \to \infty} [L_n^\alpha(f; z) - S_n^\alpha(f; z)] = 0, \qquad z \in E_1(R, \alpha),$$

where

$$E_\ell(R, \alpha) := \left\{ z \in \mathbf{C} : |z + \alpha| < (R + \alpha) \left(\frac{R + \alpha}{1 + R\alpha} \right)^\ell \right\}, \qquad \ell = 1, 2, \ldots.$$

Moreover for any integer $\ell \geq 1$,

$$\lim_{n \to \infty} [L_n^\alpha(f; z) - S_{\ell,n}(f; z)] = 0 \quad \text{for} \quad z \in E_\ell(R, \alpha),$$

where

$$S_{\ell,n}^\alpha(f; z) := \sum_{k=0}^{n} \sum_{j=0}^{\ell-1} \sum_{m=0}^{j(n+1)} \binom{j(n+1)}{m} \alpha^{j(n+1)-m} (1 - \alpha^2)^m a_{k+m} (z + \alpha)^k$$

$$- \alpha \sum_{k=0}^{n} \sum_{j=0}^{\ell-1} \sum_{m=0}^{n-k+j(n+1)} \binom{n - k + j(n+1)}{m}$$

$$\times (1 - \alpha^2)^m \alpha^{n-k+j(n+1)-m} a_{k+m} (1 + \alpha z)^k$$

(see Brück [21]). From Example 2 in [3, p. 132], we see that if $f_0(z) = \frac{1}{\rho - z}$, then the polynomial of best uniform approximation to $f_0(z)$ of degree at most $n - 1$ on the disc $|z| \leq 1$ is given by

$$\hat{p}_{n-1}(f_0; z) = \frac{1}{\rho - z} - \frac{cz^{n-1}(\rho z - 1)}{\rho - z}, \quad \text{where} \quad c = \frac{1}{\rho^{n-1}(\rho^2 - 1)}.$$

Then simplifying, we have

$$\hat{p}_{n-1}(f_0; z) = \frac{1}{\rho - z} - \frac{z^{n-1}(\rho z - 1)}{(\rho - z)(\rho^2 - 1)\rho^{n-1}}$$

$$= \frac{1}{\rho - z} - \frac{z^{n-1}[\rho(z - \rho) + \rho^2 - 1]}{(\rho - z)\rho^{n-1}(\rho^2 - 1)}$$

$$= \frac{1}{\rho - z} + \frac{z^{n-1}}{\rho^{n-2}(\rho^2 - 1)} - \frac{z^{n-1}}{\rho^{n-1}(\rho - z)}$$

$$= \frac{\rho^{n-1} - z^{n-1}}{(\rho - z)\rho^{n-1}} + \frac{z^{n-1}}{\rho^{n-2}(\rho^2 - 1)}.$$

Also for $|z| = 1$

$$f_0(z) - \hat{p}_{n-1}(f_0; z) = \left| \frac{z^{n-1}(\rho z - 1)}{\rho^{n-1}(\rho - z)} \right| = \left| \frac{\rho z - 1}{z - \rho} \right| = 1,$$

and because of the maximum principle here the left hand side is ≤ 1 in $|z| < 1$.

CHAPTER 2

HERMITE AND HERMITE-BIRKHOFF
INTERPOLATION AND WALSH EQUICONVERGENCE

2.1. Hermite Interpolation

As a generalization of Lagrange interpolation, we introduce the notion of Hermite interpolation. Let $f(z)$ be a function analytic in an open domain D. Further let n and r be positive integers, and $z_0, \ldots, z_{n-1}$ pairwise different points from D. We shall denote the (unique) Hermite polynomial interpolant of degree $rn - 1$ and order r, of $f(z)$ in these n zeros by $h_{r,rn-1}(f; z)$. This polynomial is defined by the properties

$$h_{r,rn-1}^{(j)}(f; z_k) = f^{(j)}(z_k), \qquad k = 0, \ldots, n-1; \ j = 0, \ldots, r-1.$$

With the notation $\omega_n(z) := \prod_{k=0}^{n-1}(z - z_k)$, this polynomial can be represented in the form

$$h_{r,rn-1}(f; z) = \frac{1}{2\pi i} \int_C \frac{\omega_n(t)^r - \omega_n(z)^r}{\omega_n(t)^r} \frac{f(t)}{t - z} \, dt, \tag{1.0}$$

where C may be any rectifiable Jordan curve included with its interior in D and containing the points $z_1, \ldots, z_n$ and z in its interior. Indeed, (1.0) is a polynomial of degree at most $rn - 1$, and by Cauchy's theorem

$$h_{r,rn-1}^{(j)}(f; z_k) = \frac{(-1)^j j!}{2\pi i} \int_C \frac{f(t)}{(t - z_k)^{j+1}} \, dt, \quad k = 0, \ldots, n-1; \ j = 0, \ldots, r-1.$$

The uniqueness follows from the fundamental theorem of algebra: if we had two different polynomials with the above properties, then their not identically zero difference, a polynomial of degree at most $rn - 1$, together with its first $r - 1$ derivatives, would vanish at the points $z_0, \ldots, z_{n-1}$. Counting multiplicities, this would result in rn roots for a polynomial of degree at most $rn - 1$, which is impossible.

Now let $f \in A_\rho$ $(\rho > 1)$ and let $h_{r,rn-1}(f; z)$ denote the Hermite interpolant to f in the zeros of $(z^n - 1)^r$. Then $h_{r,rn-1}(f; z) \in \pi_{rn-1}$ and it satisfies the conditions

$$h_{rn-1}^{(j)}(f; w^k) = f^{(j)}(w^k), \quad j = 0, 1, \ldots, r-1; \quad k = 0, 1, \ldots, n-1 \tag{1.1}$$

25

where w is a primitive n^{th} root of unity. It is known that

$$h_{r,rn-1}(f; z) = \frac{1}{2\pi i} \int_{\Gamma_R} \frac{f(t)}{t-z} \cdot \frac{(t^n-1)^r - (z^n-1)^r}{(t^n-1)^r} \, dt \qquad (1.2)$$

where $\Gamma_R = \{t : |t| = R\}$, $\quad R < \rho$.

If $f(z) = \sum\limits_{k=0}^{\infty} a_k z^k$, then from (3.3) in Chapter 1

$$p_{rn-1,0}(f; z) := \sum_{k=0}^{rn-1} a_k z^k = \frac{1}{2\pi i} \int_{\Gamma_R} \frac{f(t)}{t-z} \cdot \frac{t^{rn} - z^{rn}}{t^{rn}} \, dt \qquad (1.3)$$

is the Taylor series of f of degree $rn - 1$ about the origin. From (1.2) and (1.3), we see that

$$\Delta_{rn-1,1}(f; z) := h_{r,rn-1}(f; z) - p_{rn-1,0}(f; z) = \frac{1}{2\pi i} \int_{\Gamma_R} \frac{f(t) K_n(t, z)}{t-z} \, dt \ (1.4)$$

where

$$K_n(t, z) := \frac{z^{rn}}{t^{rn}} - \frac{(z^n-1)^r}{(t^n-1)^r} \cdot \qquad (1.5)$$

In order to examine the behaviour of $K_n(t, z)$ for large n, we shall prove

LEMMA 1. *For all t with $|t| > 1$, the following identity holds:*

$$\frac{z^r}{t^r} - \frac{(z-1)^r}{(t-1)^r} = \frac{t-z}{t^r} \sum_{s=1}^{\infty} \frac{\beta_{s,r}(z)}{t^s} \qquad (1.6)$$

where $\beta_{s,r}(z)$ is a polynomial in z of degree $r - 1$ given by

$$\beta_{s,r}(z) := \sum_{k=0}^{r-1} \binom{r+s-1}{k} (z-1)^k, \quad s = 1, 2, \ldots . \qquad (1.7)$$

PROOF. From (1.7), we can easily check that

$$\begin{cases} \beta_{s+1,r}(z) - z\beta_{s,r}(z) & = -\binom{r+s-1}{s} (z-1)^r \\ \beta_{1,r}(z) & = z^r - (z-1)^r. \end{cases} \qquad (1.8)$$

From (1.6), on expanding the left side in powers of $1/t$, we see that on using (1.8)

$$\frac{z^r}{t^r} - \frac{(z-1)^r}{(t-1)^r} = \frac{z^r}{t^r} - (z-1)^r \left[\frac{1}{t^r} + \frac{1}{t^r} \sum_{k=1}^{\infty} \binom{r+k-1}{k} \frac{1}{t^k} \right]$$

$$= \frac{z^r - (z-1)^r}{t^r} - \frac{1}{t^r} \sum_{k=1}^{\infty} \frac{\binom{r+k-1}{k}(z-1)^r}{t^k}$$

$$= \frac{\beta_{1,r}(z)}{t^r} + \frac{1}{t^r} \sum_{k=1}^{\infty} \frac{\beta_{k+1,r}(z) - z\beta_{k,r}(z)}{t^k}$$

$$= \frac{t-z}{t^{r+1}} \beta_{1,r}(z) + \sum_{k=2}^{\infty} \frac{\beta_{k,r}(z)}{t^{r+k}} - z \sum_{k=2}^{\infty} \frac{\beta_{k,r}(z)}{t^{r+k}}$$

$$= \frac{t-z}{t^r} \sum_{k=1}^{\infty} \frac{\beta_{k,r}(z)}{t^k}$$

after shifting the summation index in the second sum. This completes the proof.
□

The following two lemmas will give some estimates on the polynomials $\beta_{j,r}(z)$. From (1.7), we see that

$$\beta_{j,r}(z) = \binom{r+j-1}{r} r \int_0^1 t^{j-1}(z-t)^{r-1} dt, \quad j = 1, 2, \dots . \qquad (1.9)$$

This is easily done on expanding the right side in powers of $(z-1)$ and then using Euler's formula for the Beta-function. We shall prove

LEMMA 2. *There exists a constant c_0 depending only on r such that*

$$|\beta_{j,r}(z)| \leq c_0 j^{r-1} \max\{1, |z|^{r-1}\}, \quad j = 1, 2, \dots . \qquad (1.10)$$

PROOF. If $|z| > 1$, then $|z - t|^{r-1} \leq (|z| + 1)^{r-1} \leq 2^{r-1}|z|^{r-1}$ for each $t \in [0, 1]$. If $|z| < 1$, $|z - t|^{r-1} \leq 2^{r-1}$ for $0 \leq t \leq 1$. These two inequalities give the result.
□

LEMMA 3. (a) *If $|z| > 1$, then there exist constants $c_1 = c_1(r) > 0$ and $N_1 = N_1(r, z)$ such that*

$$c_1 j^{r-1} |z|^{n(r-1)} \leq |\beta_{j,r}(z^n)|, \quad n \geq N_1. \qquad (1.11)$$

(b) *If $|z| < 1$, then there are constants $c_2 = c_2(r) > 0$ and $N_2 = N_2(r, j, z)$ such that*

$$c_2 j^{r-1} \leq |\beta_{j,r}(z^n)| \quad for \quad n \geq N_2. \qquad (1.12)$$

(c) *If $\beta_{j,r}(z)$ has no zero on the unit circle, then there is a constant $c_3 = c_3(r) > 0$ such that*

$$c_3 \leq |\beta_{j,r}(z)| \quad \text{when} \quad |z| = 1. \tag{1.13}$$

PROOF. (a) We get (1.11) from (1.9) and the fact that for each z with $|z| > 1$, we have

$$\int_0^1 t^{j-1}(1 - tz^{-n})^{r-1} dt \rightarrow \frac{1}{j} \quad \text{as} \quad n \rightarrow \infty.$$

(b) From (1.9), we have

$$\beta_{j,r}(0) = (-1)^{r-1} \binom{r+j-2}{r-1} \neq 0$$

when $|z| < 1$, $z^n \rightarrow 0$ and $n \rightarrow \infty$ which proves (1.12).

(c) If $\beta_{j,r}(z)$ has no zero on the unit circle then the constant c_3 in (1.13) can be taken to be $\inf_{|w|=1} |\beta_{j,r}(w)| > 0$.

There are values of j and r for which $\beta_{j,r}(z)$ has zeros on $|z| = 1$. Thus when $j = 1$, $\beta_{1,r}(z) = z^r - (z-1)^r$ which has a zero on $|z| = 1$ when r is a multiple of 6.

For any positive integer $\ell \geq 1$, we set

$$\Delta_{rn-1,\ell}(f;z) := h_{r,rn-1}(f;z) - h_{r,rn-1,0}(f;z) - \sum_{j=1}^{\ell-1} h_{r,rn-1,j}(f;z) \tag{1.14}$$

where

$$h_{r,rn-1,0}(f;z) := p_{rn-1,0}(f;z) = \sum_{k=0}^{rn-1} a_k z^k$$

and with $\Gamma_R = \{z \,|\, |z| = R\}$,

$$h_{r,rn-1,j}(f;z) := \beta_{j,r}(z^n) \sum_{k=0}^{n-1} a_{k+n(r+j-1)} z^k,$$

$$= \beta_{j,r}(z^n) \frac{1}{2\pi i} \int_{\Gamma_R} \frac{f(t)}{t-z} \frac{t^n - z^n}{t^{n(r+j)}} dt \quad j = 1, 2, \ldots,$$

so that

$$\Delta_{rn-1,\ell}(f;z) = \frac{1}{2\pi i} \int_{\Gamma_R} \frac{f(t)K(t,z)}{t-z} dt, \tag{1.15}$$

where

$$K(t,z) := \frac{z^{rn}}{t^{rn}} - \frac{(z^n-1)^r}{(t^n-1)^r} - \frac{t^n - z^n}{t^{rn}} \sum_{j=1}^{\ell-1} \frac{\beta_{j,r}(z^n)}{t^{jn}}. \tag{1.16}$$

We shall now prove

THEOREM 1. *If $f \in A_\rho$ ($\rho > 1$), then for any fixed integer $\ell \geq 1$,*

$$\lim_{n \to \infty} \Delta_{rn-1,\ell}(f; z) = 0 \quad \text{for} \quad |z| < \rho^{1+\frac{\ell}{r}}, \tag{1.17}$$

the convergence being uniform and geometric for all $|z| \leq R < \rho^{1+\frac{\ell}{r}}$. Moreover the result of (1.17) is best possible.

PROOF. Using (1.6), we see that the kernel $K(t,z)$ in (1.16) can be written as

$$K(t,z) = \frac{t^n - z^n}{t^{rn}} \sum_{j=\ell}^{\infty} \frac{\beta_{j,r}(z^n)}{t^{jn}}. \tag{1.18}$$

From (1.15) and (1.18) we see that

$$|\Delta_{rn-1,\ell}(f; z)| \leq \frac{1}{2\pi} M 2\pi R \frac{|z|^n - R^n}{R^{rn}(|z| - R)} \sum_{j=\ell}^{\infty} \frac{|\beta_{j,r}(z^n)|}{R^{jn}}$$

$$\leq \frac{MR(|z|^n - R^n)}{(|z| - R)R^{rn}} \sum_{j=\ell}^{\infty} \frac{c_1 j^{r-1}|z|^{n(r-1)}}{R^{jn}} \quad (|z| < R)$$

where the last inequality is obtained on using Lemma 2.

Since $\sum\limits_{j=\ell}^{\infty} j^{r-1} R^{-jn} < \infty$ for $R > 1$, we see that

$$|\Delta_{rn-1,\ell}(f, z)| \leq c \frac{|z|^{rn}}{R^{rn+\ell n}}$$

from which we get (1.17). $\qquad\qquad\qquad\qquad\qquad\qquad\qquad\qquad\qquad\square$

REMARK. If $\ell \geq 1$, the operator $\Delta_{rn-1,\ell}(f; z)$ can be also interpreted as the Hermite interpolant in the zeros of $(z^n - 1)^r$ of the difference $f(z) - p_{(\ell+r-1)n-1}(f; z)$ where, as in Ch. 1, $p_j(f; z)$ is the Taylor expansion of degree j of f. If we set $h_{r,rn-1}(f; 1, z) := h_{r,rn-1}(f; z)$ when we want to emphasize the interpolation in the zeros of $(z^n - 1)^r$, we can write

$$\Delta_{rn-1,\ell}(f; z) = h_{r,rn-1}(f; 1, z) - h_{r,rn-1}\big(p_{(\ell+r-1)n-1}(f); 1, z\big).$$

Since

$$p_{(\ell+r-1)n-1}(f; z) = \frac{1}{2\pi i} \int_{\Gamma_R} \frac{f(t)}{t - z} \cdot \frac{t^{(\ell+r-1)n} - z^{(\ell+r-1)n}}{t^{(\ell+r-1)n}} \, dt,$$

we see that in (1.15), $\frac{K(t;z)}{t-z}$ is the Hermite interpolant to the difference

$$\frac{(t^n - 1)^r - (z^n - 1)^r}{(t - z)(t^n - 1)^r} - \frac{t^{(\ell+r-1)n} - z^{(\ell+r-1)n}}{(t - z)t^{(\ell+r-1)n}} = \Big[\frac{z^{(\ell+r-1)n}}{t^{(\ell+r-1)n}} - \frac{(z^n - 1)^r}{(t^n - 1)^r}\Big]/(t-z).$$

Since

$$z^{n(\ell+r-1)} = (z^n - 1 + 1)^{\ell+r-1} = \sum_{j=0}^{r-1} \binom{\ell+r-1}{j} (z^n - 1)^j + (z^n - 1)^r q(z)$$

$$= \beta_{\ell,r}(z^n) + (z^n - 1)^r q(z)$$

where $q(z)$ is a polynomial of degree $\ell - 1$ in z^n, we see that

$$K(t,z) = \frac{(t^n - 1)^r \beta_{\ell,r}(z^n) - (z^n - 1)^r \beta_{\ell,r}(t^n)}{t^{(\ell+r-1)n}(t^n - 1)^r}.$$

This expression for $K(t,z)$ when compared with (1.18) shows that the following identity holds:

$$\frac{t^n - z^n}{t^{rn}} \sum_{j=\ell}^{\infty} \frac{\beta_{j,r}(z^n)}{t^{jn}} = \frac{(t^n - 1)^r \beta_{\ell,r}(z^n) - (z^n - 1)^r \beta_{\ell,r}(t^n)}{t^{(\ell+r-1)n}(t^n - 1)^r}$$

$$+ \frac{(z^n - 1)^r (q(z) - q(t))}{t^{(\ell+r-1)n}}. \tag{1.19}$$

This identity can be easily proved directly when $\ell = 1$, since the left side in (1.19) because of (1.6) in Lemma 1, becomes

$$\frac{z^{rn}}{t^{rn}} - \frac{(z^n - 1)^r}{(t^n - 1)^r}. \tag{1.20}$$

Also for $\ell = 1$, the right side becomes, on using (1.8),

$$\frac{(t^n - 1)^r (z^{nr} - (z^n - 1)^r) - (z^n - 1)^r (t^{nr} - (t^n - 1)^r)}{t^{rn}(t^n - 1)^r}$$

which simplifies easily into (1.20) and completes the verification for $\ell = 1$ since $q(z) = q(t) =$ constant.

2.2. Generalizations of Theorem 1

We shall first find the Hermite interpolant to a monomial

$$g(z) = z^{j+(r+s)n}, \quad 0 \le j \le n - 1$$

in the zeros of $(z^n - 1)^r$. Since

$$g(z) = z^j (z^n - 1 + 1)^{r+s} = z^j \sum_{k=0}^{r-1} \binom{r+s}{k} (z^n - 1)^k + (z^n - 1)^r q(z)$$

where $q(z)$ is a polynomial, it follows that

$$h_{r,rn-1}(g, 1, z) = z^j \sum_{k=0}^{r-1} \binom{r+s}{k} (z^n - 1)^k = z^j \beta_{s+1,r}(z^n).$$

Similarly, if we find the Hermite interpolant to $g(z)$ in the zeros of $(z^n - \alpha^n)^r$, where $\alpha \in D_\rho$, then

$$h_{r,rn-1}(g; \alpha, z) = z^j \alpha^{n(r+s)} \beta_{s+1,r}(z^n/\alpha^n). \tag{2.1}$$

If $\alpha, \beta \in D_\rho$, α, β both not equal to zero, then for any $f \in A_\rho$ $(\rho > 1)$, we define

$$\Delta_{pn,rn}^{\alpha,\beta}(f; z) := h_{p,pn-1}(f; \alpha, z) - h_{p,pn-1}(L_{rn-1}(f, \beta, z); \alpha, z), \tag{2.2}$$

where $m, p, n, m \geq pn$ are positive integers. We shall now prove

THEOREM 2. *If $f \in A_\rho$ $(\rho > 1)$ and if $\alpha, \beta \in D_\rho$ $(\rho > 1)$, $|\alpha| + |\beta| \neq 0$, then*

$$\lim_{n \to \infty} \Delta_{pn,rn}^{\alpha,\beta}(f; z) = 0 \quad for \quad |z| < \sigma \tag{2.3}$$

where

$$\sigma = \rho / \max \left\{ \left(\frac{|\alpha|}{\rho} \right)^{\frac{r+1-p}{p}}, \left(\frac{|\beta|}{\rho} \right)^{r/p} \right\}, \tag{2.4}$$

the convergence being uniform and geometric in $|z| \leq \tau < \sigma$ and the result is best possible in the same sense as in Theorem 1.

PROOF. Since

$$L_{rn-1}(f; \beta, z) = \frac{1}{2\pi i} \int_{\Gamma_R} \frac{f(t)}{t - z} \cdot \frac{t^{rn} - z^{rn}}{t^{rn} - \beta^{rn}} dt$$

it follows that

$$\Delta_{rn,pn}^{\alpha,\beta}(f; z) = \frac{1}{2\pi i} \int_{\Gamma_R} f(t) K(t, z) dt \tag{2.5}$$

where $K(t, z)$ is the Hermite interpolant of $\Lambda(z)$ in the zeros of $(z^n - \alpha^n)^p$, i.e., $K(t, z) = h_{p,pn-1}(\Lambda(\cdot); \alpha, z)$. Here $\Lambda(z)$ is given by

$$\Lambda(z) = \frac{(t^n - \alpha^n)^p - (z^n - \alpha^n)^p}{(t^n - \alpha^n)^p (t - z)} - \frac{t^{rn} - z^{rn}}{t^{rn} - \beta^{rn}} \cdot \frac{1}{t - z}$$

$$= \left[\frac{z^{rn} - \beta^{rn}}{t^{rn} - \beta^{rn}} - \frac{(z^n - \alpha^n)^p}{(t^n - \alpha^n)^p} \right] / (t - z) = \Lambda_{1,n}(z) - \Lambda_{2,n}(z),$$

where we set

$$\Lambda_{1,n}(z) = \frac{z^{rn}(t^n - \alpha^n)^p - (z^n - \alpha^n)^p t^{rn}}{(t - z)(t^{rn} - \beta^{rn})(t^n - \alpha^n)^p},$$

$$\Lambda_{2,n}(z) = \frac{\beta^{rn}[(t^n - \alpha^n)^p - (z^n - \alpha^n)^p]}{(t - z)(t^{rn} - \beta^{rn})(t^n - \alpha^n)^p}.$$

Since $\Lambda_{2,n}(z) \in \pi_{np-1}$, we have

$$h_{p,pn-1}(\Lambda_{2,n}; \alpha, z) = \Lambda_{2,n}(z). \tag{2.6}$$

From (2.1) we may replace z^{rn} by its interpolant in the zeros of $(z^n - \alpha^n)^p$, i.e., by $\alpha^{nr}\beta_{r-p+1,p}(z^n/\alpha^n)$ in order to find its Hermite interpolant and so

$$h_{p,pn-1}(\Lambda_{1,n}; \alpha, z) =$$

$$= \frac{\alpha^{rn}\beta_{r-p+1,p}\left(\frac{z^n}{\alpha^n}\right)(t^n - \alpha^n)^p - \alpha^{rn}\beta_{r-p+1,p}\left(\frac{t^n}{\alpha^n}\right)(z^n - \alpha^n)^p}{(t - z)(t^{rn} - \beta^{rn})(t^n - \alpha^n)^p}. \tag{2.7}$$

From (2.6) and (2.7) we see that

$$K(t, z) = h_{p,pn-1}(\Lambda_{1,n}; \alpha, z) - \Lambda_{2,n}(z). \tag{2.8}$$

Since $\Lambda_{2,n}(z) = \frac{|z|^{np}|\beta|^{rn}}{|t|^{rn+np}} + O(1)$ for large n, it follows that $\Lambda_{2,n}(z)$ tends to zero as $n \to \infty$ if

$$|z| \leq \rho/\left(\frac{|\beta|}{\rho}\right)^{r/p}. \tag{2.9}$$

By Lemma 2, $\beta_{r-p+1,p}(t^n/\alpha^n) \leq C\left(\frac{|t|^n}{|\alpha|^n}\right)^{p-1}$ so that

$$\left|\frac{\alpha^{rn}\beta_{r-p+1,p}(t^n/\alpha^n)(z^n - \alpha^n)^p}{(t^{rn} - \beta^{rn})(t^n - \alpha^n)^p}\right| \leq |\alpha|^{rn}\frac{|t|^{n(p-1)}}{|\alpha|^{n(p-1)}}\frac{|z|^{np}}{|t|^{rn+np}} \tag{2.10}$$

and the right side will tend to zero as $n \to \infty$, if

$$|z| < \frac{\rho}{\left(\frac{|\alpha|}{\rho}\right)^{\frac{r-p+1}{p}}}. \tag{2.11}$$

Similarly,

$$\left|\frac{\alpha^{rn}\beta_{r-p+1,p}(z^n/\alpha^n)(t^n - \alpha^n)^p}{(t^n - \alpha^n)^p(t^{rn} - \beta^{rn})}\right| \leq C|\alpha|^{rn}\frac{|z|^{n(p-1)}}{|\alpha|^{n(p-1)}|t|^{rn}} \tag{2.12}$$

and this will tend to zero as $n \to \infty$, if

$$|z| < \frac{\rho}{\left(\frac{|\alpha|}{\rho}\right)^{\frac{r-p+1}{p-1}}}. \tag{2.13}$$

Combining (2.7), (2.10) and (2.11), we see that $h_{p,pn-1}(\Lambda_{1,n};\alpha,z)$ tends to zero as $n \to \infty$, if

$$|z| \le \min \left(\frac{\rho}{\left(\frac{|\alpha|}{\rho}\right)^{\frac{r-p+1}{p-1}}} , \frac{\rho}{\left(\frac{|\alpha|}{\rho}\right)^{\frac{r-p+1}{p}}} \right) = \frac{\rho}{\left(\frac{|\alpha|}{\rho}\right)^{\frac{r-p+1}{p}}} . \qquad (2.14)$$

Combining (2.9) and (2.14), we get (2.4) which completes the proof. □

REMARK 1. Instead of (2.2), one may consider a slightly more general operator, viz.,

$$\Delta_{pn,m}^{\alpha,\beta}(f;z) := h_{p,pn-1}(f;\alpha,z) - h_{p,pn-1}(L_{m-1}(f;\beta,z);\alpha;z) \qquad (2.15)$$

where $m = rn + q$, $p \le r$, $q = ns + O(1)$, $0 \le s < 1$. In this case no special difficulties arise and the result is slightly more general. In this case the expression for σ in (2.4) becomes

$$\rho/ \max \left\{ \left(\frac{|\alpha|}{\rho}\right)^{\frac{r+1-p}{p}} , \left(\frac{|\beta|}{\rho}\right)^{\frac{r+s}{p}} \right\}. \qquad (2.16)$$

If in (2.2), we put $\beta = 1$, $\alpha = 0$, then $h_{p,pn-1}(f;0,z)$ is the Taylor section of degree $pn - 1$ for f about the origin and we get the Corollary in Chapter 1.

2.3. Mixed Hermite Interpolation

If $1 \le p < r$ are integers, then we consider the operator

$$H_{pn,rn}^{\alpha,\beta}(f;z) := h_{p,pn-1}(f;\alpha,z) - h_{p,pn-1}(h_{rn-1}(f;\beta,z);\alpha,z). \qquad (3.1)$$

As in the previous section, we can see that

$$H_{pn,rn}^{\alpha,\beta}(f;z) = \frac{1}{2\pi i} \int_{\Gamma_R} f(t)K(t,z)dt \qquad (3.2)$$

where $K(t,z)$ is the Hermite interpolant in the zeros of $(z^n - \alpha^n)^p$ of the polynomial $K_1(t,z)$, where

$$K_1(t,z) = \frac{(t^n - \alpha^n)^p - (z^n - \alpha^n)^p}{(t-z)(t^n - \alpha^n)^p} - \frac{(t^n - \beta^n)^r - (z^n - \beta^n)^r}{(t-z)(t^n - \beta^n)^r}$$

$$= \left[\frac{(z^n - \beta^n)^r}{(t^n - \beta^n)^r} - \frac{(z^n - \alpha^n)^p}{(t^n - \alpha^n)^p} \right] \frac{1}{t-z} .$$

Since

$$(z^n - \beta^n)^r = \sum_{j=0}^{r} \binom{r}{j} (z^n - \alpha^n)^j (\alpha^n - \beta^n)^{r-j}$$

with a similar expansion for $(t^n - \beta^n)^r$, we see that

$$K_1(t, z) = \left(((t^n - \alpha^n)^p \sum_{j=0}^{r} \binom{r}{j} (z^n - \alpha^n)^j (\alpha^n - \beta^n)^{r-j} \right.$$

$$\left. - (z^n - \alpha^n)^p \sum_{j=0}^{r} \binom{r}{j} (t^n - \alpha^n)^j (\alpha^n - \beta^n)^{r-j} \right)$$

$$\bigg/ \left((t - z)(t^n - \beta^n)^r (t^n - \alpha^n)^p \right).$$

From this it is clear that

$$K(t, z) = \sum_{j=0}^{p-1} \binom{r}{j} \frac{(\alpha^n - \beta^n)^{r-j}}{(t^n - \beta^n)^r} \frac{\{(t^n - \alpha^n)^{p-j} - (z^n - \alpha^n)^{p-j}\}}{(t^n - \alpha^n)^{p-j}(t - z)} (z^n - \alpha^n)^j$$

$$(3.3)$$

since the right side in (3.3) is a polynomial of degree $pn - 1$ and the values of $K(t, z)$ and of $K_1(t, z)$ on the zeros of $(z^n - \alpha^n)^p$ are the same.

From (3.3) we see that for $|t| = R < \rho$, we have

$$|K(t, z)| \leq C \max \left[\max_{1 \leq j \leq p-1} \left\{ \frac{\max (|\alpha|^n, |\beta|^n)^{r-j} |z|^{nj}}{R^{nr}} \right\}, \right.$$

$$\left. \max_{1 \leq j \leq p-1} \left\{ \frac{|z|^{np} \max (|\alpha|^n, |\beta|^n)^{r-j}}{R^{nr+n(p-j)}} \right\} \right].$$

From this inequality, we see that the right side will tend to zero when

$$|z| < \frac{R^{r/j}}{(\max (|\alpha|, |\beta|))^{(r-j)/j}}, \quad j = 1, \ldots, p - 1$$

and

$$|z| \leq \frac{R^{\frac{r+p-j}{p}}}{\max (|\alpha|, |\beta|)^{\frac{r-p-j}{p}}}.$$

Therefore (3.2) will tend to zero when $n \to \infty$, provided

$$|z| < \min \left\{ \frac{RR^{(r-p+1)/p}}{\max (|\alpha|, |\beta|)^{\frac{r-p+1}{p}}}, \min_{1 \leq j \leq p-1} \frac{R^{r/j}}{\max (|\alpha|, |\beta|)^{\frac{r-j}{j}}} \right\}$$

$$= \min \left\{ \frac{R}{\max \left(\frac{|\alpha|}{R}, \frac{|\beta|}{R} \right)^{\frac{r-p+1}{p}}}, \min_{1 \leq j \leq p-1} \frac{R}{\max \left(\frac{|\alpha|}{R}, \frac{|\beta|}{R} \right)^{\frac{r-j}{j}}} \right\}.$$

Letting $R \to \rho$, we see that

$$\lim_{n \to \infty} H_{pn, rn}^{\alpha, \beta}(f; z) = 0 \quad \text{for} \quad |z| < \frac{\rho}{\left[\max \left(\frac{|\alpha|}{\rho}, \frac{|\beta|}{\rho} \right) \right]^{\frac{r-p+1}{p}}}. \quad (3.4)$$

We have thus proved

THEOREM 3. *If* $\alpha \neq \beta$, $\alpha, \beta \in D_\rho$ *and if* $f \in A_\rho$ $(\rho > 1)$ *then the polynomial* $H^{\alpha,\beta}_{pn,rn}(f;z)$, $1 \leq p < r$, *defined by* (3.1) *tends to zero as* $n \to \infty$, *when*

$$|z| < \rho \Big/ \max \left[\frac{|\alpha|}{\rho}, \frac{|\beta|}{\rho} \right]^{\frac{r-p+1}{p}}.$$

A different, but similar result is the following. If r, ℓ are positive integers, $m = nq + c$, and $q \geq r$ is a given positive integer, we set

$$\Delta^{\alpha,\beta}_{rn-1,\ell,m}(f;z) := h_{r,rn-1}(f;\alpha,z) - h_{r,rn-1}\big(h_{\ell,\ell m-1}(f;\beta,z);\alpha,z\big). \qquad (3.5)$$

We shall now prove

THEOREM 4. *Let* r, ℓ, m *be positive integers,* $m = nq + c$, $q \geq r$ *and* c *a constant. Then*

$$\lim_{n\to\infty} \Delta^{\alpha,\beta}_{rn-1,\ell,m}(f;z) = 0 \quad for \quad z \in D_\sigma, \qquad (3.6)$$

where

$$\sigma := \rho / \max \left\{ \left(\frac{|\beta|}{\rho} \right)^{\frac{\ell q}{r}}, \; \left(\frac{|\beta|}{\rho} \right)^{\frac{(\ell-1)q}{r}} \left(\frac{|\alpha|}{\rho} \right)^{\frac{q+1-r}{r}}, \; \left(\frac{|\alpha|}{\rho} \right)^{\frac{\ell q+1-r}{r}} \right\}. \qquad (3.7)$$

More precisely, for any $\rho \leq \mu < \sigma$, we have

$$\overline{\lim_{n\to\infty}} \left\{ \max_{|z|=\rho} |\Delta^{\alpha,\beta}_{rn-1,\ell,m}(f;z)| \right\}^{1/rn} \leq \frac{\mu}{\sigma}.$$

Taking $\alpha = 1$, $|\beta| < 1$, it is easy to see that $\sigma = \rho^{\frac{\ell q+1}{r}}$.

If $\alpha = 1$, $\beta = 0$, $\ell = 1$, $m = nq$ and $q = \ell' + r - 1$, then we obtain from (3.11) the following result:

$$\lim_{n\to\infty} \left[h_{r,rn-1}(f;1,z) - \sum_{j=1}^{\ell'-1} \beta_{j,r}(z^n)p_{n-1,r+j-1}(f;z) \right] = 0, \quad |z| < \rho^{1+\frac{\ell'}{r}}, \qquad (3.8)$$

where $p_{n-1,j}(f;z)$ is given by (3.3) of Ch. 1.

The proof of Theorem 4 will be based on the following:

LEMMA 4. *For any* $\alpha, \beta \in \mathbb{C}$, *we have*

$$(z^m - \beta^m)^\ell = (z^n - \alpha^n)^r Q_{\ell m - rn}(z) + R_{\ell, rn-1}(z)$$

where $Q_{\ell m - rn}(z) \in \pi_{\ell m - rn}$ *and* $R_{\ell, rn-1}(z) \in \pi_{rn-1}$ *is given by*

$$R_{\ell, rn-1}(z) = (-\beta^m)^\ell + \sum_{n=1}^{\ell} \binom{\ell}{k} (-\beta^m)^{\ell-k} z^{kc} \alpha^{nqk} \beta_{qk-r+1,r}(z^n/\alpha^n) \quad (3.9)$$

with $\beta_{s,r}(z)$ *defined in* (1.7).

PROOF. The binomial theorem gives

$$(z^m - \beta^m)^\ell = (-\beta^m)^\ell + \sum_{k=1}^{\ell} \binom{\ell}{k} (-\beta^m)^{\ell-k} z^{mk}$$

and since $z^{mk} = z^{k(nq+c)} = z^{kc}(z^n)^{qk}$ and $q > r$, we see that

$$z^{nqk} = (z^n - \alpha^n + \alpha^n)^{qk} = \sum_{j=0}^{r-1} \binom{qk}{j} (z^n - \alpha^n)^j (\alpha^n)^{qk-j} + (z^n - \alpha^n)^r q_k(z)$$

$$= \alpha^{nqk} \beta_{qk-r+1,r}(z^n/\alpha^n) + (z^n - \alpha^n)^r q_k(z)$$

where $q_k(z) \in \pi_{(k-r)n}$. From this, observation (3.9) is immediate. $\quad\square$

LEMMA 5. *The operator* $\Delta_{rn-1,\ell,m}^{\alpha,\beta}(f;z)$ *in* (3.5) *has the integral representation*

$$\Delta_{rn-1,\ell,m}^{\alpha,\beta}(f;z) = \frac{1}{2\pi i} \int_{\Gamma_R} f(t) K(t,z) dt \quad (1 < R < \rho)$$

where

$$K(t,z) = \frac{(t^n - \alpha^n)^r R_{\ell,rn-1}(z) - (z^n - \alpha^n)^r R_{\ell,rn-1}(t)}{(t^m - \beta^m)^\ell (t^n - \alpha^n)^r (t-z)}. \quad (3.10)$$

PROOF. Since $h_{r,rn-1}$ is linear and reproduces polynomials of degree $\le rn - 1$, we see that

$$\Delta_{rn-1,\ell,m}^{\alpha,\beta}(f;z) = h_{r,rn-1}(g;\alpha,z)$$

where

$$g(z) = h_{r,rn-1}(f;\alpha,z) - h_{\ell,\ell m-1}(f;\beta,z) = \frac{1}{2\pi i} \int_{\Gamma_R} f(t) K_1(t,z) \, dt$$

and

$$K_1(t,z) = \left[\frac{(t^n - \alpha^n)^r - (z^n - \alpha^n)^r}{(t^n - \alpha^n)^r (t-z)} - \frac{(t^m - \beta^m)^\ell - (z^m - \beta^m)^\ell}{(t^m - \beta^m)^\ell (t-z)} \right] dt. \quad (3.11)$$

So in order to find $K(t, z)$, it is enough to find the Hermite interpolant of the kernel $K_1(t, z)$ of (3.11) in the zeros of $(z^n - \alpha^n)^r$. Observe that on simplifying, we have

$$K_1(t, z) = [\frac{(z^m - \beta^m)^\ell}{(t^m - \beta^m)^\ell} - \frac{(z^n - \alpha^n)^r}{(t^n - \alpha^n)^r}]/(t - z)$$

$$= \frac{(t^n - \alpha^n)^r(z^m - \beta^m)^\ell - (z^n - \alpha^n)^r(t^m - \beta^m)^\ell}{(t^m - \beta^m)^\ell(t^n - \alpha^n)^r(t - z)}.$$

By Lemma 3, $(z^m - \beta^m)^\ell = R_{\ell,rn-1}(z) + Q_{\ell m-rn}(z)(z^n - \alpha^n)^r$, $(t^m - \beta^m)^\ell = R_{\ell,rn-1}(t)$
$+ Q_{\ell m-rn}(z)(t^n - \alpha^n)^r$. Since $K(t, z)$ in (3.10) is a polynomial in z of degree $\leq rn - 1$, and since it coincides with $K_1(t, z)$ in the zeros of $(z^n - \alpha^n)^r$, we see that $K(t, z)$ in (3.10) is the required kernel.   □

PROOF OF THEOREM 4. Using the expression (3.9) for $R_{\ell,rn-1}(z)$, we see that $K_1(t, z) = K_2(t, z) + K_3(t, z)$ where

$$K_2(t, z) = \frac{1}{(t^m - \beta^m)^\ell(t - z)}$$

$$\times \left\{(-\beta^m)^\ell + \sum_{k=1}^{\ell} \binom{\ell}{k} (-\beta^m)^{\ell-k} z^{kc} \alpha^{nqk} \beta_{qk-r+1,,r}(z^n/\alpha^n)\right\}$$

$$K_3(t, z) = - \frac{(z^n - \alpha^n)^r}{(t^m - \beta^m)^\ell(t^n - \alpha^n)^r(t - z)}$$ (3.12)

$$\times \left\{(-\beta^m)^\ell + \sum_{k=1}^{\ell} \binom{\ell}{k} (-\beta^m)^{\ell-k} t^{kc} \alpha^{nqk} \beta_{qk-r+1,r}(t^n/\alpha^n)\right\}.$$

We see from the expression for $K_3(t, z)$ in (3.12) on using Lemma 2 that $K_3(t, z)$ will tend to zero as $n \to \infty$, if

(A) $\dfrac{|z|^{nr}|\beta|^{m\ell}}{R^{m\ell+nr}}$ and $\dfrac{|z|^{nr}|\beta|^{m(\ell-k)}|\alpha|^{nqk}\left(\frac{|R|^n}{|\alpha|^n}\right)^{r-1}}{R^{\ell m+nr}}$, $1 \leq k \leq \ell$

tend to zero as $n \to \infty$. Similarly $K_2(t, z)$ will tend to zero if

(B) $\left(\dfrac{|z|}{|\alpha|}\right)^{n(r-1)} \dfrac{(|\beta|)^{m(\ell-k)}|\alpha|^{nqk}}{R^{\ell m}}$, $1 \leq k \leq \ell$

tends to zero as $n \to \infty$. Since $m = nq + c$ (c a constant), we see that the terms in (A) will tend to zero if

$$|z| < \min_{1 \leq k \leq \ell} \left[\frac{R}{\left(\frac{|\beta|}{R}\right)^{\frac{\ell q}{r}}}, \frac{R}{\left(\frac{|\beta|}{R}\right)^{\frac{q(\ell-k)}{r}}} \left(\frac{|\alpha|}{R}\right)^{\frac{qk+1-r}{r}}\right], \quad 1 \leq k \leq \ell.$$

The terms in (B) will tend to zero if

$$|z| \leq \min_{1 \leq k \leq \ell} \left[\frac{R}{\left(\frac{|\beta|}{R} \right)^{\frac{q(\ell-k)}{r-1}} \left(\frac{|\alpha|}{R} \right)^{\frac{qk+1-r}{r-1}}} \right].$$

Letting $R \to \rho$, we see that $K(t,z) \to 0$ as $n \to \infty$ when $|z| < \min(\sigma_1, \sigma_2)$ where

$$\sigma_1 = \rho / \max \left[\left(\frac{|\beta|}{\rho} \right)^{\frac{\ell q}{r}}, \left(\frac{|\beta|}{\rho} \right)^{\frac{q(\ell-1)}{r}} \left(\frac{|\alpha|}{\rho} \right)^{\frac{q+1-r}{r}}, \left(\frac{|\alpha|}{\rho} \right)^{\frac{\ell q+1-r}{r}} \right]$$

and

$$\sigma_2 = \rho / \max \left[\left(\frac{|\beta|}{\rho} \right)^{\frac{q(\ell-1)}{r-1}} \left(\frac{|\alpha|}{\rho} \right)^{\frac{q+1-r}{r-1}}, \left(\frac{|\alpha|}{\rho} \right)^{\frac{\ell q+1-r}{r-1}} \right].$$

Since $\sigma = \min(\sigma_1, \sigma_2)$, the result follows.

$\square$

REMARK. 1. If in Theorem 4, where we suppose that $m = nq + c$, $q \geq r$, we also require that c is not a constant, but that $c = sn + O(1)$, $0 < s < \frac{1}{\ell}$, then the details of the proof remain the same with minor changes and the value of σ depends upon s. Then it is easy to see that in this case

$$\sigma = \rho / \max \left\{ \left(\frac{|\beta|}{\rho} \right)^{\frac{\ell(q+s)}{r}}, \left(\frac{|\beta|}{\rho} \right)^{(\ell-1)(q+s)/r} \left(\frac{|\alpha|}{\rho} \right)^{\frac{q+1-r}{r}}, \left(\frac{|\alpha|}{\rho} \right)^{\frac{\ell q+1-r}{r}} \right\}.$$

2. If we take $\alpha = 0$, $r = 1$, then

$$\Delta_{n-1,\ell,m}^{0,\beta}(f;z) := S_{n-1}(f;z) - S_{n-1}(h_{\ell,\ell m-1}(f;\beta,\cdot);z).$$

2.4. Mixed Hermite and ℓ_2-Approximation

Let $f \in A_\rho$ ($\rho > 1$) and let $0 \neq \alpha \in D_\rho$. If $m > n$ are fixed positive integers, let ω be a primitive m^{th} root of unity. We shall consider the problem of finding the polynomial $P_{rm+n}(f;\alpha,z) \in \pi_{rm+n}$ for each $f \in A_\rho$ which satisfies the Hermite interpolation condition

$$P_{rm+n}^{(\nu)}(f;\alpha,\alpha\omega^k) = f^{(\nu)}(\alpha\omega^k), \quad k = 0,1,\ldots,m-1, \ \nu = 0,1,\ldots,r-1, \ (4.1)$$

and also minimizes

$$\sum_{k=0}^{m-1} |P_{rm+n}^{(r)}(f;\alpha,\alpha\omega^k) - f^{(r)}(\alpha\omega^k)|^2, \quad r \geq 0 \tag{4.2}$$

over all polynomials in π_{rm+n} which satisfy (4.1). If $L_{rm+n}(f;\beta,z)$ is the Lagrange interpolant to f in the zeros of $z^{rm+n} - \beta^{rm+n}$, we shall prove

THEOREM 5. *For each $f \in A_\rho$ ($\rho > 1$) and for each non-negative integer r, let $P_{rm+n}(f; \alpha, z)$ satisfy (4.1) and (4.2). If $0 \le q := \lim\limits_{n \to \infty} \frac{n}{m} \le 1$ and if $\alpha, \beta \in D_\rho$, then*

$$\lim_{n \to \infty} [P_{rm+n}(f; \alpha, z) - L_{rm+n}(f; \beta, z)] = 0 \tag{4.3}$$

for all

$$|z| < \sigma := \frac{\rho}{\max\left\{ \frac{|\beta|}{\rho}, \frac{|\alpha|}{\rho} \right\}^{\frac{1}{r+q}}}. \tag{4.4}$$

It is clear from (4.1) that $P_{rm+n}(f; \alpha, z)$ will have the form

$$P_{rm+n}(f; \alpha, z) = h_{rm-1}(f; \alpha, z) - (z^m - \alpha^m)^r Q_n(z)$$

where $Q_n(z)$ will be determined by the requirement (4.2). Observe that

$$\frac{d^r}{dz^r} (z^m - \alpha^m)^r]_{z=\alpha\omega^k} = \alpha^{mr-r} \omega^{-kr} \sum_{\nu=0}^{r} (-1)^{r-\nu} \binom{r}{\nu} (m\nu)_r$$

$$= \alpha^{mr-r} \omega^{-kr} \cdot m^r \cdot r! \tag{4.5}$$

where $(x)_0 = 1$ and $(x)_k = x(x-1)\ldots(x-k+1)$ for any positive integer k. Then the problem of minimizing (4.2) reduces to finding the polynomial $Q_n(z) \in \pi_n$ such that

$$\sum_{k=0}^{m-1} |g(\alpha\omega^k) - Q_n(\alpha\omega^k)|^2 = \min_{p \in \pi_n} \sum_{k=0}^{m-1} |g(\alpha\omega^k) - p(\alpha\omega^k)|^2 \tag{4.6}$$

where

$$g(z) := \frac{z^r \cdot \{h_{rm-1}^{(r)}(f; \alpha, z) - f^{(r)}(z)\}}{\alpha^{mr} m^r \cdot r!}. \tag{4.7}$$

We shall prove

LEMMA 6. *The polynomial $Q_n(z) \in \pi_n$ satisfying (4.6) is explicitly given by*

$$Q_n(z) = -\frac{1}{2\pi i} \int_{\Gamma_R} \frac{f(t)t^{m-n-1}(t^{n+1} - z^{n+1})}{(t-z)(t^m - \alpha^m)^{r+1}} dt \tag{4.8}$$

where $1 < R < \rho$ and $\Gamma_R = \{z : |z| = R\}$.

PROOF. The polynomial $Q_n(z)$ which minimizes the right side in (4.6) is obtained by truncating the Lagrange interpolant to $g(z)$ on the zeros of $z^m - \alpha^m$. In order to find $L_{m-1}(g; \alpha, z)$, we shall calculate $g(\alpha\omega^k)$. To do this we observe that

$$h_{r, rn-1}(f; \alpha, z) - f(z) = -\frac{1}{2\pi i} \int_{\Gamma_R} \frac{f(t)}{(t^m - \alpha^m)^r} K(t, z) dt$$

where $K(t,z) = \frac{(z^m - \alpha^m)^r}{t - z}$. Since

$$z^r \frac{d^r}{dz^r} K(t,z)]_{z=\alpha\omega^k} = \frac{m^r \cdot r! \alpha^{mr}}{(t - \alpha\omega^k)}, \qquad k = 0, 1, \ldots, m-1$$

we see that

$$g(\alpha\omega^k) = [z^r \{h^{(r)}_{rn-1}(f; \alpha, z) - f^{(r)}(z)\}]_{z=\alpha\omega^k}/\alpha^{mr} m^r \cdot r!$$

$$= \frac{1}{2\pi i} \int_{\Gamma_R} \frac{f(t)}{(t^m - \alpha^m)^r} [z^r \frac{d^r}{dz^r} K(t,z)]_{z=\alpha\omega^k} dt$$

$$= \frac{1}{2\pi i} \int \frac{f(t)}{(t^m - \alpha^m)^r} \frac{1}{t - \alpha\omega^k} dt,$$

Hence it follows that

$$L_{m-1}(g; \alpha, z) = \frac{1}{2\pi i} \int_{\Gamma_R} \frac{f(t)(t^m - z^m)}{(t - z)(t^m - \alpha^m)^{r+1}} dt$$

so that $Q_n(z)$ is as given by (4.8). $\qquad \square$

PROOF OF THEOREM 5. From Lemma 5 we see that

$$P_{rm+n}(f; \alpha, z) - L_{rm+n}(f; \beta, z) = \frac{1}{2\pi i} \int_{\Gamma_R} \frac{f(t)}{t - z} K_1(t,z) dt$$

where after some simplification we have

$$K_1(t,z) = -\frac{z^{rm+n+1} - \beta^{rm+n+1}}{t^{rm+n+1} - \beta^{rm+n+1}} + \left(\frac{z^m - \alpha^m}{t^m - \alpha^m}\right)^r \cdot \frac{\alpha^m - t^{m-n-1}z^{n+1}}{t^m - \alpha^m}.$$

We write $K_1(t,z) = K_2(t,z) + K_3(t,z) + K_4(t,z)$ where

$$K_2(t,z) = \frac{z^{rm+n+1} - \beta^{rm+n+1}}{t^{rm+n+1} - \beta^{rm+n+1}} - \frac{z^{rm+n+1}}{t^{rm+n+1}}$$

$$= \frac{\beta^{rm+n+1}(z^{rm+n+1} - t^{rm+n+1})}{t^{rm+n+1}(t^{rm+n+1} - \beta^{rm+n+1})},$$

$$K_3(t,z) = \left\{ \left(\frac{z}{t}\right)^{rm} - \left(\frac{z^m - \alpha^m}{t^m - \alpha^m}\right)^r \right\} \frac{z^{n+1}}{t^{n+1}},$$

$$K_4(t,z) = \left(\frac{z^m - \alpha^m}{t^m - \alpha^m}\right)^r \cdot \left\{ \frac{z^{n+1}}{t^{n+1}} + \frac{\alpha^m - t^{m-n-1}z^{n+1}}{t^m - \alpha^m} \right\}$$

$$= \left(\frac{z^m - \alpha^m}{t^m - \alpha^m}\right)^r \cdot \frac{\alpha^m(t^{n+1} - z^{n+1})}{t^{n+1}(t^m - \alpha^m)}.$$

From the expressions for $K_\ell(t,z)$ ($\ell = 2, 3, 4$), we see that as $n \to \infty$, $|K_2(t,z)|$ $\to 0$ if $|z| < R/(\frac{|\beta|}{R})$.

From (1.6), we easily have

$$\left(\frac{z^m}{t^m}\right)^r - \left(\frac{z^m - \alpha^m}{t^m - \alpha^m}\right)^r = \frac{t^m - z^m}{t^{(r+1)m}} \cdot \alpha^{rm} \sum_{s=0}^{\infty} \frac{\beta_{s+1,r}(z^m/\alpha^m)\alpha^{sm}}{t^{sm}},$$

and from Lemma 2, we have

$$\beta_{s+1,r}(z^m/\alpha^m) \le c \cdot s^{r-1} \left(\frac{|z|}{|\alpha|}\right)^{(r-1)m}.$$

Thus $K_3(t,z) \to 0$ as $n \to \infty$, if $\frac{|z|^{rm+n}|\alpha|^m}{R^{(r+1)m+n}} \to 0$, and so we require

$$|z| < \frac{R}{\left(\frac{|\alpha|}{R}\right)^{1/(r+q)}}.$$

Lastly, we see from the expression for $K_4(t,z)$ that $K_4(t,z) \to 0$ as $n \to \infty$ if

$$\frac{|z|^{rm+n+1}|\alpha|^m}{R^{mr+m+n+1}} \to 0, \quad \text{i.e., if} \quad |z| < \frac{R}{\left(\frac{|\alpha|}{R}\right)^{\frac{m}{rm+n+1}}}.$$

Since $\lim \frac{n}{m} = q$, on combining the above three and on letting $R \to \rho$, we see that $P_{rm+n}(f;\alpha,z) - L_{rm+n}(f;\beta,z) \to 0$ if

$$|z| \le \min \left\{ \frac{\rho}{\left(\frac{|\beta|}{\rho}\right)}, \frac{\rho}{\left(\frac{|\alpha|}{\rho}\right)^{1/r}}, \frac{\rho}{\left(\frac{|\alpha|}{\rho}\right)^{\frac{1}{r+q}}} \right\}.$$

From this (4.4) is immediate.   □

2.5. A Lemma and Its Applications

We shall consider here the following problem and see how it extends the result of Theorem 5 on Walsh equiconvergence.

PROBLEM. Let $f_0(z),\ldots\ldots,f_{r-1}(z)$ be r given functions in A_ρ ($\rho > 1$). Let $\{p_{\ell,j}\}_{j=0}^{n-1}$ ($\ell = 0, 1, \ldots, r-1$) be r sets of given real numbers. For each ℓ and a set of real numbers $\{p_{\ell,j}\}_{j=0}^{n-1}$ we define a linear operator $\mathcal{L}_\ell$ on the space of polynomials of degree $\le n-1$ such that

$$\text{if} \quad Q_n(z) = \sum_{j=0}^{n-1} c_j z^j, \quad \text{then} \quad \mathcal{L}_\ell(Q_n) = \sum_{j=0}^{n-1} c_j p_{\ell,j} z^j. \tag{5.1}$$

The problem is to find the polynomial $P_{m,n,r}(z)$ which minimizes the sum

$$\sum_{\ell=0}^{r-1} \sum_{k=0}^{m-1} |f_\ell(\omega^k) - \mathcal{L}_\ell Q_n(\omega^k)|^2, \quad \omega^m = 1, \quad m > n, \tag{5.2}$$

over all polynomials $Q_n \in \pi_{n-1}$. Let

$$L_{m-1}(f_\ell; z) := \sum_{j=0}^{m-1} \alpha_{\ell,j}^{(m)} z^j.$$

Then we can easily prove

LEMMA 7. *The polynomial* $P_{m,n,r}(z)$ *which minimizes* (5.2) *is given by* $\sum_{j=0}^{n-1} c_j z^j$, *where*

$$c_j = \frac{1}{K_j} \sum_{\ell=0}^{r-1} p_{\ell,j} \alpha_{\ell,j}^{(m)}, \quad K_j = \sum_{\ell=0}^{r-1} (p_{\ell,j})^2, \quad j = 0, \ldots, n-1. \quad (5.3)$$

PROOF. Since

$$|f_\ell(\omega^k) - \mathcal{L}_\ell Q_n(\omega^k)|^2 = |L_{m-1}(f_\ell; \omega^k) - \mathcal{L}_\ell Q_n(\omega^k)|^2 = |\sum_{j=0}^{m-1} d_{\ell j} \omega^{kj}|^2$$

where

$$d_{\ell,j} = \begin{cases} \alpha_{\ell,j}^{(m)} - p_{\ell,j} c_j, & j = 0, 1, \ldots, n-1, \\ \alpha_{\ell,j}^{(m)}, & j = n, \ldots, m-1, \end{cases} \quad (5.4)$$

we see on using the properties of roots of unity that

$$\sum_{\ell=0}^{r-1} \sum_{k=0}^{m-1} |f_\ell(\omega^k) - \mathcal{L}_\ell Q_n(\omega^k)|^2 = m \sum_{\ell=0}^{r-1} \sum_{j=0}^{m-1} |d_{\ell,j}|^2.$$

If we put

$$\begin{cases} c_j := \rho_j e^{i\theta_j}, & j = 0, 1, \ldots, n-1 \\ \alpha_{\ell,j}^{(m)} := \sigma_{\ell,j} e^{i\phi_{\ell,j}}, & j = 0, 1, \ldots, m-1; \quad \ell = 0, 1, \ldots, r-1 \end{cases} \quad (5.5)$$

then from (5.4), (5.5) we get

$$|d_{\ell,j}|^2 = \begin{cases} |\sigma_{\ell,j} e^{i\phi_{\ell,j}} - p_{\ell,j} \rho_j e^{i\theta_j}|^2, & j = 0, 1, \ldots, n-1 \\ |\sigma_{\ell,j}|^2, & j = n, \ldots, m-1 \end{cases}$$

and the problem of finding the minimum of (5.2) is equivalent to finding that of

$$\sum_{\ell=0}^{r-1} \sum_{j=0}^{n-1} (p_{\ell,j})^2 \rho_j^2 + \sum_{j=0}^{m-1} \sigma_{\ell,j}^2 - 2 \sum_{j=0}^{n-1} p_{\ell,j} \rho_j \sigma_{\ell,j} \cos(\theta_j - \phi_{\ell,j})$$

where ρ_j runs over the reals and $0 \le \theta_j < 2\pi$.

Elementary calculations now show that ρ_j and θ_j are determined by the following equations:

$$\begin{cases} \rho_j \sum_{\ell=0}^{r-1} (p_{\ell,j})^2 - \sum_{\ell=0}^{r-1} p_{\ell,j} \sigma_{\ell,j} \cos(\theta_j - \phi_{\ell,j}) = 0 \\ \sum_{\ell=0}^{r-1} p_{\ell,j} \sigma_{\ell,j} \sin(\theta_j - \phi_{\ell,j}) = 0 \end{cases} \quad (j = 0, \ldots, m-1).$$

Thus we have

$$\begin{cases} \frac{\sin \theta_j}{A_j} = \frac{\cos \theta_j}{B_j} = \frac{1}{\sqrt{A_j^2 + B_j^2}} \,, \\[2mm] A_j = \sum_{\ell=0}^{r-1} p_{\ell,j} \sigma_{\ell,j} \sin \phi_{\ell,j}, \quad B_j = \sum_{\ell=0}^{r-1} p_{\ell,j} \sigma_{\ell,j} \cos \phi_{\ell,j} \end{cases} \quad (j = 0, \ldots, m-1). \tag{5.6}$$

Therefore

$$\rho_j = \frac{\sqrt{A_j^2 + B_j^2}}{\sum_{\ell=0}^{r-1} (p_{\ell,j})^2} = \frac{B_j \cos \theta_j + A_j \sin \theta_j}{K_j}, \quad j = 0, \ldots, n-1. \tag{5.7}$$

From (5.5), (5.6) and (5.7), we get (5.3). $\qquad \square$

We now choose $f_j(z) := f^{(j)}(z)$, where $f(z) \in A_\rho$. If we set

$$\mathcal{L}_\ell(Q_n) := \sum_{j=0}^{n-1} (j)_\ell c_j z^{j-\ell},$$

then $\mathcal{L}_\ell(Q_n) = Q_n^{(\ell)}$ and the problem of minimizing (5.2) is equivalent to minimizing the sum

$$\sum_{\ell=0}^{r-1} \sum_{k=0}^{m-1} |\omega^{k\ell} f^{(\ell)}(\omega^k) - Q_n^{(\ell)}(\omega^k)|^2, \quad Q_n \in \pi_{n-1}. \tag{5.8}$$

We can then prove

LEMMA 8. *If* $f(z) = \sum_{\nu=0}^{\infty} a_\nu z^\nu \in A_\rho$, *then the unique polynomial which minimizes* (5.8) *is given by*

$$P_{m,n,r}(z) := \sum_{j=0}^{n-1} c_j z^j, \quad c_j = \frac{1}{A_{0,j}(r)} \sum_{\lambda=0}^{\infty} A_{\lambda,j}(r) a_{j+\lambda m}, \tag{5.9}$$

where $A_{\lambda,j}(r) = \sum_{\ell=0}^{r-1} (j)_\ell (j + \lambda m)_\ell, \quad \lambda = 0, 1, \ldots$.

PROOF. In this case $p_{\ell,j} = (j)_\ell$. Since

$$z^\ell f^{(\ell)}(z) = \sum_{\nu=0}^{\infty} (\nu)_\ell a_\nu z^\nu \,,$$

it follows that the Lagrange interpolant to $z^\ell f^{(\ell)}(z)$ in the mth roots of unity is

$$\sum_{j=0}^{m-1} \alpha_{\ell,j}^{(m)} z^j, \quad \text{where} \quad \alpha_{\ell,j}^{(m)} = \sum_{\lambda=0}^{\infty} (j + \lambda m)_\ell a_{j+\lambda m}.$$

The result (5.9) is now a direct application of (5.3) in Lemma 7. $\qquad \square$

If we set

$$S_{n,\lambda,r}(f;z) := \sum_{j=0}^{n-1} \frac{A_{\lambda,j}(r)}{A_{0,j}(r)} z^j a_{j+\lambda m}, \quad \lambda = 0, 1, 2, \ldots \tag{5.10}$$

then the following theorem is easily proved:

THEOREM 6. *Let* $f(z) \in A_\rho$ $(\rho > 1)$ *and let* m, n *be positive integers such that* $\lim\limits_{n \to \infty} \frac{m}{n} = q > 1$. *Suppose that for a given positive integer* r, $P_{m,n,r}(f; z)$ *is the polynomial which minimizes* (5.8) *on the* m^{th} *roots of unity. Then for any integer* $\mu \geq 1$

$$\lim_{n \to \infty} [P_{m,n,r}(f; z) - \sum_{\lambda=0}^{\mu-1} S_{n,\lambda,r}(f; z)] = 0 \quad for \quad |z| < \rho^{1+\mu q}.$$

As another application of Lemma 8, we consider the polynomial $\widetilde{P}_{sm+n}(f; \alpha, z) \in \pi_{sm+n-1}$ analogous to $P_{sm+n}(f; \alpha, z)$ in Theorem 5. We now require that $\widetilde{P}_{sm+n}(f; \alpha, z)$ satisfies

$$\widetilde{P}_{sm+n}^{(\nu)}(f; \alpha, \alpha\omega^k) = f^{(\nu)}(\alpha\omega^k), \quad k = 0, 1, \ldots, m-1 \, ; \, \nu = 0, 1, \ldots, s-1 \quad (5.11)$$

for some $\alpha \in D_\rho$ and also minimizes

$$\sum_{k=0}^{m-1} \sum_{\nu=0}^{s+1} |\widetilde{P}_{sm+n}^{(\nu)}(f; \alpha, \omega^k) - f^{(\nu)}(\alpha\omega^k)|^2 \tag{5.12}$$

over all polynomials in π_{sm+n-1} satisfying (5.11). We prove the following analogue of Theorem 5:

THEOREM 7. *For each* $f \in A_\rho$ $(\rho > 1)$ *and for each non-negative integer* s *and* $\alpha, \beta \in D_\rho$, *let* $\widetilde{P}_{sm+n}(f; \alpha, z)$ *satisfy* (5.11) *and* (5.12). *If* $\lim\limits_{n \to \infty} \frac{n}{m} = q \leq 1$ *then*

$$\lim_{n \to \infty} [\widetilde{P}_{sm+n}(f; \alpha, z) - L_{sm+n}(f; \beta, z)] = 0$$

for all z $(|z| < \sigma)$ *where* σ *is given by*

$$\sigma = \frac{\rho}{\max \left\{ \frac{|\beta|}{\rho}, \left(\frac{|\alpha|}{\rho} \right)^{\frac{1}{s+q}} \right\}}.$$

PROOF. As in the proof of Theorem 5, let

$$\widetilde{P}_{sm+n}(f; \alpha, z) = h_{s,sm-1}(f; \alpha, z) + (z^m - \alpha^m)^s Q_n(z)$$

where Q_n will be determined by (5.12). Besides (4.5), we also have

$$\frac{d^{s+1}}{dz^{s+1}} (z^m - \alpha^m)^s]_{z=\alpha\omega^k} = m^s \cdot (s+1)! \frac{1}{2} s(m-1)\alpha^{s(m-1)-1}\omega^{-k(s+1)}. \tag{5.13}$$

Then minimizing (5.12) can be shown to be equivalent to minimizing

$$m^s \cdot s! \sum_{k=0}^{m-1} \sum_{\ell=0}^{1} |\mathcal{L}_\ell(Q_n)(\omega^k) - f_\ell(\omega^k)|^2, \quad Q_n \in \pi_{n-1}$$

where $f_\ell(z) := [h_{sm-1}^{(s+\ell)}(f; \alpha, z) - f^{(s+\ell)}(z)]z^{s+\ell}/m^s \cdot s! \alpha^{ms-s-\ell}$ $(\ell = 0, 1)$ and $\mathcal{L}_0, \mathcal{L}_1$ are linear operators on the space π_{n-1}. $\mathcal{L}_0$ is the identity operator and

$$\mathcal{L}_1(Q_n)(z) = (s+1) \left(\frac{1}{2} s(m-1) + z \frac{d}{dz} \right) Q_n$$

$$= (s+1) \sum_{j=0}^{n-1} \left(\frac{1}{2} s(m-1) + j \right) c_j z^j$$

where $Q_n(z) = \sum_{j=0}^{n-1} c_j z^j$. So

$$p_{0j} = 1 \quad \text{and} \quad p_{1j} = (s+1) \left(j + \frac{1}{2} s(m-1) \right), \quad j = 0, 1, \ldots, n-1. \quad (5.14)$$

Since

$$h_{s,sm-1}(f; \alpha, z) - f(z) = -\frac{1}{2\pi i} \int_{\Gamma_R} \frac{f(t)(z^m - \alpha^m)^s}{(t^m - \alpha^m)^s(t - z)} dt,$$

it follows on using (4.5) and (5.13) that

$$\begin{cases} f_0(\alpha\omega^k) = -\frac{1}{2\pi i} \int_{\Gamma_R} \frac{f(t)}{(t^m - \alpha^m)^s} \frac{1}{t - \alpha\omega^k} dt \\[2mm] f_1(\alpha\omega^k) = -\frac{s+1}{2\pi i} \int_{\Gamma_R} \frac{f(t)}{(t^m - \alpha^m)^s} \left\{ \frac{\frac{1}{2}s(m-1)}{t - \alpha\omega^k} + \frac{\alpha\omega^k}{(t - \alpha\omega^k)^2} \right\} dt, \\[2mm] \qquad\qquad k = 0, \ldots, m-1. \end{cases} \quad (5.15)$$

From (5.15) we can find the Lagrange interpolants to f_0 and f_1 on the zeros of $z^m - \alpha^m$. Thus we have

$$L_{m-1}(f_0; \alpha, z) = -\frac{1}{2\pi i} \int_{\Gamma_R} \frac{f(t)}{(t^m - \alpha^m)^{s+1}} \frac{t^m - z^m}{t - z} dt \quad (5.16)$$

and

$$L_{m-1}(f_1; \alpha, z) = -\frac{s+1}{2\pi i} \int_{\Gamma_R} \frac{f(t)}{(t^m - \alpha^m)^s} K(t, z) dt, \quad (5.17)$$

where

$$K(t, z) = \left[\frac{\frac{1}{2}s(m-1)}{t^m - \alpha^m} \cdot \frac{t^m - z^m}{t - z} - \frac{d}{dt} \left\{ \frac{t(t^m - z^m)}{(t^m - \alpha^m)(t - z)} \right\} \right]$$

since the value of $\frac{d}{dt} \left\{ \frac{t(t^m - z^m)}{(t^m - \alpha^m)(t - z)} \right\}$ at $z = \alpha\omega^k$ is easily seen to be $-\frac{\alpha\omega^k}{(t - \alpha\omega^k)^2}$.

From (5.16) we see that the coefficient of z^j in $L_{m-1}(f_0; \alpha, z)$ is given by

$$\alpha_{0,j}^{(m)} = -\frac{1}{2\pi i} \int_{\Gamma_R} \frac{f(t) t^{m-j-1}}{(t^m - \alpha^m)^{s+1}} \, dt$$

and from (5.17) we have for the coefficient of z^j in $L_{m-1}(f_1; \alpha, z)$

$$\alpha_{1j}^{(m)} = -\frac{s+1}{2\pi i} \int_{\Gamma_R} \frac{f(t)}{(t^m - \alpha^m)^s} \left\{ \frac{1}{2} \frac{s(m-1)t^{m-1-j}}{t^m - \alpha^m} - \frac{d}{dt} \left(\frac{t^{m-j}}{t^m - \alpha^m} \right) \right\} dt$$

$$= -\frac{1}{2\pi i} \int_{\Gamma_R} \frac{f(t) t^{m-1-j}}{(t^m - \alpha^m)^{s+1}} \left\{ p_{1j} + \frac{m(s+1)\alpha^m}{t^m - \alpha^m} \right\} dt.$$

By Lemma 7, applied with $r = 2$, $\quad Q_n(z) = \sum\limits_{j=0}^{n-1} c_j z^j$ where

$$c_j = \frac{p_{0j}\alpha_{0j}^{(m)} + p_{1j}\alpha_{1j}^{(m)}}{(p_{0j})^2 + (p_{1j})^2}.$$

Using (5.14) and the values of $\alpha_{0j}^{(m)}$ and $\alpha_{1j}^{(m)}$ above, we have

$$c_j = -\frac{1}{2\pi i} \int_{\Gamma_R} \frac{f(t) t^{m-1-j}}{(t^m - \alpha^m)^{s+1}} \, dt$$

$$-\frac{m(s+1)\alpha^m}{2\pi i} \frac{p_{1j}}{1 + (p_{1j})^2} \int_{\Gamma_r} \frac{f(t) t^{m-1-j}}{(t^m - \alpha^m)^{s+2}} \, dt.$$

Therefore $Q_n(z) = -\frac{1}{2\pi i} \int_{\Gamma_R} f(t) K_1(t, z) dt$ where

$$K_1(t, z) = \frac{t^{m-n}(t^n - z^n)}{(t^m - \alpha^m)^{s+1}(t - z)} + \frac{m(s+1)\alpha^m}{(t^m - \alpha^m)^{s+2}} \sum_{j=0}^{n-1} \frac{p_{1j} t^{m-j-1} z^j}{1 + (p_{1j})^2}$$

so we have now

$$\widetilde{P}_{sm+n}(f; \alpha, z) - L_{sm+n}(f; \beta, z) = \frac{1}{2\pi i} \int_{\Gamma_R} f(t) \widetilde{K}(t, z) dt$$

where

$$\widetilde{K}(t, z) = \frac{(t^m - \alpha^m)^s - (z^m - \alpha^m)^s}{(t^m - \alpha^m)^s(t - z)} + (z^m - \alpha^m)^s K_1(t, z)$$

$$- \frac{t^{sm+n} - z^{sm+n}}{t^{sm+n} - \beta^{sm+n}} \cdot \frac{1}{t - z}$$

$$= \frac{1}{t - z} [\widetilde{K}_1(t, z) + \widetilde{K}_2(t, z) + \widetilde{K}_3(t, z)] + \widetilde{K}_4(t, z).$$

Here we have set

$$\widetilde{K}_1(t, z) := \frac{z^{sm+n} - \beta^{sm+n}}{t^{sm+n} - \beta^{sm+n}} - \frac{z^{sm+n}}{t^{sm+n}} = \beta^{sm+n} \frac{(z^{sm+n} - t^{sm+n})}{t^{sm+n}(t^{sm+n} - \beta^{sm+n})},$$

$$\widetilde{K}_2(t,z) := \frac{z^{sm+n}}{t^{sm+n}} - \frac{(z^m - \alpha^m)^s}{(t^m - \alpha^m)^s} \cdot \frac{z^n}{t^n} = \frac{z^n}{t^n} \left\{ \frac{z^{sm}}{t^{sm}} - \frac{(z^m - \alpha^m)^s}{(t^m - \alpha^m)^s} \right\},$$

$$\widetilde{K}_3(t,z) := \frac{(z^m - \alpha^m)^s}{(t^m - \alpha^m)^s} \left\{ \frac{z^n}{t^n} - \frac{t^{m-n}z^n - \alpha^m}{t^m - \alpha^m} \right\}$$

$$= \frac{(z^m - \alpha^m)^s}{(t^m - \alpha^m)^s} \left\{ \frac{\alpha^m(t^n - z^n)}{t^n(t^m - \alpha^m)} \right\},$$

$$\widetilde{K}_4(t,z) := \frac{m(s+1)(z^m - \alpha^m)^s \alpha^m}{(t^m - \alpha^m)^{s+2}} \sum_{j=0}^{n-1} \frac{p_{1j} t^{m-j-1} z^j}{1 + (p_{1j})^2} \ .$$

From these expressions for the kernels $\widetilde{K}_\nu(t,z)$, $\nu = 1,\ldots,4$, we see that as $n \to \infty$, $\widetilde{K}_1(t,z) \to 0$ if $|z| < \frac{R}{\frac{|\beta|}{R}}$. Using Lemma 2 and the identity (1.6) in Lemma 1 we see that $\widetilde{K}_2(t,z)$ tends to zero as $n \to \infty$ if $|z| < \frac{R}{\left|\frac{\alpha}{R}\right|^{\frac{1}{s+q}}}$. $\widetilde{K}_3(t,z)$ also tends to zero for the same bound for $|z|$. Since $0 < \frac{p_{1j}}{1+(p_{1j})^2} < 1$, we see that

$$|\widetilde{K}_4(t,z)| \le \frac{m(s+1)|z|^{sm}|\alpha|^m}{|R|^{m(s+2)}} R^{m-n} \frac{(|z|^n - R^n)}{|z| - R}$$

and so $\widetilde{K}_4(t,z)$ tends to zero for $|z| < \frac{R}{\left|\frac{z}{R}\right|^{\frac{1}{s+q}}}$. So

$$\widetilde{P}_{sm+n}(f; \alpha, z) - L_{sm+n}(f; \beta, z)$$

will tend to zero as $n \to \infty$, when

$$|z| < \min \left\{ \frac{R}{\frac{|\beta|}{R}}, \frac{R}{\left(\frac{|\alpha|}{R}\right)^{\frac{1}{s+q}}} \right\}.$$

On letting $R \to \rho$, we get the required result. $\qquad\square$

2.6. Birkhoff Interpolation

When we state an interpolation problem where not necessarily consecutive higher order derivatives are prescribed, then we call this *lacunary* (or Hermite-Birkhoff, or simply Birkhoff) interpolation. This type of interpolation was introduced by G. Birkhoff [5] at the beginning of the 19th century, but remained unnoticed for a long time. After the pioneering work of P. Turán in the late 50's, the theory of lacunary interpolation gained new momentum, and nowadays it is a fruitful field in approximation theory.

Since our subject is overconvergence, we will be concerned with one specific (although rather general) setup. Let $f(z) \in A_\rho$ ($\rho > 1$) and let

$$0 = m_0 < m_1 < \cdots < m_q, \quad \sum_{s=1}^{q} m_s = M_q \quad (q \ge 1) \tag{6.1}$$

be integers. For a given integer $n \geq 1$, we would like to construct polynomials $B_n(f; z)$ of degree at most $(q+1)n - 1$ such that

$$B_n^{(m_s)}(f; \omega^t) = f^{(m_s)}(\omega^t), \quad t = 1, \ldots, n; \quad s = 1, \ldots, q, \tag{6.2}$$

where ω is a primitive n^{th} root of unity.

In contrast to Hermite interpolation, the existence and uniqueness of such interpolation polynomials is highly non-trivial. Our first theorem says that for sufficiently large n's this is true.

THEOREM 8. *If*

$$n \geq \max_{1 \leq s \leq q} \frac{m_s}{s} \tag{6.3}$$

then the problem of $(0, m_1, \ldots, m_q)$ *interpolation is uniquely solvable for any* $f(z) \in A_\rho$.

The complete proof of this theorem is out of the scope of this monograph; at the appropriate places we will refer to the literature where the interested reader can complete the argument.

Sketch of proof. Consider the polynomials

$$p(z) = z^{\lambda n + k}, \quad k = 0, 1, \ldots, n - 1; \quad \lambda > q.$$

We look for the polynomial $B_n(p, z)$ in the form

$$B_n(p; z) = \sum_{j=0}^{q} \alpha_{j,\lambda,k,n} z^{jn+k} \in \pi_{(q+1)n-1} \tag{6.4}$$

with some numbers $\alpha_{j,\lambda,k,n}$. Conditions (6.2) for $f = p$ can be written as

$$B_n^{(m_s)}(p; \omega^t) = \sum_{j=0}^{q} \alpha_{j,\lambda,k,n}(jn + k)_{m_s} \omega^{t(k-m_s)} = (\lambda n + k)_{m_s} \omega^{t(k-m_s)} \tag{6.5}$$

$$(t = 0, 1, \ldots, n - 1; \quad s = 0, 1, \ldots, q)$$

where we use the notation $(a)_\nu = a(a-1) \ldots (a - \nu + 1)$ for non-negative integers a and ν. (Note that if $a < \nu$ then $(a)_\nu = 0$; further $(a)_0 = 1$.) This leads to the following system of linear equations:

$$\sum_{j=0}^{q} \alpha_{j,\lambda,k,n}(jn + k)_{m_s} = (\lambda n + k)_{m_s}, \quad s = 0, 1, \ldots, q. \tag{6.6}$$

The determinant of this system is

$$M_n(k) := \begin{vmatrix} 1 & 1 & \cdots & 1 \\ (k)_{m_1} & (n+k)_{m_1} & \cdots & (qn+k)_{m_1} \\ \vdots & \vdots & & \vdots \\ (k)_{m_q} & (n+k)_{m_q} & \cdots & (qn+k)_{m_q} \end{vmatrix}. \tag{6.7}$$

Under the condition (6.3), $M_n(k) \neq 0$ $(k = 0, \ldots, n-1)$. The interested reader may find a proof of this statement in [67, Corollary 4.18].

This means that $\alpha_{j,\lambda,k,n}$ are uniquely determined and (6.4) satisfies the conditions (6.5). (Note that if $\lambda \leq q$ then $B_n(p; z) \equiv p(z)$.) Now if

$$f(z) = \sum_{k=0}^{\infty} a_k z^k = \sum_{k=0}^{n-1} \sum_{\lambda=0}^{\infty} a_{\lambda n+k} z^{\lambda n+k} \tag{6.8}$$

then by the linearity of the operator B_n we obtain

$$B_n(f; z) = \sum_{k=0}^{(q+1)n-1} a_k z^k + \sum_{k=0}^{n-1} \sum_{\lambda=q+1}^{\infty} a_{\lambda n+k} \sum_{j=0}^{q} \alpha_{j,\lambda,k,n} z^{jn+k}. \tag{6.9}$$

This proves the existence of the operator B_n for any $f(z) \in A_\rho$. Uniqueness follows from the non-vanishing of (6.7) and the linearity of the operator.

Next we prove the convergence of the operator B_n.

THEOREM 9. *We have*

$$\overline{\lim_{n \to \infty}} \max_{|z| \in R} |f(z) - B_n(f; z)|^{\frac{1}{(q+1)n}} \leq \begin{cases} \frac{1}{\rho} & \text{if } 0 \leq R \leq 1, \\ \frac{R}{\rho} & \text{if } 1 \leq R < \rho \end{cases} \tag{6.10}$$

for any $f(z) \in A_\rho$.

The result shows that we have geometric convergence in every closed subdomain of $|z| < \rho$.

PROOF. We have to estimate the coefficients $\alpha_{j,\lambda,k,n}$ in (6.6) for sufficiently large n's. First we find a lower estimate for the determinant (6.7). With the notation $x = \frac{k}{n}$ we get

$$(jn+k)_{m_s} = n^{m_s}(x+j)\left(x+j-\frac{1}{n}\right)\cdots\left(x+j-\frac{m_s-1}{n}\right)$$
$$= n^{m_s}\left[(x+j)^{m_s} + O\left(\frac{1}{n}\right)\right] \qquad (j = 0, 1, \ldots, q; \; s = 1, \ldots, q)$$

where the constant involved in $O(1/n)$ depends only on q and $m_1, \ldots, m_q$. Hence for $M_n(k)$ we obtain

$$
M_n(k) = n^{M_q}
\begin{vmatrix}
1 & 1 & \cdots & 1 \\
x^{m_1} & (x+1)^{m_1} & \cdots & (x+q)^{m_1} \\
\vdots & \vdots & & \vdots \\
x^{m_q} & (x+1)^{m_q} & \cdots & (x+q)^{m_q}
\end{vmatrix}
+ O(n^{M_q-1}).
$$

Here the determinant is a so-called generalized Vandermonde determinant for $x \geq 0$ which is known to be postive (cf. e.g. [67, p. 51]). Hence

$$
c_1 n^{M_q} \leq M_n(k) \leq c_2 n^{M_q} \tag{6.11}
$$

where $c_1, c_2 > 0$ depend only on $q, m_1, \ldots, m_q$.

In order to estimate $\alpha_{j,\lambda,k,n}$ we apply Cramer's rule for the system (6.6). To do this, we have to establish lower and upper estimates for the determinant $M_{n,j,\lambda}(k)$ obtained from (6.7) by replacing the $(j + 1)$-st column by $[1 \ (\lambda n + k)_{m_1} \ldots (\lambda n+k)_{m_q}]^T$. Applying the same estimate to this determinant as above, we obtain

$$
M_{n,j,\lambda}(k) =
$$

$$
= n^{M_q}
\begin{vmatrix}
1 & \cdots & 1 & 1 & 1 & \cdots & 1 \\
x^{m_1} & \cdots & (x+j-1)^{m_1} & (x+\lambda)^{m_1} & (x+j+1)^{m_1} & \cdots & (x+q)^{m_1} \\
\vdots & & \vdots & \vdots & \vdots & & \vdots \\
x^{m_q} & \cdots & (x+j-1)^{m_q} & (x+\lambda)^{m_q} & (x+j+1)^{m_q} & \cdots & (x+q)^{m_q}
\end{vmatrix}
$$

$$
+ O(n^{M_q-1}) \qquad (j = 0, \ldots, q; \ k = 0, \ldots, n-1), \tag{6.12}
$$

where again $x = \frac{k}{n}$. Expanding the determinant by the $(j + 1)$-st column we get

$$
|M_{n,j,\lambda}(k)| \leq c_3 \lambda^{m_q} n^{M_q}, \quad j = 0, \ldots, q; \ k = 0, \ldots, n-1; \ \lambda > q. \tag{6.13}
$$

To obtain a lower estimate, we move the $(j + 1)$-st column to the last column. The resulting determinant (which is $(-1)^{q-j}$ times the original determinant) is again a generalized Vandermonde determinant which is positive for $0 \leq x \leq 1$, and its lower bound depends only on $\lambda, q, m_1, \ldots, m_q$. Thus

$$
|M_{n,j,\lambda}(k)| \geq c_4 n^{M_q}, \quad j = 0, \ldots, q; \quad k = 0, \ldots, n-1; \quad q < \lambda \leq q+\ell. \tag{6.14}
$$

Now (6.11), (6.13) and (6.14) imply

$$|\alpha_{j,\lambda,k,n}| = \frac{|M_{n,j,\lambda}(k)|}{M_n(k)} \begin{cases} \leq c_5 \lambda^{m_q}, & j = 0, \ldots, q; \ k = 0, \ldots, n-1; \ \lambda > q, \\ \geq c_6 > 0, & j = 0, \ldots, q; \ k = 0, \ldots, n-1; \\ & \quad q < \lambda \leq q + \ell. \end{cases}$$
(6.15)

After these preliminaries, we can prove the theorem. Taking the difference of (6.8) and (6.9), on using (6.15) we have

$$|f(z) - B_n(f; z)| = \Big| \sum_{k=0}^{n-1} \sum_{\lambda=q+1}^{\infty} a_{\lambda n+k} z^k \Big(1 - \sum_{j=0}^{q} \alpha_{j+\lambda,k,n} z^{jn}\Big)\Big|$$

$$= O\Big(\sum_{k=0}^{n-1} \sum_{\lambda=q+1}^{\infty} (\rho - \varepsilon)^{-\lambda n - k} |z|^k \max\left(1, |z|^{qn}\right) \lambda^{m_q} \Big)$$

with an arbitrary $\varepsilon > 0$. Now if $|z| < \rho - \varepsilon$ then this yields

$$O\Big(\max(1, |z|^{qn}) \sum_{\lambda=q+1}^{\infty} \frac{\lambda^{m_q}}{(\rho_\varepsilon)^{\lambda n}} \Big) = O\Big(\frac{\max(1, |z|^{qn})}{(\rho - \varepsilon)^{(q+1)n}} \Big)$$

which, since $\varepsilon > 0$ was arbitrary, results in (6.10). $\qquad\square$

After having settled the convergence, we now turn to the overconvergence problem. For this we need another operator

$$S_{n,\ell}(f; z) = \sum_{k=0}^{(q+1)n-1} a_k z^k + \sum_{k=0}^{n-1} \sum_{\lambda=q+1}^{q+\ell-1} a_{\lambda n+k} \sum_{j=0}^{q} \alpha_{j,\lambda,k,n} z^{jn+k}$$
(6.16)

which contains another fixed parameter $\ell \geq 1$. (In case $\ell = 1$, the second term in (6.16) does not appear, and $S_{n,1}$ is simply the Taylor section.) We will consider the difference of (6.9) and (6.16):

$$\Delta_{n,\ell}(f; z) := B_n(f; z) - S_{n,\ell}(f; z) = \sum_{k=0}^{n-1} \sum_{\lambda=q+\ell}^{\infty} a_{\lambda n+k} \sum_{j=0}^{q} \alpha_{j;\lambda,k,n} z^{jn+k}. \quad (6.17)$$

THEOREM 10. *We have for each* $f(z) \in A_\rho$

$$\overline{\lim_{n \to \infty}} \max_{|z|=R} |\Delta_{n,\ell}(f; z)|^{\frac{1}{(q+1)n}} = \begin{cases} \dfrac{1}{\rho^{\frac{q+\ell}{q+1}}} & if \ |R| \leq 1 \\[2mm] \dfrac{|z|^{\frac{q}{q+1}}}{\rho^{\frac{q+\ell}{q+1}}} & if \ 1 \leq |R| \leq \rho \\[2mm] \dfrac{|z|}{\rho^{1+\frac{\ell}{q+1}}} & if \ \rho \leq |R|. \end{cases}$$
(6.18)

The theorem shows that the exact radius of overconvergence is $\rho^{1+\frac{\ell}{q+1}}$.

PROOF. First we prove the upper estimates in (6.18). Just like in the proof of Theorem 2, we obtain from (6.17)

$$|\Delta_{n,\ell}(f;z)| = O\left(\sum_{k=0}^{n-1}\sum_{\lambda=q+\ell}^{\infty}(\rho-\varepsilon)^{-\lambda n-k}\lambda^{m_q}\max(1,|z|^{qn+k})\right)$$

$$= O\left((\rho-\varepsilon)^{-(q+\ell)n}\right)\sum_{k=0}^{n-1}(\rho-\varepsilon)^{-k}\max(1,|z|^{qn+k})$$

$$= \begin{cases} O\left((\rho-\varepsilon)^{-(q+\ell)n}\right) & \text{if } |z| \leq 1 \\[2mm] O\left(\dfrac{|z|^{q^n}}{(\rho-\varepsilon)^{(q+\ell)n}}\right) & \text{if } 1 \leq |z| \leq \rho \\[2mm] O\left(\dfrac{|z|^{(q+1)n}}{(\rho-\varepsilon)^{(q+\ell+1)n}}\right) & \text{if } \rho \leq |z| \end{cases}$$

which proves the upper estimate since $\varepsilon > 0$ is arbitrary.

Now we turn to proving the equality in (6.18). Let first $R \leq 1$, and suppose there exists $f(z) \in A_\rho$ and $0 < \delta < 1$ such that

$$\max_{|z|=R}|\Delta_{n,\ell}(f;z)| \leq \left(\frac{\delta}{\rho^{q+\ell}}\right)^n, \tag{6.19}$$

contrary to (6.18). Dividing (6.17) by z^{t+1}, $0 \leq t \leq q+\ell-1$, and integrating over $|z| = R \leq 1$ we get

$$\frac{1}{2\pi i}\int_{|z|=R}\frac{\Delta_{n,\ell}(f;z)}{z^{t+1}}dt = \alpha_{0,q+\ell,t,n}a_{(q+\ell)n+t} + \sum_{\lambda=q+\ell+1}^{\infty}\alpha_{0,\lambda,t,n}a_{\lambda n+t}.$$

Hence using (6.19), (6.15), as well as

$$|a_{\lambda n+t}| = O\left((\rho-\varepsilon)^{-\lambda n}\right) \tag{6.20}$$

with an arbitrary $\varepsilon > 0$ we get

$$a_{(q+\ell)n+t} = O\left(\left(\frac{\delta}{\rho^{q+\ell}}\right)^n + \sum_{\lambda=q+\ell+1}^{\infty}\frac{\lambda^{m_q}}{(\rho-\varepsilon)^{\lambda n}}\right) \tag{6.21}$$

$$= O\left(\left(\frac{\delta}{\rho^{q+\ell}}\right)^n + \frac{1}{(\rho-\varepsilon)^{(q+\ell+1)n}}\right) = O\left(\left(\frac{\delta}{\rho^{q+\ell}}\right)^n\right), \quad 0 \leq t \leq q+\ell-1,$$

provided $\varepsilon > 0$ is small enough. Since the numbers $(q+\ell)n+t$ with $n = 0,1,\ldots$ and $t = 0,\ldots,q+\ell-1$ exhaust all nonnegative integers, we conclude that

$$|a_n| = O\left(\left(\frac{\delta}{\rho}\right)^n\right), \quad 0 < \delta < 1 \tag{6.22}$$

which contradicts $f \in A_\rho$.

Now let $1 \le R \le \rho$, and suppose there exist $f \in A_\rho$ and $0 < \delta < 1$ such that

$$\max_{|z|=R} |\Delta_{n,\ell}(f; z)| \le \left(\frac{\delta R^q}{\rho^{q+\ell}}\right)^n, \tag{6.23}$$

contrary to (6.18). Dividing (6.17) by z^{qn+t+1}, $0 \le t \le q+\ell-1$, and integrating over $|z| = R$ we get

$$\frac{1}{2\pi i} \int_{|z|=R} \frac{\Delta_{n,\ell}(f; z)}{z^{qn+\ell+1}} \, dz = \alpha_{q,q+\ell,t,n} a_{(q+\ell)n+t} + \sum_{\lambda=q+\ell+1}^{\infty} \alpha_{q,\lambda,t,n} a_{\lambda n+t}.$$

Hence using (6.23), (6.15) and (6.20) we have the same estimate as in (6.21) and (6.22), again a contradiction.

Finally, let $\rho < R$, and suppose there exist $f \in A_\rho$ and $0 < \delta < 1$ such that

$$\max_{|z|=R} |\Delta_{n,\ell}(f; z)| \le \left(\frac{\delta R}{\rho^{q+\ell+1}}\right)^n, \tag{6.24}$$

contrary to (6.18). Dividing (6.17) by z^{n-t+1}, $0 \le t \le q + \ell$, and integrating over $|z| = R$ we get

$$\frac{1}{2\pi i} \int_{|z|=R} \frac{\Delta_{n,\ell}(f; z)}{z^{n-t+1}} dz = \alpha_{0,q+\ell,n-t,n} a_{(q+\ell+1)n-t}$$
$$+ \sum_{\lambda=q+\ell+1}^{\infty} \alpha_{0,\lambda,n-t,n} a_{(\lambda+1)n-t}.$$

Hence using (6.24), (6.15) and (6.20) we obtain

$$a_{(q+\ell+1)n-t} = O\left(\left(\frac{\delta}{\rho^{q+\ell+1}}\right)^n + \sum_{\lambda=q+\ell+1}^{\infty} \frac{\lambda^{m_q}}{(\rho-\varepsilon)^{(\lambda+1)n}}\right)$$
$$= O\left(\left(\frac{\delta}{\rho^{q+\ell+1}}\right)^n + \frac{1}{(\rho-\varepsilon)^{(q+\ell+2)n}}\right) = O\left(\left(\frac{\delta}{\rho^{q+\ell+1}}\right)^n\right), \quad 0 \le t \le q+\ell,$$

provided $\varepsilon > 0$ is small enough. Since the numbers $(q + \ell + 1)n - t$ with $n = 1, 2, \dots$ and $t = 0, \dots, q + \ell$ cover all positive integers, we conclude that (6.22) holds, a contradiction.   □

2.7. Historical Remarks

(a) Theorem 1 is due to Cavaretta, Sharma and Varga [30]. Theorems 2 and 3 are due to Akhlagi, Jakimovski and Sharma [2].

Theorem 4, even in the more general case when c is not necessarily a constant but $c = sn + O(1)$, $0 < s \leq 1/\ell$, was proved by Lou Yuanren [74]. It generalizes the special case (3.8) due to Ivanov and Sharma [50]. Theorem 5, when $\alpha = 1, \beta = r = 0$, reduces to a result of Rivlin [88] on ℓ_2-approximation on the mth roots of unity. The more general case $\alpha = 1$, $\beta = 0$, $r \geq 1$ was proved by Cavaretta, Sharma and Varga [30]. The special case $\alpha = 1$, $\beta = 0$ of Theorem 7 can be found in Cavaretta, Dikshit and Sharma [24].

Theorem 8 can be found in [67], [86] or [94]. Theorems 9 and 10 are new. Note that the results of Section 2.6 applied with $m_s = s$, $s = 1, \ldots, q$ lead to the results on Hermite interpolation in this chapter (where different methods were used).

(b) As in the case of Lagrange interpolation, it is possible to get overconvergence results for the Hermite interpolation on the roots of the Möbius transform mentioned in Section 1.6. Let $h_{n,p,r}^{\alpha}(f;z)$ be the Hermite interpolation polynomial of order p on the roots

$$z_{k,n}^{\alpha}(r) = \frac{rw_{k,n} - \alpha}{1 - rw_{k,n}}, \qquad k = 0, \ldots, n, \; 0 < \alpha < 1$$

of the polynomial

$$\omega_{n,r}^{\alpha}(z) := \frac{(z+\alpha)^{n+1} - r^{n+1}(1+\alpha z)^{n+1}}{1 - (r\alpha)^{n+1}},$$

where $w_{k,n}$ are the roots of $z^{n+1} - 1$. Sharma and Ziegler [99] proved that if $f(z)$ is analytic in $\{z \in \mathbf{C} : |z + \alpha| < R + \alpha\}$, $R > 1$, then for $0 < r, s < 1$

$$\lim_{n \to \infty} [h_{n,p,r}(f;z) - h_{n,p,s}(f;z)] = 0,$$

provided

$$z \in \left\{ z \in \mathbf{C} : |z + \alpha| < \frac{R + \alpha}{(\max(r,s))^{1/p}} \left(\frac{R + \alpha}{1 + R\alpha} \right)^{1/p} \right\}.$$

For further details refer to Brück [21].

CHAPTER 3

A GENERALIZATION OF THE TAYLOR SERIES TO RATIONAL FUNCTIONS AND WALSH EQUICONVERGENCE

In this chapter we consider least square approximation when polynomials are replaced by rational functions with given poles. Equiconvergence of the rational functions appearing in these considerations will also be investigated. Finally, we will consider a discrete analogue of the least square approximation for rational functions.

3.1. Rational Functions with a Minimizing Property

If $f(z) = \sum_{\nu=0}^{\infty} a_\nu z^\nu$ is analytic in and on the circle $\Gamma = \{z \mid |z| = 1\}$, then the polynomial $\sum_{\nu=0}^{n-1} a_\nu z^\nu$ is characterized by the property that it minimizes the least square difference

$$\oint_\Gamma |f(z) - p(z)|^2 |dz|, \quad \Gamma = \{z \mid |z| = 1\}$$

over all polynomials $p(z) \in \pi_{n-1}$. It also interpolates $f^{(\nu)}(z)$ $(\nu = 0, 1, \ldots, n-1)$ at the origin. We now give a simple generalization of the Taylor series and its two properties for rationals with given poles.

THEOREM 1. *Let $f(z)$ be analytic in and on Γ. If $\alpha_1, \ldots, \alpha_n$ are n given numbers, $|\alpha_k| > 1$ $(k = 1, \ldots, n)$, and $R_n(z)$ is the unique rational function of the form*

$$R_n(z) := \frac{Q_n(z)}{\prod_{k=1}^{n}(z - \alpha_k)}, \quad Q_n(z) \in \pi_n \tag{1.1}$$

which minimizes

$$\oint_\Gamma |f(z) - R_n(z)|^2 |dz| \tag{1.2}$$

over all $Q_n \in \pi_n$, then $R_n(z)$ interpolates $f(z)$ in the points $0, \frac{1}{\bar{\alpha}_1}, \ldots, \frac{1}{\bar{\alpha}_n}$.

REMARKS. If one of the factors $z - \alpha_k$ in the denominator has multiplicity λ_k, then the node $\frac{1}{\bar{\alpha}_k}$ has also to have the same multiplicity for interpolation.

If one of the poles α_k is infinite, then it is dropped out in (1.1). This shows that if for any integer $m \geq -1$, we consider rational functions of the form

$$Q_{n+m}(z)\Big/\prod_{k=1}^{n}(z-\alpha_k),$$

then

$$\min_{Q_{n+m}\in\pi_{n+m}} \oint_{\Gamma}\left|f(z) - \frac{Q_{n+m}(z)}{\prod\limits_{k=1}^{n}(z-\alpha_k)}\right|^2 |dz|$$

is attained when $Q_{n+m}(z)$ interpolates $f(z)\prod\limits_{k=1}^{n}(z-\alpha_k)$ at the nodes $0, \frac{1}{\bar{\alpha}_1}, \ldots,$
$\frac{1}{\bar{\alpha}_n}$.

PROOF OF THEOREM 1. Rational functions of the form (1.1) can be interpreted as linear combinations of the linearly independent functions

$$1, \frac{1}{z-\alpha_1}, \ldots, \frac{1}{z-\alpha_n}. \tag{1.3}$$

Taking proper linear combinations of these functions, we can obtain an orthogonal system $r_0(z) \equiv 1, \ldots, r_n(z)$ on Γ. By a theorem of Walsh [115, Ch. 6, Theorem 1], $R_n(z) = \sum\limits_{k=0}^{n} c_k r_k(z)$ will be the solution of the minimization problem (1.1) if and only if

$$\oint_{\Gamma} [f(z) - R_n(z)]\,\overline{r_k(z)}\,|dz| = 0 , \quad k = 0, \ldots, n. \tag{1.4}$$

It is well-known that each function in (1.3) can be expressed as a linear combination of $r_0, \ldots, r_n$. Therefore (1.4) is equivalent to

$$\oint_{\Gamma} [f(z) - R_n(z)]\,\frac{|dz|}{\bar{z}-\bar{\alpha}_k} = 0 , \quad k = 1, \ldots, n, \tag{1.5}$$

and

$$\oint_{\Gamma} [f(z) - R_n(z)]\,|dz| = 0. \tag{1.6}$$

We have to show that if $R_n(z)$ is the interpolating rational function of the form (1.1) to $f(z)$ at $0, \frac{1}{\bar{\alpha}_1}, \ldots, \frac{1}{\bar{\alpha}_n}$ then (1.5) is satisfied. But (1.5) is equivalent to

$$\oint_{\Gamma} \frac{f(z) - R_n(z)}{1 - \bar{\alpha}_k z}\,dz = 0 , \quad k = 1, \ldots, n ,$$

and this is true because of the interpolating property the integrand is an analytic function. (1.6) is also fulfilled because of the interpolation property at $z = 1$.

So far we assumed that the α_k's are pairwise different. The proof easily extends to the general case by using the corresponding Hermite interpolation polynomials.   □

From Theorem 1, it is easy to see that

$$f(z) - R_n(z) = \frac{1}{2\pi i} \oint_\Gamma f(t) K(t, z) dt \tag{1.7}$$

where

$$K(t, z) = \frac{z \prod_{\nu=1}^{n} (\bar{a}_\nu z - 1) \prod_{\nu=1}^{n} (t - a_\nu)}{t(t - z) \prod_{\nu=1}^{n} (z - a_\nu) \prod_{\nu=1}^{n} (\bar{a}_\nu t - 1)}. \tag{1.8}$$

If $f(z)$ is analytic in $|z| < \rho$, $\rho > 1$ and if the poles α_k are $\geq \sigma > 1$, then from (1.7) and (1.8) we can show that the sequence $\{R_n(z)\}$ converges to the limit $f(z)$ for

$$|z| < \frac{\sigma^2 \rho + \rho + 2\sigma}{2\sigma\rho + \sigma^2 + 1} \quad (> \min(\sigma, \rho)). \tag{1.9}$$

For taking the circle $\Gamma_{\rho'}$, with $1 < \rho' < \rho < \rho$, we have

$$\left| \frac{t - \alpha_k}{\bar{\alpha}_k t - 1} \right| < \frac{\rho' + \sigma}{1 + \rho'\sigma} \quad \text{and} \quad \left| \frac{\bar{\alpha}_k z - 1}{z - \alpha_k} \right| < \frac{\sigma R - 1}{\sigma - R}, \quad R = |z| < \sigma.$$

For n sufficiently large, and for $R < \rho'$, the kernel $K(t, z)$ will tend to zero if

$$\frac{\rho' + \sigma}{1 + \rho'\sigma} \cdot \frac{\sigma R - 1}{\sigma - R} < 1.$$

Simplifying this we get (1.9) on letting $\rho' \to \rho$.

If $\rho = \sigma$, then the right side in (1.7) becomes

$$\frac{\sigma^3 + 3\sigma}{3\sigma^2 + 1}.$$

If $f(z)$ is analytic at every finite point of the z-plane, then letting $\rho \to \infty$, we get the region of convergence to be

$$\frac{\sigma^2 + 1}{2\sigma}.$$

But if we let $\sigma \to \infty$, then the region of convergence is only the circle $|z| < \rho$ in which $f(z)$ is analytic and $R_n(z)$ is the Taylor section. Thus Theorem 1 is a generalization of Taylor's series.

If we take $f(z) = \frac{1}{z-\rho}$, and let $\alpha_1 = \alpha_2 = \cdots = \alpha_n = \sigma$, then

$$R_n(z) = \frac{-z(\rho-\sigma)^n(z-\frac{1}{\sigma})^n + \rho(\rho-\frac{1}{\sigma})^n(z-\sigma)^n}{\rho(\rho-\frac{1}{\sigma})^n(z-\sigma)^n(z-\rho)}$$

so that $(z-\sigma)^n R_n(z)$ is a polynomial of degree $\leq n$ which interpolates $\frac{1}{z-\rho}$ at 0 and at $\frac{1}{\sigma}$ (multiplicity n). This shows the sharpness of the bound (1.9), because

$$f(z) - R_n(z) = \frac{z(z-\frac{1}{\sigma})^n(\rho-\sigma)^n}{\rho(\rho-\frac{1}{\sigma})^n(z-\sigma)^n(z-\rho)} .$$

If z is on the unit circle, then $\left|\frac{z\sigma-1}{z-\sigma}\right| = 1$ and $f(z) - R_n(z)$ will tend to zero if

$$\frac{\rho-\sigma}{\rho\sigma - 1} < 1.$$

REMARK. In Theorem 1, if in (1.1), $Q_n(z) \in \pi_{n-1}$, then it can be seen easily that in order to minimize (1.2), the polynomial $Q_n(z)$ will interpolate $f(z)$ in the nodes $\frac{1}{\bar{\alpha}_1}, \ldots, \frac{1}{\bar{\alpha}_n}$.

3.2. Interpolation on roots of $z^n - \sigma^n$

We shall now consider a special case of Theorem 1 and take a rational function $r_{n+m,n}(z)$ of the form

$$r_{n+m,n}(z) = \frac{P_{n+m}(z)}{z^n - \sigma^n}, \quad \sigma > 0, \quad P_{n+m}(z) \in \pi_{n+m} \quad (m \geq -1) \qquad (2.1)$$

which minimizes the integral

$$\oint_\Gamma |f(z) - r(z)|^2 |dz|, \qquad (2.2)$$

$r(z)$ the form (2.1). In this case by Theorem 1 we know that $P_{n+m}(z)$ interpolates $f(z)(z^n - \sigma^n)$ in the zeroes of $z^{m+1}(z^n - \sigma^{-n})$ and we can write

$$f(z) - r_{n+m,n}(z) = \frac{1}{2\pi i} \oint_{\Gamma_\tau} \frac{z^{m+1}(z^n - \sigma^{-n})(t^n - \sigma^n)f(t)}{(z^n - \sigma^n)(t^n - \sigma^{-n})t^{m+1}(t-z)} dt \qquad (2.3)$$

where Γ_τ is the circle $|t| = \tau$, $1 < \tau < \rho$ and $f(z) \in A_\rho$.

If $m = -1$, it follows from the remark in Section 3.1, that $r_{n-1,n}(z)$ will interpolate $f(z)(z^n - \sigma^n)$ in the zeros of $(z^n - \sigma^{-n})$. When $m = -\mu$, $2 \leq \mu \leq n$, then we shall show that there is a close connection between the rationals $r_{n-\mu,n}(z)$ and $r_{n-1,n}(z)$. More precisely, we have

LEMMA 1. *Let $\rho > 1$, $\sigma > 1$ and let μ be a fixed integer, $2 \le \mu \le n$. Let $P_{n-\mu,n}(z)$ denote the polynomial $\in \pi_{n-\mu}$ which attains the minimum*

$$\min_{Q \in \pi_{n-\mu}} \oint_{\Gamma} \left| f(z) - \frac{Q(z)}{z^n - \sigma^n} \right|^2 |dz|, \quad f(z) \in A_\rho. \tag{2.4}$$

Then $P_{n-\mu,n}(z)$ is obtained by truncating the polynomial $P_{n-1,n}(z)$. Indeed, we have

$$P_{n-\mu,n}(z) = \frac{1}{2\pi i} \oint_{\Gamma_\tau} \frac{f(t) t^{\mu-1}(t^{n-\mu+1} - z^{n-\mu+1})(-\sigma^n)}{(t-z)(t^n - \sigma^{-n})} \, dt, \quad 1 < \tau < \rho. \tag{2.5}$$

PROOF. The problem of minimizing (2.4) is equivalent to the problem of finding the following minimum

$$\min_{a_j} \oint_{\Gamma} \left| f(z) - \sum_{j=1}^{n-\mu} a_j f_j(z) \right|^2 |dz| \tag{2.6}$$

where $f_j(z) = z^j/(z^n - \sigma^n)$. This minimum will be attained if

$$\oint_{\Gamma} \left\{ f(z) - \sum_{j=0}^{n-\mu} a_j f_j(z) \, \overline{f_\ell(z)} \right\} |dz| = 0, \quad \ell = 0, \dots, n - \mu. \tag{2.7}$$

Since

$$f_\ell(z) = \frac{1}{n} \sum_{k=0}^{n-1} \frac{1}{\sigma^{n-\ell-1}} \frac{\omega^{k\ell+k}}{z - \sigma\omega^k}, \quad \omega^n = 1,$$

and since

$$\frac{1}{2\pi i} \oint_{\Gamma} \frac{f(z)}{z - \sigma\omega^{-k}} |dz| = \frac{1}{2\pi i} \oint_{\Gamma} \frac{f(z)}{z - \sigma\omega^{-k}} \frac{dz}{iz}$$

$$= \frac{1}{2\pi} \oint_{\Gamma} \frac{f(z)}{z - \frac{\omega^k}{\sigma}} \, dz \, \frac{1}{\sigma\omega^{-k}}$$

$$= i \frac{\omega^k}{\sigma} f\left(\frac{\omega^k}{\sigma}\right),$$

it follows that

$$\frac{1}{2\pi i} \oint_{\Gamma} f(z) \, \overline{f_\ell(z)} \, |dz| = \frac{1}{n} \sum_{k=0}^{n-1} \frac{1}{\sigma^{n-\ell-1}} \oint_{\Gamma} \frac{f(z) \omega^{-k\ell-k}}{z - \sigma\omega^{-k}} |dz|$$

$$= \frac{i}{n\sigma^{n-\ell}} \sum_{k=0}^{n-1} f\left(\frac{\omega^k}{\sigma}\right) \omega^{-k\ell}.$$

Similarly

$$
\frac{1}{2\pi i} \oint_\Gamma f_j(z)\, \overline{f_\ell(z)}\, |dz| = \frac{i}{n\sigma^{n-\ell}} \sum_{k=0}^{n-1} \frac{(\frac{\omega^k}{\sigma})^j \omega^{-k\ell}}{\frac{1}{\sigma^n} - \sigma^n}
$$

$$
= \frac{i}{n\sigma^{n-\ell}} \frac{\sigma^{-j}}{(\sigma^{-n} - \sigma^n)} \sum_{k=0}^{n-1} \omega^{k(j-\ell)}
$$

$$
= \begin{cases} \frac{i}{(1-\sigma^{2n})}, & j = \ell \\ 0, & j \neq \ell. \end{cases}
$$

So from (2.7) we get

$$
\frac{a_\ell}{1-\sigma^{2n}} = \frac{1}{n\sigma^{n-\ell}} \sum_{k=0}^{n-1} f\left(\frac{\omega^k}{\sigma}\right) w^{-k\ell} = \frac{1}{n\sigma^{n-\ell}} \frac{1}{2\pi i} \oint_\Gamma \sum_{k=0}^{n-1} \frac{f(t)}{t - \frac{\omega^k}{\sigma}} w^{-k\ell} dt
$$

$$
= \frac{1}{n\sigma^{n-\ell}} \frac{1}{2\pi i} \oint_\Gamma f(t) \sum_{k=0}^{n-1} \frac{\omega^{-k\ell}}{t - \frac{\omega^k}{\sigma}} .
$$

Since

$$
\frac{t^{n-1-\ell}}{t^n - \sigma^{-n}} = \frac{1}{n} \sum_{k=0}^{n-1} \frac{(\frac{\omega^k}{\sigma})^{n-1-\ell}}{(\frac{\omega^k}{\sigma})^{n-1}(t - \frac{\omega^k}{\sigma})} = \frac{\sigma^\ell}{n} \sum_{k=0}^{n-1} \frac{\omega^{-k\ell}}{t - \frac{\omega^k}{\sigma}} ,
$$

so

$$
\frac{a_\ell}{1-\sigma^{2n}} = \frac{1}{\sigma^n} \frac{1}{2\pi i} \oint_\Gamma \frac{f(t) t^{n-1-\ell}}{t^n - \sigma^{-n}} \, dt. \tag{2.8}
$$

Then the polynomial

$$
P_{n-\mu,n}(z) = \sum_{\ell=0}^{n-\mu} a_\ell z^\ell
$$

$$
= \frac{\sigma^{-n} - \sigma^n}{2\pi i} \oint_\Gamma \frac{f(t) t^{\mu-1}(t^{n-\mu+1} - z^{n-\mu+1})}{(t - z)(t^n - \sigma^{-n})} \, dt.
$$

Writing $\sigma^{-n} - \sigma^n = \sigma^{-n} - t^n + t^n - \sigma^n$ and splitting the above into two integrals, we get (2.5). From (2.3)

$$
r_{n-1,n}(t) = \frac{1}{2\pi i} \oint_\Gamma \frac{(\sigma^n - \sigma^{-n})(z^n - t^n)f(t)}{(z^n - \sigma^n)(t^n - \sigma^{-n})(t - z)} \, dt
$$

and it is easily seen that (2.5) is the truncation of this.   □

REMARK. If we find the polynomial $P_{n-\mu,n}(z) \in \pi_{n-\mu}$, $2 \leq \mu \leq n$ which minimizes

$$
\min_{Q \in \pi_{n-\mu}} \oint_\Gamma \left| f(z) - \frac{Q(z)}{\omega(z)} \right|^2 |dz|, \quad \omega(z) = \prod_{j=1}^{n} |z - \alpha_j| \tag{2.9}
$$

then the polynomial $P_{n-\mu,n}(z) = \sum\limits_{j=0}^{n-\mu} a_j f_j(z)$ is easily obtained by solving the equations

$$\sum_{j=0}^{n-\mu} a_j \oint_\Gamma f_j(z)\,\overline{f_\ell(z)}\,|dz| = \oint_\Gamma f(z)\,\overline{f_\ell(z)}\,|dz|,$$

where we have set $f_\ell(z) = z^\ell/\omega(z)$ ($\ell = 0,1,\ldots,n-\mu$). It can be seen by a simple example, that the interesting property of Lemma 1 does not hold in general when $z^n - \sigma^n$ is replaced by $\omega(z) = \prod\limits_{\nu=1}^{n}(z-\alpha_\nu)$, $|\alpha_\nu| > 1$. In other words, it is not true in general that $P_{n-\mu,n}(z)$ is obtained by truncating $P_{n-1,n}(z)$. Thus for example

$$\min_{a,b} \oint_\Gamma \left| f(z) - \frac{a}{z-\alpha_1} - \frac{b}{z-\alpha_2} \right|^2 |dz| \qquad (2.10)$$

is attained when

$$R_2(z) = \begin{vmatrix} \frac{1}{z-\alpha_1} & \frac{1}{z-\alpha_2} & 0 \\ \frac{1}{\frac{1}{\overline{\alpha_1}}-\alpha_1} & \frac{1}{\frac{1}{\overline{\alpha_1}}-\alpha_2} & f(\frac{1}{\overline{\alpha_1}}) \\ \frac{1}{\frac{1}{\overline{\alpha_2}}-\alpha_1} & \frac{1}{\frac{1}{\overline{\alpha_2}}-\alpha_2} & \frac{1}{\overline{\alpha_2}}f(\frac{1}{\overline{\alpha_2}}) \end{vmatrix} \div \begin{vmatrix} \frac{1}{\frac{1}{\overline{\alpha_1}}-\alpha_1} & \frac{1}{\frac{1}{\overline{\alpha_1}}-\alpha_2} \\ \frac{1}{\frac{1}{\overline{\alpha_2}}-\alpha_1} & \frac{1}{\frac{1}{\overline{\alpha_2}}-\alpha_2} \end{vmatrix}. \qquad (2.11)$$

Similarly,

$$\min_a \oint_\Gamma \left| f(z) - \frac{a}{(z-\alpha_1)(z-\alpha_2)} \right|^2 |dz|$$

is attained when

$$a = \oint_\Gamma \frac{f(z)}{(\overline{z}-\overline{\alpha}_1)(\overline{z}-\overline{\alpha}_2)}\,|dz| \Big/ \oint_\Gamma \frac{1}{(z-\alpha_1)(z-\alpha_2)} \cdot \frac{1}{(\overline{z}-\overline{\alpha}_1)(\overline{z}-\overline{\alpha}_2)}\,|dz|.$$

Simple calculation shows that

$$a = \frac{\frac{1}{\overline{\alpha}_1}f(\frac{1}{\overline{\alpha}_1}) - \frac{1}{\overline{\alpha}_2}f(\frac{1}{\overline{\alpha}_2})}{\frac{1}{\overline{\alpha}_1(\frac{1}{\overline{\alpha}_1}-\alpha_1)(\frac{1}{\overline{\alpha}_1}-\alpha_2)} - \frac{1}{\overline{\alpha}_2(\frac{1}{\overline{\alpha}_2}-\alpha_1)(\frac{1}{\overline{\alpha}_2}-\alpha_2)}}.$$

The value of a obtained from (2.11) on putting $z = 0$ is different from the value of a above except when $a_1 = \sigma$ and $a_2 = -\sigma$.

3.3. Equiconvergence of $R^\alpha_{n+m,n}(z)$ and $r_{n+m,n}(z)$ for $m \geq -1$

Let $R^\alpha_{n+m,n}(z)$ denote the rational function $\frac{B_{n+m}(z)}{z^n-\sigma^n}$, $B_{n+m}(z) \in \pi_{n+m}$, which interpolates $(z^n - \sigma^n)f(z) \in A_\rho$ in the zeros of $z^{n+m+1} - \alpha^{n+m+1}$ where $0 \neq \alpha \in D_\tau$, $\tau = \min(\sigma,\rho)$. Then we can write

$$f(z) - R^\alpha_{n+m,n}(z) = \frac{1}{2\pi i} \oint_{\Gamma_\tau} \frac{(z^{n+m+1}-\alpha^{n+m+1})(t^n-\sigma^n)f(t)}{(z^n-\sigma^n)(t^{n+m+1}-\alpha^{n+m+1})(t-z)}\,dt \qquad (3.1)$$

where $1 < \tau < \rho$ and $|z| < \tau$. From (3.1), it follows that if $K = \{z : |z| < \tau\}$, then for m fixed

$$\varlimsup_{n \to \infty} |f(z) - R^\alpha_{n+m,n}(z)|^{1/n} \leq \frac{1}{\tau} \max \{1, |z|\} < 1, \quad z \in K.$$

From (3.1), we get

$$R^\alpha_{n+m,n}(z) = \frac{1}{2\pi i} \oint_{\Gamma_\tau} \frac{(t^{n+m+1} - z^{n+m+1})(t^n - \sigma^n)}{(t^{n+m+1} - \alpha^{n+m+1})(z^n - \sigma^n)(t-z)} f(t) \, dt \qquad (3.2)$$

so that if $\rho > \sigma$, and $f(z) = \sum\limits_{k=0}^{\infty} a_k z^k$, then for $|z| > \sigma$, we have

$$\lim_{n \to \infty} R^\alpha_{n+m,n}(z) = \begin{cases} 0 & \text{if} \quad m = -1 \\ \frac{1}{2\pi i} \oint_{\Gamma_\tau} \frac{t^{m+1} - z^{m+1}}{t-z} \cdot \frac{f(t)}{t^{m+1}} \, dt = \sum\limits_{k=0}^{m} a_k z^k, & m > -1. \end{cases}$$
$$(3.3)$$

The following theorem shows that the difference $R^\alpha_{n+m,n}(z) - r_{n+m,n}(z)$ converges to zero in a larger region. The proof (in a more general setting; see Theorem 3 or 4) will be given later.

THEOREM 2. *Let $\rho, \sigma > 1$ and let $|\alpha| < \min(\rho, \sigma)$. If $f \in A_\rho$, and if*

$$\Delta^\alpha_{n+m,n}(z; f) := R^\alpha_{n+m,n}(z) - r_{n+m,n}(z), \qquad (3.4)$$

then

$$\lim_{n \to \infty} \Delta_{n+m,n}(z; f) = 0 \begin{cases} \forall \, |z| < \rho_1, & \text{if} \quad \sigma \geq \rho_1 \\ \\ \forall \, |z| \neq \sigma & \text{if} \quad \sigma < \rho_1 \end{cases} \qquad (3.5)$$

where $\rho_1 := \rho^2/|\alpha|$.

For fixed integers $m \geq -1$, $n > 1$, set $N = n+m+1$, let $\frac{1}{\sigma} < |\alpha| < \min(\sigma, \rho)$, and

$$\alpha_{n,m}(z) := \alpha^{n+m+1} - z^{m+1}\sigma^{-n}, \quad \beta_{n,m}(z) := z^{m+1}(z^n - \sigma^{-n}). \qquad (3.6)$$

Let $S_{N,\nu}(z) \in \pi_{(\nu+1)N-1}$ which interpolates $(\alpha_{n,m}(z))^\nu (z^n - \sigma^n) f(z)$ in the Hermite sense in the zeros of $(\beta_{n,m}(z))^{\nu+1}$. We shall now prove

LEMMA 2. *If $f(z) \in A_\rho$, then for each n sufficiently large,*

$$\lim_{\nu \to \infty} \frac{S_{N,\nu}(z)}{(\alpha_{n,m}(z))^\nu} = (z^n - \sigma^n) f(z) \qquad (3.7)$$

uniformly in $|z| \leq 1$. *Also*

$$S_{N,\nu}(z) - \alpha_{n,m}(z)S_{N,\nu-1}(z) = \left(\beta_{n,m}(z)\right)^\nu P^\alpha_{n+m}(z;\nu), \quad \nu = 1, 2, \dots \quad (3.8)$$

where $P^\alpha_{n+m}(z;\nu) \in \pi_{N-1}$. *Consequently for* $|z| \leq 1$,

$$(z^n - \sigma^n)f(z) = \sum_{\nu=0}^\infty \left(\frac{\beta_{n,m}(z)}{\alpha_{n,m}(z)}\right)^\nu P^\alpha_{n+m}(z;\nu). \quad (3.9)$$

REMARK. When $\nu = 0$, $P^\alpha_{n+m}(z;0)$ interpolates $(z^n - \sigma^n)f(z)$ in the zeros of $\beta_{n,m}(z)$ and so from the polynomial $P^\alpha_{n+m}(z;0)$ we get

$$r_{n+m,n}(z) := \frac{P^\alpha_{n+m}(z;0)}{z^n - \sigma^n}$$

and $P^\alpha_{n+m}(z;1)$ interpolates $\quad \dfrac{(z^n-\sigma^n)f(z)-P^\alpha_{n+m}(z,0)}{\beta_{n,m}(z)} \alpha_{n,m}(z) \quad$ in the zeros of $\beta_{n,m}(z)$.

PROOF OF LEMMA 2. From the interpolation properties of $S_{N,\nu}(z)$, we see that $S_{N,\nu-1}(z)$ interpolates $\quad (z^n - \sigma^n)f(z)(\alpha_{n,m}(z))^{\nu-1} \quad$ in the zeros of $\left(\beta_{n,m}(z)\right)^\nu$, while $S_{N,\nu}(z)$ interpolates $(z^n - \sigma^n)f(z)\left(\alpha_{n,m}(z)\right)^\nu$ in the zeros of $\left(\beta_{n,m}(z)\right)^{\nu+1}$. It follows that

$$S_{n,\nu}(z) - \alpha_{n,m}(z)S_{n,\nu-1}(z) = \left(\beta_{n,m}(z)\right)^\nu P^\alpha_{n+m}(z,\nu)$$

where $P^\alpha_{n+m}(z;\nu)$ is a polynomial of degree $\leq N(\nu+1) - 1 - (N\nu - 1 + m + 1) = n + m$. This proves (3.8).

In order to prove (3.7) we observe, from the definition of $S_{N,\nu}(z)$, that

$$S_{N,\nu}(z) = \frac{1}{2\pi i} \oint_{\Gamma_\tau} \frac{\left(\alpha_{n,m}(t)\right)^\nu (t^n - \sigma^n)f(t)}{(t-z)\left(\beta_{n,m}(t)\right)^{\nu+1}} \{\left(\beta_{n,m}(t)\right)^{\nu+1} - \left(\beta_{n,m}(z)\right)^{\nu+1}\}dt \quad (3.10)$$

so that

$$E_\nu(z) := (z^n - \sigma^n)f(z) - \frac{S_{N,\nu}(z)}{\left(\alpha_{n,m}(z)\right)^\nu}$$

$$= \frac{1}{2\pi i} \oint_{\Gamma_\tau} \left\{\frac{\beta_{n,m}(z)}{\beta_{n,m}(t)}\right\}^{\nu+1} \cdot \left\{\frac{\alpha_{n,m}(t)}{\alpha_{n,m}(z)}\right\}^\nu \frac{(t^n - \sigma^n)f(t)}{t-z} dt$$

where $|\alpha| < \tau < \rho$. Then it follows that

$$\overline{\lim_{\nu \to \infty}} \|E_\nu(z)\|^{1/\nu}_{|z|\leq 1} < \frac{(1+\sigma^{-n})(|\alpha|^{n+m+1} + \tau^{m+1}\sigma^{-n})}{\tau^{m+1}(\tau^n - \sigma^{-n})(1 - \sigma^{-n})} < 1$$

for n sufficiently large.

(3.9) follows from (3.7) and (3.8). $\qquad\square$

Since $B^\alpha_{n+m}(z) \in \pi_{n+m}$ and it interpolates $(z^n - \sigma^n)f(z)$ in the zeros of $z^{n+m+1} - \alpha^{n+m+1}$, we would like to express $B^\alpha_{n+m}(z)$ in terms of the polynomials $P^\alpha_{n+m}(z;\nu)$. This is given by

COROLLARY 1. *For $n > n_0\,(m, \rho, \sigma)$, we have*

$$B_{n+m}^{\alpha}(z) = \sum_{\nu=0}^{\infty} P_{n+m}^{\alpha}(z; \nu). \qquad (3.11)$$

PROOF. If $\omega^{n+m+1} = 1$, then from (3.8)

$$B_{n+m}^{\alpha}(\alpha\omega) = (z^n - \sigma^n)f(z)]_{z=\alpha\omega}$$

$$= \sum_{\nu=0}^{\infty} \left\{ \frac{\beta_{n,m}(\alpha\omega)}{\alpha_{n,m}(\alpha\omega)} \right\}^{\nu} P_{n+m}^{\alpha}(\alpha\omega; \nu) = \sum_{\nu=0}^{\infty} P_{n+m}^{\alpha}(\alpha\omega; \nu)$$

since $z^{n+m+1} - \alpha^{n+m+1} = \beta_{n,m}(z) - \alpha_{n,m}(z)$ vanishes for $z = \alpha\omega$ and so $\beta_{n,m}(\alpha\omega) = \alpha_{n,m}(\alpha\omega)$.

(3.11) follows from the uniqueness of Lagrange interpolation. □

We set $r_{n+m,n}^{\alpha}(z, \nu) := \frac{P_{n+m}^{\alpha}(z;\nu)}{z^n - \sigma^n}$, and for any integer $\ell \geq 1$, put

$$\Delta_{n+m,n}^{\alpha,\ell}(z; f) := R_{n+m,n}^{\alpha}(z) - \sum_{\nu=0}^{\ell-1} r_{n+m,n}^{\alpha}(z; \nu).$$

We can now prove

THEOREM 3. *Let $\rho > 1$, $\sigma > 1$ and an integer $m \geq -1$, be given. If $f(z) \in A_\rho$ and if for any given integer $\ell \geq 1$, and $|\alpha| < \min\,(\rho, \sigma)$ we put*

$$\Delta_{n+m,n}^{\alpha,\ell}(f; z) := R_{n+m,n}^{\alpha}(z) - \sum_{\nu=0}^{\ell-1} r_{n+m,n}^{\alpha}(z; \nu)$$

then

$$\lim_{n\to\infty} \Delta_{n+m,n}^{\alpha}(f; z) = 0 \quad \begin{cases} for & |z| < \rho_\ell, & if & \sigma \geq \rho_\ell \\ for & |z| \neq \sigma & if & \sigma < \rho_\ell \end{cases} \qquad (3.12)$$

where

$$\rho_\ell := \rho^{\ell+1}/(\max\,(|\alpha|, \sigma^{-1}))^{\ell}.$$

The convergence in (3.12) is uniform and geometric on compact subsets of the above regions. Moreover, the result is sharp.

PROOF. From (3.11), we see that

$$\Delta_{n+m,n}^{\alpha,\ell}(f; z) = \sum_{\nu=\ell}^{\infty} \frac{P_{n+m}^{\alpha}(z; \nu)}{z^n - \sigma^n}. \qquad (3.13)$$

In order to obtain an integral representation for $\Delta_{n+m,n}^{\alpha,\ell}(f;z)$, we observe that by definition

$$S_{N,\nu}(z) = \frac{1}{2\pi i} \oint_{\Gamma_\tau} \frac{(t^n - \sigma^n)f(t)\big(\alpha_{n,m}(t)\big)^\nu \{\big(\beta_{n,m}(t)\big)^{\nu+1} - \big(\beta_{n,m}(z)\big)^{\nu+1}\}}{(t-z)\big(\beta_{n,m}(t)\big)^{\nu+1}} \, dt$$

$$(3.14)$$

where $\tau < \min(\rho,\sigma)$, so that from (3.7), we have

$$P_{n+m}^\alpha(z;\nu) = \frac{S_{N,\nu}(z) - \alpha_{n,m}(z)S_{n,\nu-1}(z)}{\big(\beta_{n,m}(z)\big)^\nu} .$$

Using (3.14) and observing that

$$\frac{1}{2\pi i} \oint_{\Gamma_\tau} f(t)(t^n - \sigma^n)(\alpha_{n,m}(t))^{\nu-1} \frac{\alpha_{n,m}(t) - \alpha_{n,m}(z)}{t - z} \, dt = 0$$

we see that

$$P_{n+m}^\alpha(z,\nu) = \frac{1}{2\pi i} \oint_{\Gamma_\tau} \frac{f(t)(t^n - \sigma^n)}{t - z} K_\nu(t,z) dt$$

where

$$K_\nu(t,z) = \frac{\alpha_{n,m}(z)\beta_{n,m}(t) - \alpha_{n,m}(t)\beta_{n,m}(z)}{\alpha_{n,m}(t)\beta_{n,m}(t)} \left\{ \frac{\alpha_{n,m}(t)}{\beta_{n,m}(t)} \right\}^\nu .$$

Hence from (3.13), we see that

$$\Delta_{m+n,n}^{\alpha,\ell}(f;z) = \frac{1}{2\pi i} \oint_{\Gamma_\tau} \frac{f(t)}{t - z} K(t,z) dt \qquad (3.15)$$

where for $|\alpha| < \min(\rho,\sigma)$ and n sufficiently large

$$K(t,z) = \frac{t^n - \sigma^n}{z^n - \sigma^n} \cdot \sum_{\nu=\ell}^\infty K_\nu(t,z) \qquad (3.16)$$

$$= \frac{t^n - \sigma^n}{z^n - \sigma^n} \cdot \frac{\alpha_{n,m}(z)\beta_{n,m}(t) - \alpha_{n,m}(t)\beta_{n,m}(z)}{\alpha_{n,m}(t)\beta_{n,m}(t)} \sum_{\nu=\ell}^\infty \left(\frac{\alpha_{n,m}(t)}{\beta_{n,m}(t)} \right)^\nu$$

$$= \frac{t^n - \sigma^n}{z^n - \sigma^n} \cdot \frac{\alpha_{n,m}(z)\beta_{n,m}(t) - \alpha_{n,m}(t)\beta_{n,m}(z)}{\alpha_{n,m}(t)\beta_{n,m}(t)} \left(\frac{\alpha_{n,m}(t)}{\beta_{n,m}(t)} \right)^\ell \times$$

$$\times \frac{\beta_{n,m}(t)}{t^{n+m+1} - \alpha^{n+m+1}}$$

$$= \frac{t^n - \sigma^n}{z^n - \sigma^n} \cdot \frac{\alpha_{n,m}(z)\beta_{n,m}(t) - \alpha_{n,m}(t)\beta_{n,m}(z)}{\beta_{n,m}(t)(t^{n+m+1} - \alpha^{n+m+1})} \left(\frac{\alpha_{n,m}(t)}{\beta_{n,m}(t)} \right)^{\ell-1} .$$

Denoting $\gamma := \min\left(|\alpha|, \sigma^{-1}\right)$, we obtain by (3.6)

$$|\alpha_{n,m}(z)| \le c\gamma^n, \quad |\beta_{n,m}(z)| \le c \max\left(|z|, \sigma^{-1}\right)^n \quad \text{for all} \quad z \in \mathbb{C},$$

and

$$|\beta_{n,m}(t)| \ge c|t|^n \quad \text{if} \quad |t| > \sigma^{-1}.$$

Hence using term-by-term estimates,

$$|\alpha_{n,m}(z)\beta_{n,m}(t) - \alpha_{n,m}(t)\beta_{n,m}(z)| \le c\gamma^n\left[\max\left(|t|, \sigma^{-1}\right)^n + \max\left(|z|, \sigma^{-1}\right)^n\right]$$
$$\le c\gamma^n \max\left(|z|, |t|\right) \quad \text{if} \quad |t| > \sigma^{-1}.$$

Thus from (3.16) we get

$$|K(t,z)| \le c \frac{\max\left(|t|, \sigma\right)^n}{\max\left(|z|, \sigma\right)^n} \frac{\gamma^n \max\left(|z|, |t|\right)^n}{|t|^n} \cdot \frac{\gamma^{n(\ell-1)}}{|t|^{n\ell}}$$
$$\le c \left[\frac{\max\left(|t|, \sigma\right) \max\left(|z|, |t|\right) \cdot \gamma^\ell}{\max\left(|z|, \sigma\right) \cdot |t|^{\ell+1}}\right]^n \quad (|z| \ne \sigma, |t| > 1). \tag{3.17}$$

Now if $\sigma \ge \frac{\rho^{\ell+1}}{\gamma^\ell}$ then for $|z| < \frac{\rho^{\ell+1}}{\gamma^\ell}$ we get $|K(t,z)| \le c\left[\frac{\sigma}{\sigma}|z|\frac{\gamma^\ell}{|t|^{\ell+1}}\right]^n \to 0$ if $|t| < \rho$ is close enough to ρ.

On the other hand, if $\sigma < \frac{\rho^{\ell+1}}{\gamma^\ell}$ then we distinguish two cases:

Case 1: $\rho \le \sigma$. Then from (3.17)

$$|K(t,z)| \le c \left[\frac{\max\left(|t|, \sigma\right)\gamma^\ell}{|t|^{\ell+1}}\right]^n \le c\left(\frac{\sigma\gamma^\ell}{|t|^{\ell+1}}\right)^n \to 0 \quad (n \to \infty)$$

again if $|t| < \rho$ and $|t|$ is close enough to ρ.

Case 2: $\rho > \sigma$. Then assuming $|t| > \sigma$ we get for $|z| \le |t|$

$$|K(t,z)| \le c \left[\frac{|t|^2 \cdot \gamma^\ell}{\sigma|t|^{\ell+1}}\right]^n < c\left(\frac{\sigma}{|t|}\right)^{(\ell-1)n} \to 0,$$

while for $|z| > |t|$

$$|K(t,z)| \le c \left[\frac{|t|\gamma^\ell}{|t|^{\ell+1}}\right]^n = c\left(\frac{\gamma}{|t|}\right)^{\ell n} \to 0$$

provided $|t| < \rho$ is close enough to ρ. This completely proves Theorem 3. □

A similar result can be proved if we set $B(z) := z^{\ell(m+1)}(z^{\ell n} - \beta^{\ell n})$, where $|\beta| < \rho$, $|\beta| \ne |\alpha|$ and where $L_{\ell(m+n)}(f_\sigma, B, z)$ denotes the Hermite-Lagrange interpolant of degree $\ell(m+n+1) - 1$ to the function $f_\sigma(z) := (z^n - \sigma^n)f(z)$ on the zeros of $B(z)$. Let $L_{m+n}(f_\sigma, \alpha, z)$ denote the Lagrange interpolant to f_σ at the zeros of $z^{m+n+1} - \alpha^{m+n+1}$. We can then prove

THEOREM 4. *Let $f(z) \in A_\rho$ ($\rho > 1$) and let $m \geq -1$ be a fixed integer. If $\alpha \neq \beta$ ($|\alpha|, |\beta| < \min(\rho, \sigma)$) are given and if*

$$\Delta_{m,n}^{\alpha,\beta}(f, z) := L_{m+n}(f_\sigma, \alpha, z) - L_{m+n}\big(L_{\ell(m+n)}(f_\sigma, \beta, z), \alpha, z\big)$$

then

$$\lim_{n \to \infty} \frac{\Delta_{m,n}^{\alpha,\beta}(f, z)}{z^n - \sigma^n} = 0$$

for $|z| < \sigma_1$ if $\sigma > \sigma_1 := \rho^{\ell+1} / \big(\max(|\alpha|, |\beta|)\big)^\ell$ and for $|z| \neq \sigma$ if $\sigma < \sigma_1$.

For $\ell = 1$, $\beta = \sigma^{-1}$, we get Theorem 2.

PROOF. Since

$$L_{\ell(m+n)}(f_\sigma, \beta, z) = \frac{1}{2\pi i} \oint_{\Gamma_\tau} f_\sigma(t) K(t, z) dt,$$

where $1 < \tau < \rho$ and $K(t, z) = \frac{B(t) - B(z)}{(t-z)B(t)}$, in order to find an integral representation for $\Delta_{m,n}^{\alpha,\beta}(f, z)$, it is enough to find the Lagrange interpolant of $K(t, z)$ in the zeros of $z^{m+n+1} - \alpha^{m+n+1}$.
Then we obtain

$$L_{m+n}\big(K(t, \cdot), \alpha, z\big) = \frac{1}{B(t)} \left[\frac{t^{\ell(m+n+1)} - \alpha^{\ell(m+n+1)}}{t^{m+n+1} - \alpha^{m+n+1}} \cdot \frac{t^{m+n+1} - z^{m+n+1}}{t - z} \right.$$

$$\left. - \beta^{\ell n} \frac{\big(t^{\ell(m+1)} - z^{\ell(m+1)}\big)}{t - z} \right].$$

Then

$$\frac{\Delta_{m,n}^{\alpha,\beta}(f, z)}{z^n - \sigma^n} = \frac{1}{2\pi i} \oint_{\Gamma_\tau} f_\sigma(t) \frac{K_1(z, t)}{z^n - \sigma^n} dt$$

where

$$K_1(t, z) = \frac{t^{m+n+1} - z^{m+n+1}}{(t - z)(t^{m+n+1} - \alpha^{m+n+1})} - L_{m+n}\big(K(t, z), \alpha, z\big)$$

$$= \frac{t^{m+n+1} - z^{m+n+1}}{(t - z)(t^{m+n+1} - \alpha^{m+n+1})} \cdot \frac{\alpha^{\ell(m+n+1)} - t^{\ell(m+1)}\beta^{\ell n}}{B(t)}$$

$$+ \frac{\beta^{\ell n}\big(t^{\ell(m+1)} - z^{\ell(m+1)}\big)}{(t - z)B(t)}.$$

Hence with the notation $\gamma := \max(|\alpha|, |\beta|)$ we obtain

$$\left| \frac{t^n - \sigma^n}{z^n - \sigma^n} K_1(t, z) \right| \leq c \left\{ \frac{\max(|t|, \sigma)}{\max(|z|, \sigma)} \left[\frac{\max(|t|, |z|)\gamma^\ell}{|t|^{\ell+1}} + \frac{|\beta|^\ell}{|t|^\ell} \right] \right\}^n$$

$$\leq c \left[\frac{\max(|t|, \sigma) \max(|t|, |z|)\gamma^\ell}{\max(|z|, \sigma)|t|^{\ell+1}} \right]^n \quad (|z| \neq \sigma).$$

The rest of the analysis is essentially the same as in Theorem 3.   □

So far we considered $m \geq -1$. We shall now show that if $m = -\mu$, $\mu \geq 2$, then also equiconvergence holds. We can prove

THEOREM 5. *Let ρ, $\sigma > 1$ and let $\mu \geq 2$ be fixed. If $f \in A_\rho$, suppose the rational function $R_{n-\mu,n}^\alpha(z) = B_{n-\mu}(z)/(z^n - \sigma^n)$ interpolates $f(z)$ in the zeros of $z^{n-\mu+1} - \alpha^{n-\mu+1}$ and the rational function $r_{n-\mu,n}(z)$ with the denominator $z^n - \sigma^n$, minimizes*

$$\oint_\Gamma |f(z) - r_{n-\mu,n}(z)|^2 |dz|.$$

Then

$$\lim_{n\to\infty} \left(R_{n-\mu,n}(z) - r_{n-\mu,n}(z)\right) = 0 \quad \begin{cases} \forall\, |z| < \rho_1, & if\ \ \sigma > \rho_1 \\ \forall\, |z| \neq \sigma & if\ \ \sigma < \rho_1 \end{cases}$$

where $\rho_1 := \rho^2/\max(|\alpha|, \sigma^{-1})$.

PROOF. In this case, we see that using (2.5) and (3.2)

$$R_{n-\mu,n}(z) - r_{n-\mu,n}(z) =$$
$$= \frac{1}{2\pi i} \oint_{\Gamma_\tau} \frac{f(t)(t^n - \sigma^n)(t^{n-\mu+1} - z^{n-\mu+1})(t^{\mu-1}\alpha^{n-\mu+1} - \sigma^{-n})}{(z^n - \sigma^n)(t - z)(t^n - \sigma^{-n})(t^{n-\mu+1} - \alpha^{n-\mu+1})} dt$$

and the result follows as in the proof of Theorem 4.   □

In order to bring out the sharpness of the result we take $\hat{f}(z) = \frac{1}{z-\rho}$. Then for $n \geq 2(\mu - 1)$, we can verify that

$$B_{n-\mu,n}(\hat{f}, z) = \frac{(z^{n-\mu+1} - \alpha^{n-\mu+1})(\rho^{\mu-1}\alpha^{n-\mu+1} - \sigma^n)}{(\rho - z)(\rho^{n-\mu+1} - \alpha^{n-\mu+1})} + \frac{\sigma^n - z^{\mu-1}\alpha^{n-\mu+1}}{\rho - z}$$

and

$$P_{n-\mu,n}(\hat{f}, z) = \frac{(\sigma^n - \sigma^{-n})\rho^{\mu-1}}{\rho^n - \sigma^{-n}} \cdot \frac{(\rho^{n-\mu+1} - z^{n-\mu+1})}{\rho - z} .$$

From the difference of $B_{n-\mu,n}(\hat{f}, z) - P_{n-\mu,n}(\hat{f}, z)$, we can verify that when $\sigma > \rho_1$ then for $z_0 = \rho_1 = \frac{\rho^2}{\max(|\alpha|, \sigma^{-1})}$, we get

$$\lim_{n\to\infty} \left| \frac{B_{n-\mu,n}(\hat{f}, z_0) - P_{n-\mu,n}(\hat{f}, z_0)}{z_0^n - \sigma^n} \right| = \frac{1}{\rho - \rho_1} = \frac{1}{\rho - \rho^2/(\max(|\alpha|, \sigma^{-1}))} ,$$

and when $\sigma = \frac{\rho^2}{|\alpha|}$ then

$$\lim_{n\to\infty} \left| \frac{B_{n-\mu,n}(\hat{f}, z) - P_{n-\mu,n}(\hat{f}, z)}{z^n - \sigma^n} \right| = \frac{|z|^{\mu-1}}{|\rho - z|} \quad (|z| > \sigma).$$

3.4. Hermite Interpolation

We shall see how some of the above theorems hold also for Hermite interpolation. For fixed integers m, r, s $(m \geq -1,\ 1 \leq r \leq s)$ and $N = n + m + 1$, let $R_{sN-1}(f, \alpha, z) := C_{sN-1}(f, \alpha, z)/(z^n - \sigma^n)^r$, where $C_{sN-1}(f, z)$ is a polynomial which interpolates $(z^n - \sigma^n)^r f(z)$ in the zeros of $(z^N - \alpha^N)^s$, where $|\alpha| < \rho$ and $f(z) \in A_\rho$. Suppose $S_{sN-1}(f, \beta, z) = Q_{sN-1}(f, \beta, z)/(z^n - \sigma^n)^s$ where $Q_{sN-1}(f, \beta, z)$ interpolates $f(z)(z^n - \sigma^n)^r$ in the zeros of $z^{sN-rn}(z^n - \beta^n)^r$, where $\beta \neq \alpha$, and $\max\ (|\alpha|, |\beta|) < \rho$. If we set

$$\Delta^{\alpha,\beta}_{N,r,s}(f, z) := R_{rN-1}(f, \alpha, z) - S_{sN-1}(f, \beta, z) \tag{4.1}$$

then we can prove

THEOREM 6. *If $f \in A_\rho$ $(\rho > 1)$ and $\alpha \neq \beta$ $(\gamma := \max\ (|\alpha|, |\beta|) < \rho)$, then*

$$\lim_{n \to \infty} \Delta^{\alpha,\beta}_{N,r,s}(f; z) = 0 \tag{4.2}$$

in the following situations:
  (a) *For $|z| < \rho_1 := \dfrac{\rho^{1 + \frac{1}{s}}}{\gamma^{\frac{1}{s}}}$, when $\sigma > \rho_1$*
  (b) *For $|z| < \rho_2 := \{ \dfrac{\rho^{s+1}}{\sigma^r \gamma} \}^{\frac{1}{s-r}}$, $|z| \neq \sigma$, when $\rho < \sigma < \rho_1$.*
  (c) *For $|z| < (\dfrac{\rho^{s-r+1}}{\gamma})^{\frac{1}{s-r}}$, $|z| \neq \sigma$ when $1 < \sigma < \rho$.*
The convergence is geometric in every compact subset of the corresponding regions.

We shall prove a slightly more general result.

THEOREM 7. *If $f(z) \in A_\rho$ $(\rho > 1)$ and if $\alpha \neq \beta$ $(\gamma < \rho)$ are any complex numbers then for any integer $\ell \geq 1$, there exist polynomials $P_{sN-1,j}(f, z)$ of degree $sN - 1$ $(j = 1, 2, \ldots, \ell - 1)$ such that*

$$\lim_{n \to \infty} [\Delta^{\alpha,\beta}_{N,r,s}(f, z) - \sum_{j=1}^{\ell-1} P_{sN-1,j}(f, z)/(z^n - \sigma^n)^r] = 0 \tag{4.3}$$

for $z \in D$, where the region D is given below:
  (a) $D = \{z|\ |z| < \rho/(\frac{\gamma}{\rho})^{\ell/s}\}$, *if* $\sigma \geq \rho^{1 + \ell/s}/\gamma^{\ell/s}$.
  (b) $D = \{z|\ |z| < \rho(\frac{\rho}{\sigma})^{\frac{r}{s-r}}/(\frac{\gamma}{\rho})^{\frac{\ell}{s-r}},\ |z| \neq \sigma\}$ *if* $1 < (\frac{\sigma}{\rho})^s < (\frac{\rho}{\gamma})^\ell$.
  (c) $D = \{z|\ |z| < \rho/(\frac{\gamma}{\rho})^{\frac{\ell}{s-r}},\ \ |z| \neq \sigma\}$, *if* $1 < \sigma < \rho$.
The convergence is uniform and geometric in every compact subset of D.

PROOF. It is easy to see that

$$\Delta^{\alpha,\beta}_{N,r,s}(f, z) = \frac{1}{2\pi i} \oint_{\Gamma_\tau} \frac{f(t)}{t - z} \left(\frac{t^n - \sigma^n}{z^n - \sigma^n} \right)^r K(t, z) dt, \tag{4.4}$$

where $1 < \tau < \rho$, and

$$K(t,z) = \left(\frac{z}{t}\right)^{sN-rn} \left(\frac{z^n - \beta^n}{t^n - \beta^n}\right)^r - \left(\frac{z^N - \alpha^N}{t^N - \alpha^N}\right)^s = K_1(t,z) - K_2(t,z)$$

on setting

$$K_1(t,z) := \left(\frac{z}{t}\right)^{sN} - \left(\frac{z^N - \alpha^N}{t^N - \alpha^N}\right)^s, \tag{4.5}$$

$$K_2(t,z) := \left(\frac{z}{t}\right)^{sN-rn} \left[\left(\frac{z}{t}\right)^{rn} - \left(\frac{z^n - \beta^n}{t^n - \beta^n}\right)^r\right]. \tag{4.6}$$

From Lemma 1 of Chapter 2, we know that

$$\begin{cases} K_1(t,z) = \dfrac{t^N - z^N}{t^{sN}} \displaystyle\sum_{s=1}^{\infty} \alpha^{(j+s-1)N} \dfrac{\beta_{j,s}(z^N \alpha^{-N})}{t^{jN}}, \\[4mm] K_2(t,z) = \left(\dfrac{z}{t}\right)^{sN-rn} \dfrac{t^n - z^n}{t^{rn}} \displaystyle\sum_{j=1}^{\infty} \dfrac{\beta_{j,r}(z^n \beta^{-n})}{t^{jn}} \beta^{(j+r-1)n}. \end{cases} \tag{4.7}$$

From (4.4), (4.5), (4.6) and (4.7), we see that

$$\Delta_{N,r,s}^{\alpha,\beta}(z,f) = \sum_{j=1}^{\infty} P_{sN-1,j}(f,z)/(z^n - \sigma^n)^r$$

where $P_{sN-1,j}(f,z)$ are polynomials of degree $\leq sN-1$ for each j and are given by

$$P_{sN-1,j}(f,z) := \frac{1}{2\pi i} \oint_{\Gamma_\tau} f(t)(t^n - \sigma^n)^r M_j(t,z)dt. \tag{4.8}$$

$M_j(t,z)$ is a polynomial of degree $sN - 1$ and

$$M_j(t,z) = M_{j,1}(t,z) - M_{j,2}(t,r)$$

where

$$M_{j,1}(t,z) := \left(\frac{\alpha}{t}\right)^{(j+s-1)N} \frac{\beta_{js}(z^N \alpha^{-N})}{t^N} \frac{t^N - z^N}{t - z} \in \pi_{sN-1}, \tag{4.9}$$

$$M_{j,2}(t,z) := \left(\frac{\beta}{t}\right)^{(j+r-1)n} \frac{\beta_{j,r}(z^n \beta^{-n})}{t^n} \cdot \frac{t^n - z^n}{t - z} \left(\frac{z}{t}\right)^{sN-rn} \in \pi_{sN-1}. \tag{4.10}$$

For any positive integer $\ell \geq 1$, we have

$$\Delta_{N,r,s}^{\alpha,\beta}(f,z) - \sum_{j=1}^{\ell-1} P_{sN-1,j}(f,z)/(z^n - \sigma^n)^r = \sum_{j=\ell}^{\infty} P_{sN-1,j}(f,z)/(z^n - \sigma^n)^r. \tag{4.11}$$

Using Lemma 2 from Chapter 2, we see that for $|z| > \rho$ and $|t| < \rho$, we have

$$|M_{j,1}(t,z)| \leq |\alpha|^{(j+s-1)N} \frac{\max(1, |z|^{N(s-1)}|\alpha|^{-N(s-1)})|z|^N}{|t|^{sN+jN}} j^{s-1}$$

$$\leq |\alpha|^{jN}|z|^N \max(|\alpha|^{(s-1)N}, |z|^{N(s-1)})/|t|^{sN+jN} j^{s-1}$$

$$\leq \frac{|\alpha|^{jN}|z|^{Ns}}{|t|^{(s+j)N}} j^{s-1}.$$

Similarly from (4.10), we obtain

$$|M_{j,2}(t,z)| \leq |\beta|^{jn} \frac{\max(|\beta|^{n(r-1)}, |z|^{n(r-1)})}{|t|^{jn+sN}} |z|^{sN-rn+n} j^{r-1}$$

$$\leq \frac{\beta^{jn}|z|^{sN}}{|t|^{jn+sN}} j^{r-1}.$$

If $|t| = R < \rho$ and $|z| > \rho$, then using (4.8), we see that

$$\left|(z^n - \sigma^n)^{-r}\right| \left|\sum_{j=\ell}^{\infty} P_{sn-1,j}(z,f)\right|$$

$$\leq C \left|\frac{R^n + \sigma^n}{|z|^n - \sigma^n}\right|^r \sum_{j=\ell}^{\infty} \left\{\frac{|\alpha|^{jN}|z|^{sN}}{R^{jN+sN}} j^{s-1} + \frac{|\beta|^{jn}|z|^{sN}}{R^{jn+sN}} j^{r-1}\right\}$$

$$\leq C \left|\frac{R^n + \sigma^n}{|z|^n - \sigma^n}\right|^r \frac{|z|^{sn}|z|^{s(m+1)}\gamma^n}{R^{(s+\ell)n}R^{(s+\ell)(m+1)}} \times$$

$$\times \sum_{j=0}^{\infty} \left\{(j+\ell)^{s-1}\left(\frac{|\alpha|}{R}\right)^{jN} + (j+\ell)^{r-1}\left(\frac{|\beta|}{R}\right)^{jn}\right\}$$

$$\leq C \left|\frac{R^n + \sigma^n}{|z|^n - \sigma^n}\right|^r \frac{|z|^{sn}\gamma^n}{R^{(s+\ell)n}}.$$

From (4.11) and the above, we see that (4.3) holds when $|z| < \rho_1$, if $\sigma > \rho_1 := \left(\frac{R^{s+\ell}}{\gamma^s}\right)^{1/s}$. If $\rho < \sigma < \rho_1$, then (4.3) holds if

$$|z| < \rho_2 := \left(\frac{R^{s+\ell}}{\sigma^r \gamma^\ell}\right)^{1/(s-r)} \quad \text{and} \quad |z| \neq \sigma.$$

If $1 < \sigma < \rho$, then (4.3) holds if

$$|z| < \rho_3 := \left(\frac{R^{s+\ell-r}}{\gamma^\ell}\right)^{1/(s-r)} \quad \text{and} \quad |z| \neq \sigma.$$

These are the regions given in the theorem when $R \to \rho$ and completes the proof of the theorem.

We observe that $\rho_1 > \rho$ and $\rho_3 > \rho$, but $\rho_2 > \rho$ only if

$$\left(\frac{\rho}{\gamma}\right)^\ell > \left(\frac{\sigma}{\rho}\right)^r$$

and $\rho_2 > \sigma$, if

$$\left(\frac{\sigma}{\rho}\right)^s < \left(\frac{\rho}{\gamma}\right)^\ell.$$

$\square$

REMARK. When $s = r$, it can be seen from the proof that convergence holds for $|z| \neq \sigma$, when $\sigma < \rho_1$.

3.5. A Discrete Analogue of Theorem 1

As before let A_ρ, $1 < \rho < \infty$ be the set of functions $f(z)$ analytic in $|z| < \rho$ but not in $|z| \leq \rho$. If q is a fixed positive integer and $m = qn + c$, $0 \leq c \leq q - 1$ let $P_{n-1,m}(z, f)$, $m \geq n$ denote the polynomial of least square approximation $f(z)$ on the mth roots of unity. Then

PROPOSITION 1. *If q is a fixed integer ≥ 2, $m = qn + c$, $0 \leq c \leq q - 1$, $f \in A_\rho$, $1 < \rho < \infty$, and if $S_{n-1}(z, f)$ is the Taylor section of $f(z)$ about the origin of degree $\leq n - 1$, then*

$$\lim_{n \to \infty} \{P_{n-1,m}(z; f) - S_{n-1}(z, f)\} = 0 \quad \text{for} \quad |z| < \rho^{q+1}, \qquad (5.1)$$

the convergence being uniform and geometric in $|z| \leq r < \rho^{q+1}$. Moreover the result is sharp in the sense that (5.1) fails to be true for every $|z| = \rho^{q+1}$ for an $f \in A_\rho$.

We will not prove this statement, but give an extension to the case of rational functions with denominator $z^n - \sigma^n$, $\sigma > 1$. In order to do this we consider the following problem:

PROBLEM (P1). *Let $m \geq -1$, $q \geq 2$ be fixed integers and let $\omega = \exp\left(\frac{2\pi i}{qn}\right)$ for $f \in A_\rho$. We want to minimize*

$$\sum_{k=0}^{qn-1} \left(f(\omega^k) - R(\omega^k, f)\right)^2 \qquad (5.2)$$

over all rational functions of the form $\frac{p(z)}{z^n - \sigma^n}$, $p(z) \in \pi_{n+m}$.

We shall denote this rational minimizing function by

$$R_{n+m,n}(z,f) = \frac{P_{n+m,n}(z)}{z^n - \sigma^n}, \quad P_{n+m,n}(z,f) = \pi_{n+m},$$

and replace the Taylor section in (5.1) by the rational function $r_{n+m,n}(z,f) = \frac{P_{n+m,n}(z,f)}{z^n - \sigma^n}$ which interpolates $f(z)$ in the zeros of $(z^n - \sigma^n)z^{m+1}$.

We shall prove the following:

THEOREM 8. *Let $m \geq -1$, $q \geq 2$ be fixed integers and let $\sigma > 1$. If $f \in A_\rho$ ($\rho > 1$), then*

$$\lim_{n \to \infty} \{R_{n+m,n}(z,f) - r_{n+m,n}(z,f)\} = 0 \begin{cases} |z| < \rho^{1+q}, & \text{if } \sigma > \rho^{1+q} \\ |z| \neq \sigma, & \text{if } \sigma < \rho^{1+q} \end{cases}$$

the convergence being uniform and geometric in any compact subset of the regions described above. Moreover the result is sharp in the sense that for each $|z| = \rho^{1+q}$ if $\sigma > \rho^{1+q}$, there is an $f \in A_\rho$ for which (5.2) does not hold.

The reason for the choice of $r_{n+m,n}(z,f)$ is Theorem 1 in Section 3.1.

We shall need the following lemmas to prove the result.

LEMMA 3. *Let $m \geq -1$, $q \geq 2$ be fixed integers, $\sigma > 1$, and let $g(z) = \frac{P(z)}{z^n - \sigma^n}$, where $P(z) = \sum_{j=0}^{n+m} c_j z^j \in \pi_{n+m}$. Then the Lagrange interpolant of $g(z)$ on the $(qn)^{\text{th}}$ roots of unity is given by*

$$L_{qn-1}(z,g) = \sum_{\nu=0}^{qn-1} A_{\nu,j} z^\nu$$

$$= \sum_{j=)}^{m} A_{0,j} z^j + \sum_{\nu=0}^{q-1} \sum_{j=m+1}^{n-1} A_{\nu n+j} z^j + \sum_{\nu=1}^{q-1} \sum_{j=0}^{m} A_{\nu n+j} z^j$$

where

$$A_{\nu,j} = A_{\nu n+j} = \begin{cases} \lambda_1 c_j + \lambda_q c_{j+n}, & \nu = 0, \quad 0 \leq j \leq m \\ \lambda_{\nu+1} c_j, & 0 \leq \nu \leq q-1, \quad m+1 \leq j \leq n+1 \\ \lambda_{\nu+1} c_j + \lambda_\nu c_{j+1}, & 1 \leq \nu \leq q-1, \quad 0 \leq j \leq m \end{cases}$$

(5.3)

where

$$\lambda_\nu = \frac{\sigma^{(q-\nu)n}}{1 - \sigma^{qn}}, \quad \nu = 1, \dots, q.$$

PROOF. The Lagrange interpolant of $(z^n - \sigma^n)^{-1}$ in the $(qn)^{\text{th}}$ roots of unity is given by

$$L_{qn-1}\big(z_1(z^n - \sigma^n)^{-1}\big) = \frac{z^{qn} - \sigma^{qn}}{(z^n - \sigma^n)(1 - \sigma^{qn})} = \sum_{\nu=1}^{q} \lambda_\nu z^{(\nu-1)n}.$$

Then

$$\frac{P(\omega^k)}{\omega^{kn} - \sigma^n} = \lambda_1 \sum_{j=0}^{n+m} c_j \omega^{kj} + \lambda_2 \sum_{j=n}^{2n+m} c_{j-n}\omega^{kj} + \cdots + \lambda_q \sum_{j=(q-1)n}^{qn+m} c_{j-(q-1)n}\omega^{kj}.$$

(We make the convention that $\lambda_{q+1} = \lambda_1$.)

In each of the q summations above, there are $m+1$ distinct powers of ω^k which also appear in the preceding sums. If we group the terms involving identical powers of ω^k in separate summations and then rearrange them, then we obtain

$$\frac{P(\omega^k)}{\omega^{kn} - \sigma^n} = \sum_{j=0}^{m}(\lambda_1 c_j + \lambda_q c_{j+n})\omega^{kj} + \sum_{\nu=0}^{q-1}\sum_{j=m+1}^{n-1}\lambda_{\nu+1}c_j\omega^{(\nu n+j)k}$$

$$+ \sum_{\nu=1}^{q}\sum_{j=0}^{m}(\lambda_{\nu+1}c_j + \lambda_\nu c_{j+n})\omega^{(\nu n+j)k} \qquad (5.4)$$

$$= \sum_{j=0}^{m} A_{0,j} z^{jk} + \sum_{\nu=0}^{q-1}\sum_{j=m+1}^{n-1} A_{\nu,j}\omega^{kj} + \sum_{\nu=1}^{q}\sum_{j=0}^{m} A_{\nu,j}\omega^{kj}$$

where the $A_{\nu,j}$'s are given by (5.3).

We have $L_{qn-1}(\omega^k, f) = \sum_{j=0}^{qn-1} b_j \omega^{kj} = f(\omega^k) \quad (k = 0, 1, \ldots, qn - 1)$ with

$$b_j = \frac{1}{2\pi i}\oint_{\Gamma_\tau} \frac{f(t)t^{qn-j-1}}{t^{qn} - 1}\, dt, \quad j = 0, 1, \ldots, qn - 1. \qquad (5.5)$$

Since

$$L_{qn-1}(\omega^k, g) = \sum_{j=0}^{m} A_{0,j}\omega^{kj} + \sum_{r=0}^{q-1}\sum_{j=m+1}^{n-1}\lambda_{\nu+1}c_j\omega^{(\nu n+j)k}$$

$$+ \sum_{\nu=1}^{q}\sum_{j=0}^{m}(\lambda_{n+1}c_j + \lambda_\nu c_{j+n})\omega^{(2n+j)k}$$

we have

$$\sum_{k=0}^{qn-1}|f(\omega^k) - L_{qn-1}(\omega^k, g)|^2 = \sum_{\nu=0}^{qn-1}\left|\sum_{j=0}^{qn-1}(b_j - A_j)\omega^{kj}\right|^2$$

$$= qn\sum_{j=0}^{qn-1}|b_j - A_j|^2$$

so the problem of minimizing (5.2) is equivalent to finding the minimum of

$$G = \sum_{j=0}^{qn-1} |b_j A_{\nu,j}|^2 \tag{5.6}$$

where the A_j's are given by (5.3).

We now rewrite

$$G = \sum_{j=0}^{m} |b_j - \lambda_1 c_j - \lambda_q c_{j+n}|^2 + \sum_{\nu=0}^{q-1} \sum_{j=m+1}^{n-1} |b_{j+\nu n} - \lambda_{\nu+1} c_j|^2$$

$$+ \sum_{\nu=1}^{q-1} \sum_{j=0}^{m} |b_{j+\nu n} - \lambda_{\nu+1} c_j - \lambda_\nu c_{j+n}|^2.$$

Therefore for $0 \leq j \leq m$, we have

$$\frac{\partial G}{\partial \bar{c}_j} = -(b_j - \lambda_1 c_j - \lambda_q c_{j+n})\lambda_1 - \sum_{\nu=1}^{q-1} (b_{j+\nu n} - \lambda_{\nu+1} c_j - \lambda_\nu c_{j+n})\lambda_{\nu+1} = 0$$

$$\frac{\partial G}{\partial \bar{c}_{j+n}} = -(b_j - \lambda_1 c_j - \lambda_q c_{j+n})\lambda_q - \sum_{\nu=1}^{q-1} (b_{j+\nu n} - \lambda_{\nu+1} c_j - \lambda_\nu c_{j+n})\lambda_\nu = 0$$

which on simplifying yields for $0 \leq j \leq m$,

$$\begin{cases} \alpha c_j + \beta c_{n+j} & = \sum_{\nu=0}^{q-1} \lambda_{\nu+1} b_{\nu n+j} \\[4mm] \beta c_j + \alpha c_{n+j} & = \lambda_q b_j + \sum_{\nu=1}^{q-1} \lambda_\nu b_{\nu n+j} \end{cases} \tag{5.7}$$

where we have set

$$\alpha = \lambda_1^2 + \lambda_2^2 + \cdots + \lambda_q^2 = \frac{1 + \sigma^{qn}}{1 - \sigma^{2n}} \lambda_q$$

and

$$\beta = \lambda_q \lambda_1 + \lambda_1 \lambda_2 + \cdots + \lambda_{q-1} \lambda_q = \frac{\sigma^n + \sigma^{(q-1)n}}{1 - \sigma^{2n}} \lambda_q.$$

For $m + 1 \leq j \leq n - 1$, we have

$$\frac{\partial G}{\partial \bar{c}_j} = \sum_{\nu=0}^{q-1} \lambda_{\nu+1}^2 c_j - \sum_{\nu=0}^{q-1} \lambda_{\nu+1} b_{j+\nu n} = 0$$

which yields

$$c_j = \frac{1}{\alpha} \sum_{\nu=1}^{q} \lambda_\nu b_{j+(\nu-1)n}, \quad m+1 \le j \le n-1 \tag{5.8}$$

since $\alpha^2 - \beta^2 = \frac{1-\sigma^{2(q-1)n}}{1-\sigma^{2n}} \cdot \frac{1}{(1-\sigma^{qn})^2}$, we find from (5.7)

$$c_j = -b_j\sigma^n + \frac{\sigma^{(q-1)n}(1-\sigma^{2n})}{1-\sigma^{(q-1)2n}} \sum_{\nu=1}^{q-1} \sigma^{-\nu n} b_{j+\nu n}, \qquad 0 \le j \le m,$$

$$c_{j+n} = b_{j-n} - \frac{\sigma^{2(q-1)n}(1-\sigma^{2n})}{1-\sigma^{(q-1)2n}} \sum_{\nu=1}^{q-1} \sigma^{-\nu n} b_{j+(\nu-1)n}, \quad n \le j \le n+m$$

and for $m+1 \le j \le n-1$, we have

$$c_j = \frac{\sigma^{(q-1)n}(1-\sigma^{2n})}{1+\sigma^{qn}} \sum_{\nu=0}^{q-1} \sigma^{-\nu n} b_{j+\nu n}, \quad m+1 \le j \le n-1.$$

Thus we have the

LEMMA 4. *The rational function* $R_{n+m,n}(z,f) = \frac{P_{n+m,n}(z,f)}{z^n-\sigma^n}$ *which mini-mizes* (5.2) *is given by* $P_{n+m,n}(z,f) = \sum\limits_{j=0}^{n+m} P_\nu z^j$ *where*

$$P_j = \begin{cases} -b_j\sigma^n + \frac{\sigma^{(q-1)n}(1-\sigma^{2n})}{1-\sigma^{(q-1)2n}} \sum\limits_{\nu=1}^{q-1} \sigma^{-\nu n} b_{j+\nu n}, & 0 \le j \le m, \\[3mm] \frac{\sigma^{(q-1)n}(1-\sigma^{2n})}{1+\sigma^{qn}} \sum\limits_{\nu=0}^{q-1} \sigma^{-\nu n} b_{j+\nu n}, & m+1 \le j \le n-1, \quad (5.9) \\[3mm] b_{j-n} - \frac{\sigma^{2(q-1)n}(1-\sigma^{2n})}{1-\sigma^{(q-1)n}} \sum\limits_{\nu=1}^{q-1} \sigma^{-\nu n} b_{j+(\nu+1)n}, & n \le j \le n+m. \end{cases}$$

Recall that $r_{n+m,n}(z,f) = \frac{P_{n+m,n}(z,f)}{z^n-\sigma^n}$ where $P_{n+m,n}(z,f)$ interpo-lates $f(z)(z^n - \sigma^n)$ in the zeros of $(z^n - \sigma^n)z^{m+1}$. Thus

$$P_{n+m,n}(z,f) = \frac{1}{2\pi i} \oint_{\Gamma_\tau} f(t)(t^n-\sigma^n) \frac{t^{m+1}(t^n-\sigma^n) - z^{m+1}(z^n-\sigma^{-n})}{t^{m+1}(t^n-\sigma^{-n})} \, dt \; .$$

Hence

$$r_{n+m,n}((z,f) = \frac{1}{2\pi i} \oint_{\Gamma_\tau} \frac{(t^n-\sigma^n)f(t)}{z^n-\sigma^n} \cdot \frac{t^{m+1}(t^n-\sigma^{-n}) - z^{m+1}(z^n-\sigma^{-n})}{t^{m+1}(t^n-\sigma^{-n})} \, dt$$

$$= \frac{1}{2\pi i} \oint_{\Gamma_\tau} \frac{(t^n-\sigma^n)f(t)}{z^n-\sigma^n} \sum_{j=1}^{3} A_j(t,z) \, dt$$

where we have set

$$A_1(t,z) = \frac{t^{m+1} - z^{m+1}}{t^{m+1}(t-z)}, \qquad A_2(t,z) = \frac{z^{m+1}(t^{n-m-1} - z^{n-m-1})}{(t^n - \sigma^{-n})(t-z)},$$

$$(5.10)$$

$$A_3(t,z) = \frac{z^n(t^{m+1} - z^{m+1})}{t^{m+1}(t^n - \sigma^{-n})(t-z)}.$$

Using the values of P_j's in Lemma 4, we see that

$$R_{n+m,n}(z,f) = \frac{P_1(z) + P_2(z) + P_3(z)}{z^n - \sigma^n}$$

where

$$\begin{cases} P_1(z) &= -\sigma^n \sum_{j=0}^{m} b_j z^j + \frac{\sigma^{(q-1)n}(1-\sigma^{2n})}{1-\sigma^{(q-1)2n}} \sum_{\nu=1}^{q-1} \sigma^{-\nu n} \sum_{j=0}^{m} b_{j+\nu n} z^j, \\[2ex] P_2(z) &= \frac{\sigma^{(q-1)n}(1-\sigma^{2n})}{1+\sigma^{qn}} \sum_{\nu=0}^{q-1} \sigma^{-2n} \sum_{j=m+1}^{n-1} b_{j+\nu n} z^j, \\[2ex] P_3(z) &= \sum_{j=n}^{n+m} b_{j-n} z^j - \frac{\sigma^{2(q-1)n}(1-\sigma^{2n})}{1-\sigma^{(q-1)2n}} \sum_{\nu=1}^{q-1} \sigma^{-\nu n} \sum_{j=n}^{n+m} b_{j+(\nu-1)n} z^j. \end{cases}$$

$$(5.11)$$

We use the formula for b_j in (5.5) in the above expressions for $P_1(z)$, $P_2(z)$ and $P_3(z)$. Thus

$$R_{n+m,n}(z,f) = \sum_{j=1}^{3} \frac{P_j(z)}{z^n - \sigma^n},$$

where using (5.9), we have

$$\begin{cases} P_1(z) &= \sum_{j=0}^{m} P_j z^j = -\sigma^n \sum_{j=0}^{m} b_j z^j + \\[2ex] &\quad + \sigma^{(q-1)n}(1 - \sigma^{2n})(1 - \sigma^{(q-1)2n})^{-1} \sum_{\nu=1}^{q-1} \sigma^{-\nu n} \sum_{j=0}^{m} b_{j+\nu n} z^j \\[2ex] P_2(z) &= \sigma^{(q-1)n}(1 - \sigma^{2n})(1 + \sigma^{qn})^{-1} \sum_{\nu=0}^{q-1} \sigma^{-\nu n} \sum_{j=m+1}^{n-1} b_{j+\nu n} z^j, \\[2ex] P_3(z) &= \sum_{j=n}^{n+m} b_{j-n} z^j - \sigma^{2(q-1)n}(1 - \sigma^{2n})(1 - \sigma^{(q-1)2n})^{-1} \sum_{\nu=1}^{q-1} \sigma^{-\nu n} \times \\[2ex] &\quad \times \sum_{j=n}^{n+m} b_{j(\nu-1)n} z^j. \end{cases}$$

$$(5.12)$$

Using (2.5) in the above and simplifying, we obtain

$$P_1(z) = -\frac{\sigma^n}{2\pi i} \oint_{\Gamma_\tau} \frac{f(t)t^{qn}}{t^{qn} - 1} \cdot \frac{t^{m+1} - z^{m+1}}{(t-z)t^{m+1}} + \frac{\sigma^{(q-1)n}(1-\sigma^{2n})}{1-\sigma^{(q-1)2n}} \times$$

$$\times \frac{1}{2\pi i} \oint_{\Gamma_\tau} \frac{f(t)}{t^{qn} - 1} \sum_{j=0}^{m} \sum_{\nu=1}^{q-1} z^j t^{qn-j-\nu n-1} dt$$

$$(5.13)$$

$$= \frac{1}{2\pi i} \oint_{\Gamma_\tau} \frac{f(t)}{t^{qn} - 1} A_1(t,z)\{\sigma^{-(q-2)n} B(t,\sigma) - t^{qn}\sigma^n\} dt$$

where $A_1(t,z) = \frac{t^{m+1}-z^{m+1}}{(t-z)t^{m+1}}$ and $B(t,\sigma) = \frac{t^n(t^{(q-1)n}-\sigma^{-(q-1)n})(1-\sigma^{-2n})}{(t^n-\sigma^{-n})(1-\sigma^{-2(q-1)n})}$. Similarly,

$$
\begin{aligned}
P_2(z) &= \frac{\sigma^{-n}-\sigma^n}{1+\sigma^{-qn}} \frac{1}{2\pi i} \oint_{\Gamma_\tau} \frac{f(t)}{t^{qn}-1} \cdot \frac{t^{qn}-\sigma^{-qn}}{t^n-\sigma^{-n}} \frac{z^{m+1}(t^{n-m-1}-z^{n-m-1})}{t-z} dt \\
&= \frac{1}{2\pi i} \oint_{\Gamma_\tau} \frac{f(t)}{t^{qn}-1} \cdot \frac{(t^{qn}-\sigma^{-qn})(\sigma^{-n}-\sigma^n)}{1+\sigma^{-qn}} \times \\
&\qquad \times \frac{z^{m+1}(t^{n-m-1}-z^{n-m-1})}{(t-z)(t^n-\sigma^{-n})} dt
\end{aligned}
\tag{5.14}
$$

where

$$
A_2(t,z) = \frac{z^{m+1}(t^{n-m-1}-z^{n-m-1})}{(t-z)(t^n-\sigma^{-n})}, \quad B_2(t,\sigma) = \frac{(t^{qn}-\sigma^{-qn})(\sigma^{-n}-\sigma^n)}{1+\sigma^{-qn}}.
$$

Lastly, in the same way from (5.9) and (5.11), we get

$$
\begin{aligned}
P_3(z) &= \sum_{j=n}^{n+m} b_{j-n}z^j - \frac{\sigma^{2(q-1)n}(1-\sigma^{2n})}{1-\sigma^{(q-1)2n}} \sum_{\nu=1}^{q-1} \sum_{j=n}^{n+m} \sigma^{-\nu n}z^j b_{j+(\nu-1)n} \\
&= z^n \sum_{j=0}^{m} b_j z^j - \frac{\sigma^{2(q-1)n}(1-\sigma^{2n})}{1-\sigma^{(q-1)2n}} z^n \sum_{\nu=1}^{q-1} \sum_{j=0}^{m} \sigma^{-\nu n}z^j b_{j+\nu n} \\
&= z^n \frac{1}{2\pi i} \oint_{\Gamma_\tau} \frac{f(t)}{t^{qn}-1} \sum_{j=0}^{m} t^{qn-j-1}z^j - \\
&\quad - \frac{\sigma^{2(q-1)n}(1-\sigma^{2n})}{1-\sigma^{(q-1)2n}} z^n \sum_{\nu=1}^{q-1} \sum_{j=0}^{m} \sigma^{-\nu n}z^j \frac{1}{2\pi i} \oint_{\Gamma_\tau} \frac{f(t)}{t^{qn}-1} t^{qn-\nu n-j-1} dt \\
&= \frac{z^n}{2\pi i} \oint_{\Gamma_\tau} \frac{f(t)t^{qn}}{t^{qn}-1} \frac{t^{m+1}-z^{m+1}}{t^{m+1}(t-z)} dt \\
&\quad - \frac{1-\sigma^{-2n}}{1-\sigma^{-(q-1)2n}} \sigma^{2n} \frac{z^n}{2\pi i} \oint_{\Gamma_\tau} \frac{f(t)t^{qn}}{t^{qn}-1} \cdot \frac{(t\sigma)^{-n}-(t\sigma)^{-qn}}{1-(t\sigma)^{-n}} \times \\
&\qquad \times \frac{t^{m+1}-z^{m+1}}{(t-z)t^{m+1}} dt.
\end{aligned}
$$

Putting $A_3(t,z) = z^n(t^{m+1}-z^{m+1})((t-z)(t^n-\sigma^{-n}))^{-1}$, we have

$$
\begin{aligned}
P_3(z) &= \frac{1}{2\pi i} \oint_{\Gamma_\tau} \frac{f(t)t^{qn}}{t^{qn}-1} (t^n-\sigma^{-n})) A_3(t,z) \\
&\quad - \frac{1}{2\pi i} \frac{(1-\sigma^{-2n})\sigma^n}{1-\sigma^{-(q-1)2n}} \oint_{\Gamma_\tau} \frac{t^n(t^{(q-1)n}-\sigma^{-(q-1)n})}{t^n-\sigma^{-n}} \frac{f(t)}{t^{qn}-1} A_3(t,z) dt \\
&= \frac{1}{2\pi i} \oint_{\Gamma_\tau} \frac{f(t)}{t^{qn}-1} (t^n-\sigma^{-n})\{t^{qn}-\sigma^n B(t,\sigma)\} A_3(t,z) dt
\end{aligned}
\tag{5.15}
$$

where we have set

$$B(t,\sigma) = \frac{t^n(t^{(q-1)n} - \sigma^{-(q-1)n})(1 - \sigma^{-2n})}{(t^n - \sigma^{-n})(1 - \sigma^{-(q-1)2n})}.$$

Combining (5.13), (5.14) and (5.15), we have

$$\begin{aligned}
R_{n+m,n}(z,f) &= \frac{1}{2\pi i} \oint_{\Gamma_\tau} \frac{f(t)}{t^{qn} - 1} \frac{A_1(t,z)}{z^n - \sigma^n} B_1(t,\sigma) dt \\
&+ \frac{1}{2\pi i} \oint_{\Gamma_\tau} \frac{f(t)}{t^{qn} - 1} \frac{A_2(t,z)}{t^{qn} - 1} \frac{B_2(t,\sigma)}{z^n - \sigma^n} dt \\
&+ \frac{1}{2\pi i} \oint_{\Gamma_\tau} \frac{f(t)}{t^{qn} - 1} \frac{A_3(t,z)}{t^{qn} - 1} \frac{B_3(t,\sigma)}{z^n - \sigma^n} dt,
\end{aligned}$$

where

$$\begin{aligned}
B_3(t,\sigma) &= (t^n - \sigma^{-n})\{t^{qn} - \sigma^n B(t,\sigma)\}, \\
B_1(t,\sigma) &= \sigma^{-(q-2)n} B(t,\sigma) - t^{qn}\sigma^n
\end{aligned}$$

and

$$B_2(t,\sigma) = \frac{(t^{qn} - \sigma^{-qn})(\sigma^{-n} - \sigma^n)}{1 + \sigma^{-qn}}.$$

This gives

$$R_{n+m,n}(z,f) - r_{n+m,n}(z,f) = \frac{1}{2\pi i} \oint_{\Gamma_\tau} \sum_{i=1}^{3} \frac{A_j(t,z)K_j(t,\sigma)}{(z^n - \sigma^n)(t - z)(t^{qn} - 1)} dt$$

where

$$\begin{aligned}
K_1(t,z) &= \sigma^{-(q-2)n} B(t,\sigma) - t^{qn}\sigma^n - (t^{qn} - 1)(t^n - \sigma^n) \\
&= \sigma^{-(q-2)n} B(t,\sigma) - t^{(q+1)n} + t^n - \sigma^n,
\end{aligned}$$

(5.16)

$$\begin{aligned}
K_2(t,z) &= B_2(t,\sigma) - (t^{qn} - 1)(t^n - \sigma^n) \\
&= \frac{(t^{qn} - \sigma^{-qn})(\sigma^{-n} - \sigma^n)}{1 + \sigma^{-qn}} + t^{qn}\sigma^n - t^{(q+1)n} + t^n - \sigma^n \\
&= \frac{t^{qn}\sigma^{-n}(1 + \sigma^{-(q-2)n}) - \sigma^{-qn}(\sigma^{-n} - \sigma^n)}{1 + \sigma^{-qn}} - t^{(q+1)n} + t^n - \sigma^n
\end{aligned}$$

(5.17)

and

$$K_3(t,z) = (t^n - \sigma^n)\{t^{qn} - \sigma^n B(t,\sigma)\} - (t^{qn} - 1)(t^n - \sigma^n)$$

$$= t^{qn}(t^n - \sigma^{-n}) - \frac{\sigma^n t^n (t^{(q-1)n} - \sigma^{-(q-1)n})(1 - \sigma^{-2n})}{1 - \sigma^{-2(q-1)n}}$$

$$- t^{(q+1)n} + t^n - \sigma^n + t^{qn}\sigma^n$$

$$= t^{qn}(\sigma^n - \sigma^{-n})\left(1 - \frac{1}{1 - \sigma^{-2(q-1)n}}\right) - \frac{t^n \sigma^{-qn}(\sigma^{2n} - 1)}{1 - \sigma^{-2(q-1)n}} + t^n - \sigma^n$$

$$= \frac{t^n \sigma^{-qn}(\sigma^{2n} - 1)\{t^{(q-1)n}\sigma^{-(q-1)n} - 1\}}{1 - \sigma^{-2(q-1)n}} + t^n - \sigma^n.$$

$$(5.18)$$

An analysis of the kernels $A_j(t,z)$ and $K_j(t,\sigma)$, $j = 1,2,3$ above yields the result.

To prove that the result is sharp, we consider the point $z^* = \rho^{1+q}$ and the function $\widehat{f}(z) = (z - \rho)^{-1}$. A direct computation shows that

$$R_{n+m,n}(z,\widehat{f}) - r_{n+m,n}(z,\widehat{f}) = \sum_{j=1}^{3} \frac{A_j(\rho,z)K_j(\rho,\sigma)}{(z^n - \sigma^n)(z - \rho)(\rho^{qn} - 1)}.$$

If $\sigma > \rho^{1+q}$, we get after some simple calculations that

$$\lim_{n\to\infty} R_{n+m,n}(\rho^{1+q}, f) - r_{n+m,n}(\rho^{1+q}, f) = \frac{\rho^q}{\rho^q - 1}.$$

3.6. Historical Remarks

Theorem 1 is due to Walsh [113]. The special case $\alpha = 1$ of Theorems 3 and 5 was proved by Saff and Sharma [91]. Case (a) of Theorem 6 with $\alpha = 1$, $\beta = 0$ is a result of Bokhari and Sharma [16]. Theorem 6 with $\ell = 1, \beta = \sigma^{-1}$ is Theorem 2.2 from [16]. For $\ell > 1$, Theorem 6 and Theorem 3.3 from [16] are not comparable. Proposition 1 is due to Rivlin [88].

CHAPTER 4

SHARPNESS RESULTS

4.1. Lagrange Interpolation

In Chapter 1, Section 3, we gave upper estimates, for the order of overconvergence in the corresponding domain in case of Lagrange interpolation. Now we examine how sharp these estimates are.

THEOREM 1. *If $f(z) \in A_\rho$ and if $\ell \geq 1$ is any given integer, then*

$$\varlimsup_{n \to \infty} \max_{|z|=\mu} |L_{n-1}(f;z) - \sum_{j=0}^{\ell-1} p_{n-1,j}(f;z)|^{1/n} = \begin{cases} \frac{1}{\rho^\ell} & if \quad 0 < \mu < \rho \\ \frac{\mu}{\rho^{\ell+1}} & if \quad \mu \geq \rho. \end{cases} \quad (1.1)$$

PROOF. Let

$$f(z) = \sum_{k=0}^{\infty} a_k z^k \in A_\rho, \quad \rho > 1,$$

then $a_k = O\big((\rho - \varepsilon)^{-k}\big)$ for all $\varepsilon > 0$. Denoting

$$\Delta_{n,\ell}(f;z) := L_{n-1}(f;z) - \sum_{j=0}^{\ell-1} p_{n-1,j}(f;z) = \sum_{j=\ell}^{\infty} \sum_{k=0}^{n-1} a_{k+jn} z^k, \quad (1.2)$$

let $\varepsilon > 0$ be so small that in case $\mu < \rho$, we have $\mu < \rho - \varepsilon$ as well. Then

$$\Delta_{n,\ell}(f;z) = \sum_{k=0}^{n-1} a_{\ell n+k} z^k + O\Big(\sum_{k=0}^{n-1} |z|^k (\rho - \varepsilon)^{-(\ell+1)n-k}\Big)$$

$$= \sum_{k=0}^{n-1} a_{\ell n+k} z^k + \begin{cases} O\big((\rho - \varepsilon)^{-(\ell+1)n}\big) & if \quad 0 \leq \mu < \rho \\ O\big((\mu^n (\rho - \varepsilon)^{-(\ell+2)n}\big) & if \quad \mu \geq \rho. \end{cases} \quad (1.3)$$

Hence, the first upper estimate in (1.1) follows from

$$\Big|\sum_{k=0}^{n-1} a_{\ell n+k} z^k\Big| = O\big((\rho - \varepsilon)^{-\ell n}\big), \quad \mu < \rho - \varepsilon$$

81

when letting $\varepsilon \to 0$. The second upper estimate in (1.1) would follow also easily, but this has already been proved in Theorem 7 of Chapter 1.

So we have to prove the lower estimates in (1.1). We obtain from (1.2) for $0 < \mu < \rho$

$$|a_{\ell n+k}| = \frac{1}{2\pi}\left|\int_{\Gamma_\mu} \frac{\Delta_{n,\ell}(f;z)}{z^{k+1}}\,dz\right| + O\left((\rho-\varepsilon)^{-(\ell+1)n}\right), \quad 0 \le k \le \ell - 1. \quad (1.4)$$

Since $f \in A_\rho$, there exists a sequence $m_1 < m_2 < \ldots$, $m_j = \ell n_j + k_j$, $0 \le k_j \le \ell - 1$ such that

$$a_{m_j} \ge (\rho+\varepsilon)^{-\ell n_j}$$

for sufficiently large j. Thus we obtain from (1.4)

$$\max_{|z|=\mu} |\Delta_{n_j,\ell}(f;z)| \ge c_1(\rho+\varepsilon)^{-\ell n_j} - c_2(\rho-\varepsilon)^{-(\ell+1)n_j} \ge c_3(\rho+\varepsilon)^{-\ell n_j}$$

if $\varepsilon > 0$ is small and j is large enough. Hence

$$\overline{\lim_{n\to\infty}} \max_{|z|=\mu} |\Delta_{n,\ell}(f;z)|^{1/n} \ge \overline{\lim_{j\to\infty}} \max_{|z|=\mu} |\Delta_{n_j,\ell}(f,z)|^{\frac{1}{n_j}}$$

$$\overset{\ge}{=} (\rho+\varepsilon)^{-\ell}, \quad 0 < \mu < \rho$$

and this yields the result since $\varepsilon > 0$ was arbitrary.

Finally, let $\mu \ge \rho$. Then (1.2) leads to

$$|a_{\ell n+k}| = \frac{1}{2\pi}\left|\int_{\Gamma_\mu} \frac{\Delta_{n,\ell}(f;z)}{z^{k+1}}\,dz\right| + O\left((\rho-\varepsilon)^{-(\ell+2)n}\right), \quad n-\ell-1 \le k \le n-1.$$

Choosing a sequence $m_1 < m_2 < \ldots$, $m_j = \ell n_j + k_j$, $n_j - \ell - 1 \le k_j \le n_j - 1$ such that

$$a_{m_j} \ge (\rho+\varepsilon)^{-(\ell+1)n_j}$$

for sufficiently large j, we obtain

$$\max_{|z|=\mu} |\Delta_{n_j,\ell}(f;z)| \ge C_1\mu^{n_j}(\rho+\varepsilon)^{-(\ell+1)n_j} - C_2\mu^{n_j}(\rho-\varepsilon)^{-(\ell+2)n_j}$$

$$\ge C_3\mu^{n_j}(\rho+\varepsilon)^{-(\ell+1)n_j}$$

provided $\varepsilon > 0$ is small and j is large enough. Hence

$$\overline{\lim_{n\to\infty}} \max_{|z|=\mu} |\Delta_{n,\ell}(f;z)|^{1/n} \ge \overline{\lim_{j\to\infty}} \max_{|z|=\mu} |\Delta_{n_j,\ell}(f;z)|^{\frac{1}{n_j}}$$

$$\ge \mu(\rho+\varepsilon)^{-\ell-1}, \quad \mu \ge \rho$$

which proves the second lower estimate in (1.1). $\qquad\square$

(1.1) can be reformulated as

$$\limsup_{n\to\infty} \max_{|z|=\mu} |\Delta_{n,\ell}(f;z)|^{1/n} = K_\ell(\rho,\mu) := \begin{cases} \rho^{-\ell}, & \text{if } \mu \le \rho \\ \dfrac{\mu}{\rho^{\ell+1}}, & \text{if } \mu \ge \rho. \end{cases}$$

Theorem 1 asserts that on each circle $|z| = \mu > 0$, the best possible error estimate is indeed attained. The next question which naturally arises is the following: Is it possible to get, at some points, better estimates than those stated in (1.1)? More exactly, let

$$B_\ell(f;z) := \limsup_{n\to\infty} |\Delta_{n,\ell}(f;z)|^{1/n}$$

and

$$\delta_{\ell,\rho}(f) := \{z \mid B_\ell(f;z) < K_\ell(\rho,|z|)\}.$$

The elements of the set $\delta_{\ell,\rho}(f)$ are called (ℓ,ρ)-*distinguished* or $(\ell,\rho)-$*exceptional* points.

First, note that for $z = 0$, Theorem 1 does not give any information. Indeed, (1.3) shows that

$$\Delta_{n,\ell}(f;0) = a_{\ell n} + O\big((\rho-\varepsilon)^{-(\ell+1)n}\big)$$

and evidently there exists $f \in A_\rho$ such that $a_\ell = a_{2\ell} = a_{3\ell} = \cdots = 0$ and then

$$\varlimsup_{n\to\infty} |\Delta_{n,\ell}(f;0)|^{1/n} \le (\rho-\varepsilon)^{-\ell-1} < \rho^{-\ell}$$

if $\varepsilon > 0$ is small enough, i.e., (1.1) is not true for $\mu = 0$ in this case.

In general, it is rather exceptional that the general error term in (1.1) is not attained. This will be shown in the next theorem.

THEOREM 2. *Let* $f(z) \in A_\rho$, $\rho > 1$. *Then*

$$\varlimsup_{n\to\infty} |\Delta_{n,\ell}(f;z)|^{1/n} = \frac{1}{\rho^\ell} \tag{1.5}$$

for all but at most $\ell - 1$ *points in* $0 < |z| < \rho$; *and*

$$\varlimsup_{n\to\infty} |\Delta_{n,\ell}(f;z)|^{1/n} = \frac{|z|}{\rho^{\ell+1}} \tag{1.6}$$

for all but at most ℓ *points in* $|z| > \rho$.

In other words, the set of distinguished points consists of at most $\ell - 1$ and ℓ elements in $|z| < \rho$ and $|z| > \rho$, respectively.

PROOF. Let first $0 < |z| < \rho - \varepsilon$ with some $\varepsilon > 0$, then

$$h_n(z) := \Delta_{n,\ell}(f; z) - z^\ell \Delta_{n+1,\ell}(f; z) \tag{1.7}$$

$$= \sum_{k=0}^{n-1} a_{\ell n + k} z^k - \sum_{k=0}^{n} a_{\ell n + \ell + k} z^{\ell + k} + O\big((\rho - \varepsilon)^{-(\ell+1)n}\big)$$

$$= \sum_{k=0}^{\ell-1} a_{\ell n + k} z^k + O\big(|z|^n (\rho - \varepsilon)^{-(\ell+1)n}\big).$$

Assume the contrary: we have $0 < |z_j| < \rho$, $j = 1, \ldots, \ell$ such that

$$\varlimsup_{n \to \infty} |\Delta_{n,\ell}(f; z_j)|^{1/n} < \frac{1}{\rho^\ell}, \quad j = 1, \ldots, \ell.$$

Then by (1.7) we also get

$$\varlimsup_{n \to \infty} |h_n(z_j)|^{1/n} < \frac{1}{\rho^\ell}, \quad j = 1, \ldots, \ell, \tag{1.8}$$

i.e.,

$$\sum_{k=0}^{\ell-1} a_{\ell n + k} z_j^k = \beta_{jn}, \quad j = 1, \ldots, \ell, \tag{1.9}$$

where

$$\beta_{j,n} = h_n(z_j) + O\big(|z_j|^n (\rho - \varepsilon)^{-(\ell+1)n}\big).$$

Here by (1.8), supposing $|z_j| < \rho - \varepsilon$, we get

$$\varlimsup_{n \to \infty} |\beta_{j,n}|^{1/n} < \frac{1}{\rho^\ell}, \quad j = 1, \ldots, \ell.$$

Solving the system of equations (1.9) for $a_{\ell n + k}$, we obtain (since z_j are independent of n) with $m = \ell n + k$, $0 \leq k \leq \ell - 1$,

$$\varlimsup_{m \to \infty} |a_m|^{1/m} \leq \max_{1 \leq j \leq \ell} \varlimsup_{n \to \infty} |\beta_{j,n}|^{\frac{1}{n} \cdot \frac{n}{\ell n + k}} < \frac{1}{\rho}$$

which contradicts $f \in A_\rho$.

Now let $|z| > \rho$. Then we have

$$h_n(z) = \sum_{k=0}^{n-1} a_{\ell n + k} z^k - \sum_{k=0}^{n} a_{\ell n + \ell + k} z^{\ell + k} + O\big(|z|^n (\rho - \varepsilon)^{-(\ell+2)n}\big) \tag{1.10}$$

$$= -\sum_{k=0}^{\ell} a_{(\ell+1)n + k} z^{n + k} + O\big((\rho - \varepsilon)^{-\ell n} + |z|^n (\rho - \varepsilon)^{-(\ell+2)n}\big).$$

Assuming again

$$\overline{\lim_{n\to\infty}} |\Delta_{n,\ell}(f;z_j)|^{1/n} < \frac{|z_j|}{\rho^{\ell+1}}, \quad j = 1,\ldots,\ell+1,$$

we also have

$$\overline{\lim_{n\to\infty}} |h_n(z_j)|^{1/n} < \frac{|z_j|}{\rho^{\ell+1}}, \quad j = 1,\ldots,\ell+1. \tag{1.11}$$

Thus from (1.10)

$$\sum_{k=0}^{\ell} a_{(\ell+1)n+k} z_j^k = \beta_{jn}, \quad j = 1,\ldots,\ell+1 \tag{1.12}$$

where

$$\beta_{j,n} = z_j^{-n} h_n(z_j) + O\big(|z_j|^{-n}(\rho-\varepsilon)^{-\ell n} + (\rho-\varepsilon)^{-(\ell+2)n}\big), \quad j = 1,\ldots,\ell+1.$$

Here by (1.11)

$$\overline{\lim_{n\to\infty}} |\beta_{j,n}|^{1/n} < \frac{1}{\rho^{\ell+1}}, \quad j = 1,\ldots,\ell+1$$

provided $\varepsilon > 0$ is so small that

$$\rho^{\ell+1} < \min\big(|z_j|(\rho-\varepsilon)^{\ell}, (\rho-\varepsilon)^{\ell+2}\big), \quad j = 1,\ldots,\ell+1.$$

But then from (1.10), with $m = (\ell+1)n+k$, $0 \le k \le \ell$, we get

$$\overline{\lim_{m\to\infty}} |a_m|^{1/m} \le \max_{1\le j\le \ell+1} \overline{\lim_{n\to\infty}} |\beta_{j,n}|^{\frac{1}{n}\frac{n}{(\ell+1)n+k}} < \frac{1}{\rho},$$

again a contradiction. □

We now show that distinguished points appearing in Theorem 2 indeed exist. This will be seen from an example given in the next theorem.

THEOREM 3. (i) *Given arbitrary points* $0 < |z_j| < \rho$, $j = 1,\ldots,\lambda$, $\lambda \le \ell-1$, *there exists a function* $f_1 \in A_\rho$ *such that*

$$\overline{\lim_{n\to\infty}} |\Delta_{n,\ell}(f_1;z_j)|^{1/n} \le \frac{\max(|z_j|,1)}{\rho^{\ell+1}}, \quad j = 1,\ldots,\lambda; \tag{1.13}$$

and at all other points $|z| < \rho$, *(1.5) holds.*
(ii) *Given arbitrary points* $|z_j| > \rho$, $j = 1,\ldots,\lambda$; $\lambda \le \ell$, *there exists a function* $f_2 \in A_\rho$ *such that*

$$\overline{\lim_{n\to\infty}} |\Delta_{n,\ell}(f_2;z_j)|^{1/n} \le \frac{\max(|z_j|,\rho^2)}{\rho^{\ell+2}}, \quad j = 1,\ldots,\lambda; \tag{1.14}$$

and at all other points $|z| > \rho$, (1.6) holds.

PROOF. (i) The system of equations

$$\sum_{k=1}^{\lambda} c_k z_j^k = 1, \quad j = 1, \ldots, \lambda \tag{1.15}$$

has a unique solution in c_k's, since it has a non-zero system determinant (Vandermonde). From this solution we construct

$$f_1(z) = z^{\ell-\lambda-1} \Big(\sum_{k=1}^{\lambda} c_k z^k - 1 \Big) \sum_{m=0}^{\infty} \Big(\frac{z}{\rho} \Big)^{\ell m} \in A_\rho.$$

This function has the Taylor coefficients

$$a_{\ell m + k} := \begin{cases} 0 & \text{if } 0 \le k \le \ell - \lambda - 2, \\ -\rho^{\ell m} & \text{if } k = \ell - \lambda - 1, \qquad m = 0, 1, \ldots, \\ c_{k-\ell+\lambda+1} \rho^{-\ell m} & \text{if } \ell - \lambda \le k \le \ell - 1. \end{cases} \tag{1.16}$$

Now (1.16) and (1.15) imply that

$$\sum_{k=0}^{\ell-1} a_{\ell m + k} z_j^k = \rho^{-\ell m} \Big(-z_j^{\ell-\lambda-1} + \sum_{k=\ell-\lambda}^{\ell-1} c_{k-\ell+\lambda+1} z_j^k \Big) \tag{1.17}$$

$$= \rho^{-\ell m} z_j^{\ell-\lambda-1} \Big(\sum_{k=1}^{\lambda} c_k z_j^k - 1 \Big) = 0, \quad j = 1, \ldots, \lambda.$$

For a positive integer n, let the integers r and s $(0 \le s < \ell)$ be determined such that $\ell r + s = (\ell+1)n$. Then by (1.17) we get

$$\sum_{k=0}^{n-1} a_{\ell n + k} z_j^k = z_j^{\ell(r-n)} \sum_{k=0}^{s-1} a_{\ell r + k} z_j^k + \sum_{\nu=n}^{r-1} z_j^{\ell(\nu-n)} \sum_{k=0}^{\ell-1} a_{\ell \nu + k} z_j^k$$

$$= z_j^{\ell(r-n)} \sum_{k=0}^{s-1} a_{\ell r + k} z_j^k$$

$$= O\Big(\frac{|z_j|^{\ell(r-n)}}{\rho^{\ell r}} \Big) = O\Big(\frac{|z_j|^n}{\rho^{(\ell+1)n}} \Big), \quad j = 1, \ldots, \lambda.$$

Hence by (1.3) and $|z_j| < \rho$ $(j = 1, \ldots, \lambda)$,

$$\Delta_{n,\ell}(f_1; z_j) = O\Big(\frac{|z_j|^n}{\rho^{(\ell+1)n}} + \frac{1}{(\rho-\varepsilon)^{(\ell+1)n}} \Big), \quad j = 1, \ldots, \lambda,$$

and since $\varepsilon > 0$ was arbitrary, (1.13) follows.

In order to show that at all other points $0 < |z| < \rho$, (1.5) holds, we get from (1.7) and (1.16) (denoting $c_0 = -1$)

$$\sum_{k=0}^{\ell-1} a_{\ell n+k} z^k = \rho^{-\ell n} \sum_{k=\ell-\lambda-1}^{\ell-1} c_{k-\ell+\lambda+1} z^k = h_n(z) + O\left(|z|^n (\rho - \varepsilon)^{-(\ell+1)n}\right).$$

$$(1.18)$$

Now if (1.5) does not hold then there exists an a, $0 < a < 1$ such that

$$|\Delta_{n,\ell}(f_1; z)| = O\left(\frac{a^n}{\rho^{\ell n}}\right), \quad \text{i.e.,} \quad |h_n(z)| = O\left(\frac{a^n}{\rho^{\ell n}}\right).$$

Thus (1.18) yields

$$z^{\ell-\lambda-1} \sum_{k=0}^{\lambda} c_k z^k = O\left(a^n + \frac{|z|^n \rho^{\ell n}}{(\rho - \varepsilon)^{(\ell+1)n}}\right) \to 0 \quad \text{as} \quad n \to \infty,$$

provided $\varepsilon > 0$ is so small that $0 < |z| < \frac{(\rho-\varepsilon)^{\ell+1}}{\rho^{\ell}}$. Hence

$$\sum_{k=1}^{\lambda} c_k z^k = 1,$$

and therefore by (1.15), z must be one of the z_j's.

(ii) Again, we consider the system of equations (1.15), but now with $|z_j| > \rho$ and $\lambda \le \ell$. The solution of this system defines a function

$$f_2(z) := z^{\ell-\lambda} \left(\sum_{k=1}^{\lambda} c_k z^k - 1\right) \sum_{m=0}^{\infty} \left(\frac{z}{\rho}\right)^{(\ell+1)m} \in A_\rho,$$

whose Taylor coefficients are

$$a_{(\ell+1)m+k} := \begin{cases} 0 & \text{if } 0 \le k \le \ell-\lambda-1, \\ -\rho^{-(\ell+1)m} & \text{if } k = \ell-\lambda, \\ c_{k-\ell+\lambda}\rho^{-(\ell+1)m} & \text{if } \ell-\lambda+1 \le k \le \ell, \end{cases} \quad (1.19)$$

for $m = 0, 1, 2, \ldots$.

Hence and by (1.15)

$$\sum_{k=0}^{\ell} a_{(\ell+1)m+k} z_j^k = \rho^{-(\ell+1)m} \left(-z_j^{\ell-\lambda} + \sum_{k=\ell-\lambda+1}^{\ell} c_{k-\ell+\lambda} z_j^k\right) \quad (1.20)$$

$$= \rho^{-(\ell+1)m} z_j^{\ell-\lambda} \left(\sum_{k=1}^{\lambda} c_k z_j^k - 1\right) = 0, \quad j = 1, \ldots, \lambda.$$

For a positive integer n, let the integers r and s ($0 \leq s < \ell+1$) be determined such that $\ell n + s = (\ell+1)r$. Then by (1.20)

$$\sum_{k=0}^{n-1} a_{\ell n+k} z_j^k = \sum_{k=0}^{s-1} a_{\ell n+k} z_j^k + \sum_{\nu=r}^{n-1} z_j^{(\ell+1)r-\ell n} \sum_{k=0}^{\ell} a_{(\ell+1)\nu+k} z_j^k$$

$$= \sum_{k=0}^{s-1} a_{\ell n+k} z_j^k = O(\rho^{-\ell n}), \quad j = 1, \ldots, \lambda.$$

Hence and by (1.3)

$$\Delta_{n,\ell}(f_2; z_j) = O\left(\rho^{-\ell n} + \frac{|z_j|^n}{(\rho-\varepsilon)^{(\ell+2)n}}\right) = O\left(\frac{\max(|z_j|^n, \rho^{2n})}{(\rho-\varepsilon)^{(\ell+2)n}}\right), \quad j = 1, \ldots, \lambda,$$

whence (1.14) follows, since $\varepsilon > 0$ was arbitrary.

Finally, (1.10) and (1.19) imply (denoting $c_0 = -1$)

$$\sum_{k=0}^{\ell} a_{(\ell+1)n+k} z^{k+n} = \rho^{-(\ell+1)n} \sum_{k=\ell-\lambda}^{\ell} c_{k-\ell+\lambda} z^{k+n} \qquad (1.21)$$

$$= -h_n(z) + O\left(\frac{\max(|z|^n, \rho^{2n})}{(\rho-\varepsilon)^{(\ell+2)n}}\right).$$

If (1.6) does not hold then there exists an a, $0 < a < 1$, such that

$$\Delta_{n,\ell}(f_2; z)| = O\left(\frac{a^n |z|^n}{\rho^{(\ell+1)n}}\right), \quad \text{i.e.,} \quad |h_n(z)| = O\left(\frac{a^n |z|^n}{\rho^{(\ell+1)n}}\right).$$

Thus (1.21) yields

$$z^{\ell-\lambda} \sum_{k=0}^{\lambda} c_k z^k = O\left(a^n + \frac{\rho^{(\ell+1)n}}{(\rho-\varepsilon)^{(\ell+2)n}} + \frac{\rho^{(\ell+3)n}}{|z|^n (\rho-\varepsilon)^{(\ell+2)n}}\right) \to 0 \quad \text{as} \quad n \to \infty$$

provided $|z| > \frac{\rho^{\ell+3}}{(\rho-\varepsilon)^{\ell+2}}$. Hence

$$\sum_{k=0}^{\lambda} c_k z^k = 0$$

which shows, by (1.15), that z must be one of the z_j's. $\qquad \square$

So far distinguished points on the circle $|z| = \rho$ were excluded. Next we show that any distinct $\ell + 1$ points on this circle can always be preassigned as distinguished points.

THEOREM 4. *Given any distinct $\ell+1$ points z_j, with $|z_j| = \rho$, $j = 1,\ldots,\ell+1$, there exists a function $f_3(z) \in A_\rho$ such that*

$$\overline{\lim_{n\to\infty}} |\Delta_{n,\ell}(f_3; z_j)|^{1/n} \leq \frac{1}{\rho^{\ell+1}}, \quad j = 1,\ldots,\ell+1. \tag{1.22}$$

PROOF. Consider the system of equations

$$\sum_{s=0}^{\ell} b_{(\ell+1)n+s} w_j^{n+s} = \sum_{k=0}^{\ell-1} b_{\ell n+k} w_j^k, \quad w_j = z_j \rho^{-1}, \quad j = 1,\ldots,\ell+1 \tag{1.23}$$

in the unknowns $\{b_{(\ell+1)n+s}\}_{s=0}^{\ell}$. It is uniquely solvable, since the determinant of the system is Vandermonde:

$$b_{(\ell+1)n+s} = \sum_{k=0}^{\ell-1} c_{k,s,n} b_{\ell n+k}, \quad s = 0,\ldots,\ell \tag{1.24}$$

where

$$|c_{k,s,n}| \leq M_1, \quad k = 0,\ldots,\ell-1; \quad s = 0,\ldots,\ell; \quad n = 1,2,\ldots$$

($M_1 > 0$ is independent of n). Let $n_0 = \ell + 2$ and set $b_0 = b_1 = \cdots = b_{\ell n_0 - 1} = b_{\ell n_0} = 1$. Then the next $n_0 - 1 = \ell + 1$ numbers $b_{\ell n_0 + 1},\ldots,b_{(\ell+1)n_0 - 1}$ can be determined from the system of equations

$$\sum_{k=0}^{n_0-1} b_{\ell n_0 + k} w_j^k = 0, \quad j = 1,\ldots,\ell+1. \tag{1.25}$$

Using (1.24) successively with $n = n_0, n_0 + 1,\ldots$, we determine all the b_k, $k \geq (\ell+1)n_0$, in turn uniquely.

We now show that

$$\overline{\lim_{n\to\infty}} |b_n|^{1/n} = 1. \tag{1.26}$$

For any positive integer ν, let $\nu = (\ell+1)n+k$, $0 \leq k < \ell+1$. Then from (1.24) we have

$$|b_\nu| \leq M_1 \ell \cdot \max_{\ell n \leq j < \ell(n+1)} |b_j| \leq M_1 \ell \cdot \max_{j \leq \ell n + \ell} |b_j| \leq M_1 \ell \cdot \max_{j \leq \nu\ell/(\ell+1)+\ell} |b_j|.$$

If we set $\varphi(x) = \max_{\nu \leq x} |b_\nu|$ then

$$\varphi(N) \leq M_1 \ell \cdot \max_{j \leq N\ell/(\ell+1)+\ell} |b_j| = M_1 \ell \varphi\left(\frac{\ell N}{\ell+1} + \ell\right),$$

so that

$$\varphi(N) \leq (M_1\ell)^s \varphi\left(\left(\frac{\ell}{\ell+1}\right)^s N + \ell \sum_{j=0}^{s-1} \left(\frac{\ell}{\ell+1}\right)^j\right)$$

$$\leq (M_1\ell)^s \varphi\left(\left(\frac{\ell}{\ell+1}\right)^s N + \ell(\ell+1)\right), \quad s = 1, 2, \ldots .$$

Now let

$$s = \left[\frac{\log \frac{N}{\ell(\ell+1)}}{\log \left(1 + \frac{1}{\ell}\right)}\right] + 1,$$

then the previous iterative estimate yields

$$\varphi(N) \leq (M_1\ell)^s \varphi\left(2\ell(\ell+1)\right)$$

whence

$$\varlimsup_{n \to \infty} |b_n|^{1/n} \leq 1. \tag{1.27}$$

In order to show the opposite inequality, we observe that from the first ℓ equations in (1.23) we get

$$b_{\ell n + k} = \sum_{s=0}^{\ell} d_{s,k,n} b_{(\ell+1)n+s}, \quad k = 0, \ldots, \ell-1 \tag{1.28}$$

where

$$|d_{s,k,n}| \leq M_2, \quad s = 0, \ldots, \ell; \quad k = 0, \ldots, \ell-1; \quad n = 1, 2, \ldots$$

($M_2 > 0$ is independent of n). Let $m_0 = \ell n_0$, and choose $m_1 > m_0$ such that

$$|b_{m_1}| = \max_{(\ell+1)n_0 \leq s < (\ell+1)(n_0+1)} |b_s|.$$

Then by (1.28) (with $n = n_0$, $k = 0$)

$$1 = b_{m_0} \leq M_2(\ell+1)|b_{m_1}|. \tag{1.29}$$

Proceeding inductively, in general let $m_t = \ell n_t + k_t$, $0 \leq k_t < \ell$ be defined, then m_{t+1} is defined by

$$|b_{m_{t+1}}| = \max_{(\ell+1)n_t \leq s < (\ell+1)(n_t+1)} |b_s|,$$

so that by (1.28) (with $n = n_t$, $k = k_t$)

$$|b_{m_t}| \leq M_2(\ell+1)|b_{m_{t+1}}|, \quad t = 0, 1, \ldots .$$

Hence and by (1.29),

$$b_{m_t} \geq (M_2(\ell+1))^{-t}, \quad t = 0, 1, \ldots . \tag{1.30}$$

Since $(\ell+1)n_t \leq m_{t+1}$, we get

$$n_t = \frac{m_t - k_t}{\ell} \geq \frac{(\ell+1)n_{t-1} - \ell + 1}{\ell} \geq \cdots \geq \left(\frac{\ell+1}{\ell}\right)^t n_0 - \frac{\ell-1}{\ell} \sum_{j=0}^{t-1} \left(\frac{\ell+1}{\ell}\right)^j$$

$$> \left(\frac{\ell+1}{\ell}\right)^t (n_0 - \ell + 1) > \left(\frac{\ell+1}{\ell}\right)^t, \quad t = 1, 2, \ldots .$$

Thus

$$t \leq \frac{\log n_t}{\log \frac{\ell+1}{\ell}} \leq \frac{\log \frac{m_t}{\ell}}{\log \frac{\ell+1}{\ell}},$$

and (1.30) yields

$$b_{m_t} \geq (M_2(\ell+1))^{-\log \frac{m_t}{\ell} / \log \frac{\ell+1}{\ell}}, \quad t = 1, 2, \ldots$$

whence

$$\lim_{t \to \infty} |b_{m_t}|^{1/m_t} \geq 1.$$

This together with (1.27) proves (1.26).

We now set $a_k := \rho^{-k} b_k$, $k = 0, 1, \ldots$ and $f_3(z) := \sum_0^\infty a_k z^k$. From (1.26) we have $\varlimsup_{n \to \infty} |a_n|^{1/n} = \rho^{-1}$, i.e., $f_3 \in A_\rho$. Also following the reasoning in (1.7) and (1.10), we get for $|z| = \rho$,

$$h_n(z) = \sum_{k=0}^{\ell-1} a_{\ell n + k} z^k - \sum_{k=0}^{\ell} a_{(\ell+1)n + k} z^{n+k} + O\big((\rho - \varepsilon)^{-(\ell+1)n}\big),$$

whence using (1.23), we have

$$h_n(z_j) = \Delta_{n,\ell}(f_3; z_j) - z_j^\ell \Delta_{n+1,\ell}(f_3; z_j) = O\big((\rho - \varepsilon)^{-(\ell+1)n}\big),$$
$$j = 1, \ldots, \ell + 1.$$

Thus

$$|\Delta_{n,\ell}(f_3; z_j) - z_j^{\ell n} \Delta_{2n,\ell}(f_3; z_j)| = \Big| \sum_{s=0}^{n-1} z_j^{\ell s} \big(\Delta_{n+s,\ell}(f_3, z_j) - z_j^\ell \Delta_{n+s+1,\ell}(f_s; z_j)\big)$$

$$\leq \sum_{s=0}^{n-1} |\Delta_{n+s,\ell}(f_3; z_j) - z_j^\ell \Delta_{n+s+1,\ell}(f_3; z_j)|$$

$$= O\big(n(\rho - \varepsilon)^{-(\ell+1)n}\big), \ j = 1, \ldots, \ell,$$

whence using (1.1)

$$|\Delta_{n,\ell}(f_3; z_j)| = |\Delta_{2n,\ell}(f_3; z_j) + O\big(n(\rho - \varepsilon)^{-(\ell+1)n}\big)$$
$$= O\big(\rho^{-2\ell n} + n(\rho - \varepsilon)^{-(\ell+1)n}\big), \quad j = 1, \ldots, \ell+1,$$

i.e.,

$$\varlimsup_{n \to \infty} |\Delta_{n,\ell}(f_3; z_j)|^{\frac{1}{n}} \le (\rho - \varepsilon)^{-\ell-1}, \quad j = 1, \ldots, \ell+1.$$

Since $\varepsilon > 0$ is arbitrary, this proves the theorem.

We now want to give general conditions for a given set of points $Z = U \cup V$ to be distinguished, where

$$U = \{u_1, \ldots, u_\mu\}, \quad V = \{v_1, \ldots, v_\nu\}$$

with

$$|u_j| < \rho \quad (j = 1, \ldots, \mu), \qquad |v_j| > \rho \quad (j = 1, \ldots, \nu).$$

Set

$$X = \begin{pmatrix} 1 & u_1 & \cdots & u_1^{\ell-1} \\ \vdots & \vdots & & \vdots \\ 1 & u_\mu & \cdots & u_\mu^{\ell-1} \end{pmatrix}, \quad Y = \begin{pmatrix} 1 & v_1 & \cdots & v_1^{\ell} \\ \vdots & \vdots & & \vdots \\ 1 & v_n u & \cdots & v_\nu^{\ell} \end{pmatrix}$$

and

$$M = M(X, Y) = \begin{pmatrix} X & & & & \\ & X & & & \\ & & \ddots & & \\ & & & X & \\ Y & & & & \\ & Y & & & \\ & & \ddots & & \\ & & & Y \end{pmatrix},$$

where X is repeated $\ell + 1$ times and Y is repeated ℓ times. (The Y's begin below the last row of the last X.)

THEOREM 5. *The set Z defined above is (ℓ, ρ)-distinguished if and only if*

$$\operatorname{rank} M < \ell(\ell+1).$$

As a corollary, we obtain the following reformulation of Theorem 2:

COROLLARY 1. *If either $\mu \geq \ell$ or $\nu \geq \ell+1$ (i.e., there are at least ℓ points in $|z| < \rho$ or at least $\ell+1$ points in $|z| > \rho$), then Z is not an (ℓ, ρ)-distinguished set.*

Namely, if $\mu \geq \ell$, we take the minor of M which consists of the first ℓ rows of each X in M. Its determinant is the $(\ell+1)^{\text{th}}$ power of the Vandermonian of U, thus nonzero. Similar reasoning applies for the set V if $\nu \geq \ell+1$.

The second corollary is a reformulation of Theorem 3:

COROLLARY 2. *If $\mu + \nu \leq \ell, \nu > 0$ or $\mu < \ell, \nu = 0$ then Z is an (ℓ, ρ)-distinguished set.*

Namely, the number of rows in M is $\mu(\ell+1)+\nu < \ell(\ell+1)$ so that rank $M < \ell(\ell+1)$.

Proof of Theorem 5. (a) *Sufficiency.* Suppose rank $M < \ell(\ell+1)$. Then there exists a nonzero vector $\mathbf{b} = (b_0, \ldots, b_{\ell(\ell+1)-1})$ such that

$$M \cdot \mathbf{b}^T = 0. \tag{1.31}$$

Set

$$f(z) = \sum_{N=0}^{\infty} a_N z^N = \left[1 - \left(\frac{z}{\rho} \right)^{\ell(\ell+1)} \right]^{-1} \sum_{k=0}^{\ell(\ell+1)} b_k z^k.$$

Evidently $f \in A_\rho$ and

$$a_N = b_k \rho^{-\ell(\ell+1)m}, \tag{1.32}$$

where $N = \ell(\ell+1)m+k$, $k = 0,1,\ldots,\ell(\ell+1)-1$. From (1.30) and (1.31), we have

$$\begin{cases} \sum_{k=0}^{\ell} a_{(\ell+1)m+k} z^k = 0, & z \in U, \\ \sum_{k=0}^{\ell-1} a_{\ell m+k} z^k = 0, & z \in V \end{cases} \tag{1.33}$$

for each m. For any positive integer n, we can find integers r and s such that $\ell n + s = (\ell+1)r$, $0 \leq s < \ell+1$. Then (1.32) and (1.33) give

$$\sum_{k=0}^{n-1} a_{\ell n+k} z^k = \sum_{k=0}^{s-1} z^{(\ell+1)m-\ell n} \sum_{k=0}^{\ell} a_{(\ell+1)m+k} z^k$$

$$= \sum_{k=0}^{s-1} a_{\ell n+k} z^k = O(\rho^{-\ell n}), \quad z \in V.$$

Hence

$$\Delta_{\ell n-1}(f;z) = \sum_{k=0}^{n-1} a_{\ell n+k} z^k + O((K_{\ell+1}(\rho,|z|)+\varepsilon)^n) \qquad (1.34)$$

$$= O(\rho^{\ell n} + (K_{\ell+1}(\rho,|z|)+\varepsilon)^n)$$

$$= O((K_{\ell+1}(\rho,|z|)-\varepsilon)^n), \qquad z \in V,$$

which shows that the set V is indeed (ℓ,ρ)-distinguished.

Now, if for any positive integer n, we determine r and s $(0 \le s < \ell)$ such that $\ell r + s = (\ell+1)n$, then we have from (1.32) and (1.33)

$$\sum_{k=0}^{n-1} a_{\ell n+k} z^k = z^{\ell(r-n)} \sum_{k=0}^{s-1} a_{\ell r+k} z^k + \sum_{m=n}^{r-1} z^{\ell(m-n)} \sum_{k=0}^{\ell-1} a_{\ell m+k} z^k$$

$$= z^{\ell(r-n)} \sum_{k=0}^{s-1} a_{\ell r+k} z^k = O\left(\frac{|z|^{\ell(r-n)}}{\rho^{\ell r}}\right)$$

$$= O\left(\frac{|z|^n}{\rho^{(\ell+1)n}}\right) = O\left(\rho^{-\ell n}\left(\frac{|z|}{\rho}\right)^n\right), \qquad z \in U.$$

Therefore, we obtain

$$\Delta_{\ell,n-1}(f;z) = O\left(\left(\frac{1}{\rho^\ell}-\varepsilon\right)^n + K_{\ell+1}(\rho,|z|)+\varepsilon)^n\right) \qquad (1.35)$$

$$= O\left(\left(\frac{1}{\rho^\ell}-\varepsilon\right)^n\right), \qquad z \in U,$$

which shows that the set U is also (ℓ,ρ)-distinguished.

(b) *Necessity.* Suppose $B_\ell(f;z) < K_\ell(\rho,|z|)$ for $z \in Z$ and that rank $M = \ell(\ell+1)$. We shall show that this leads to a contradiction. Set

$$h(z) := \Delta_{\ell,n-1}(f;z) - z^\ell \Delta_{\ell,n}(f;z).$$

Then

$$h(z) = \sum_{k=0}^{\ell-1} a_{\ell n+k} z^k - \sum_{k=0}^{\ell} a_{(\ell+1)n+k} z^{n+k} + O((K_{\ell+1}(\rho,|z|)+\varepsilon)^n).$$

By supposition we have

$$h(z) = \sum_{k=0}^{\ell-1} a_{\ell n+k} z^k + O\left(\left(\frac{1}{\rho}+\varepsilon\right)^{n(\ell+1)}|z|^n + \left(\frac{1}{\rho^{\ell+1}}+\varepsilon\right)^n\right)$$

$$= O\left(\left(\frac{1}{\rho^\ell}-\varepsilon\right)^n\right), \qquad z \in U,$$

so that

$$\sum_{k=0}^{\ell-1} a_{\ell n+k} z^k = O\left(\left(\frac{1}{\rho^\ell} - \varepsilon\right)^n\right), \qquad z \in U. \tag{1.36}$$

Similarly

$$h(z) = -\sum_{k=0}^{\ell} a_{(\ell+1)n+k} z^{n+k} + O\left(\left(\frac{1}{\rho} + \varepsilon\right)^{n\ell} + \left(\frac{|z|}{\rho^{\ell+2}} + \varepsilon\right)^n\right)$$

$$= O\left(\left(\frac{|z|}{\rho^{\ell+1}} - \varepsilon\right)^n\right), \qquad z \in V,$$

that is

$$\sum_{k=0}^{\ell} a_{(\ell+1)n+k} z^k = O\left(\left(\frac{1}{\rho^{\ell+1}} - \varepsilon\right)^n\right), \qquad z \in V. \tag{1.37}$$

Since (1.36) and (1.37) are true for every value of n, we put $n = (\ell+1)m + \lambda$ in (1.36) and $n = \ell m + \lambda$ in (1.37). Then we obtain

$$\sum_{k=0}^{\ell-1} a_{\ell(\ell+1)m+\ell\lambda+k} z^k = O\left(\left(\frac{1}{\rho^{\ell(\ell+1)}} - \varepsilon\right)^m\right), \; z \in U, \lambda = 0,\ldots,\ell, \; m = 0,1,\ldots$$

$$\tag{1.38a}$$

and

$$\sum_{k=0}^{\ell} a_{\ell(\ell+1)m+(\ell+1)\lambda+k} z^k = O\left(\left(\frac{1}{\rho^{\ell(\ell+1)}} - \varepsilon\right)^m\right), \quad z \in V, \lambda = 0,\ldots,\ell-1,$$

$$, m = 0,1,\ldots. \tag{1.38b}$$

The matrix of the system of equations (1.38a)-(1.38b) is M which, by hypothesis, has full rank. So the equalities (1.38a)-(1.38b) can be uniquely solved to determine the a_k's. Hence we have

$$a_{\ell(\ell+1)m+k} = O\left(\left(\frac{1}{\rho^{\ell(\ell+1)}} - \varepsilon\right)^n\right), \qquad k = 0,1,\ldots,\ell(\ell+1) - 1,$$

so that $\limsup_{N\to\infty} |a_N|^{1/N} \le 1/\rho - \varepsilon$, which contradicts the fact that $f \in A_\rho$.
□

4.2. Hermite Interpolation

In this section we generalize the sharpness results obtained for Lagrange interpolation, to the case of rth order Hermite interpolation. Although the methods to be used are similar, new ideas applied to the more complicated situation are needed.

For the $\Delta_{rn-1,\ell}(f; z)$ introduced in Chapter 2, (1.14), we first prove

THEOREM 6. *Let $r, \ell \in \mathbb{N}$ and $\rho > 1$. Then for any $f \in A_\rho$ we have*

$$\overline{\lim_{n \to \infty}} \max_{|z|=\mu} |\Delta_{rn-1,\ell}(f;z)|^{\frac{1}{rn}} = \begin{cases} \rho^{-1-\frac{\ell-1}{r}}, & \text{if } \mu \leq 1, \\ \frac{\mu^{1-1/r}}{\rho^{1+(\ell-1)/r}}, & \text{if } 1 \leq \mu \leq \rho, \\ \frac{\mu}{\rho^{1+\ell/r}}, & \text{if } \rho \leq \mu. \end{cases}$$

PROOF. It is easily seen that if $f(z) = \sum\limits_{k=0}^{\infty} a_k z^k$ then

$$h_{rn-1}(f;z) = \sum_{k=0}^{rn-1} a_k z^k + \sum_{j=1}^{\infty} \beta_{j,r}(z^n) \sum_{k=0}^{n-1} a_{(r+j-1)n+k} z^k,$$

where $\beta_{j,n}(z)$ is defined in (1.7) of Chapter 2. On the other hand, from (1.14), Chapter 2 we get

$$\Delta_{rn-1,\ell}(f;z) = \sum_{j=\ell}^{\infty} \beta_{j,r}(z^n) \sum_{k=0}^{n-1} a_{(r+j-1)n+k} z^k. \tag{2.1}$$

Hence and from Lemma 2 of Chapter 1 we obtain

$$|\Delta_{rn-1,\ell}(f;z)| = O\left(\max\{1, |z|^{n(r-1)}\} \sum_{j=\ell}^{\infty} j^{r-1} \sum_{k=0}^{n-1} |z|^k (\rho - \varepsilon)^{-(r+j-1)n-k} \right)$$

where $\varepsilon > 0$ is arbitrary. Hence, distinguishing the cases $\mu < 1$, $1 \leq \mu < \rho$ and $\mu \geq \rho$, we get

$$\overline{\lim_{n \to \infty}} \max_{|z|=\mu} |\Delta_{rn-1,\ell}(f;z)|^{\frac{1}{rn}} \leq \rho^{-1-\frac{\ell-1}{r}} \max\left\{1, \mu^{1-\frac{1}{r}}, \mu \rho^{-\frac{1}{r}}\right\} := K_{r,\ell}(\rho, \mu). \tag{2.2}$$

To prove the opposite inequality, first notice that (2.1) and (2.2) imply

$$\Delta_{rn-1,\ell}(f;z) = \beta_{\ell,r}(z^n) \sum_{k=0}^{n-1} a_{(r+\ell-1)n+k} z^k + \Delta_{rn-1,\ell+1}(f;z) \tag{2.3}$$

$$= \beta_{\ell,r}(z^n) \sum_{k=0}^{n-1} a_{(r+\ell-1)n+k} z^k + O\big((K_{r,\ell+1}(\rho, \mu) + \varepsilon)^{nr}\big)$$

with an arbitrary $\varepsilon > 0$. Dividing by z^{k+1} and integrating

$$\frac{1}{2\pi i} \int_{\Gamma_\mu} \frac{\Delta_{rn-1,\ell}(f;z)}{z^{k+1}} \, dz = \beta_{\ell,r}(0) a_{(r+\ell-1)n} + O\big(\mu^{-k}(K_{r,\ell+1}(\rho, \mu) + \varepsilon)^{nr}\big),$$

$$k = 0, \ldots, n-1.$$

Since from (4.8), we get $\beta_{\ell,r}(0) = (-1)^{r-1} \begin{pmatrix} r + \ell - 2 \\ r - 1 \end{pmatrix} \neq 0$, it follows that

$$|a_{(r+\ell-1)n+k}| \leq c_1 \mu^{-k} \max_{|z|=\mu} |\Delta_{rn-1,\ell}(f;z)| + O(\mu^{-k}(K_{r,\ell+1}(\rho,\mu) + \varepsilon)^{nr}),$$
$$k = 0, \dots, n-1. \tag{2.4}$$

If $\mu \leq 1$ then putting $m = (r + \ell - 1)n + k$, $k = 0, \dots, r + \ell - 2$, we get

$$\overline{\lim_{n \to \infty}} \max_{|z|=\mu} |\Delta_{rn-1,\ell}(f;z)|^{\frac{1}{rn}} \geq \overline{\lim_{m \to \infty}} [\mu^k |a_m| - O((K_{r,\ell+1}(\rho,\mu) + \varepsilon)^{nr})]^{\frac{1}{rn}}$$

$$= \rho^{-1 - \frac{\ell-1}{r}}.$$

If $\mu > 1$ then by (2.1) and by (4.7) of Chapter 2, we have

$$\frac{1}{2\pi i} \int_{\Gamma_r} \frac{\Delta_{rn-1,\ell}(f;z)}{z^{(r-1)n+k+1}} dz = \begin{pmatrix} r + \ell - 1 \\ r - 1 \end{pmatrix} a_{(r+\ell-1)n+k} +$$
$$+ O(\mu^{-(r-1)n-k-1}(K_{r,\ell+1}(\rho,\mu) + \varepsilon)^{nr}), \quad k = 1, \dots, n-1. \tag{2.5}$$

Now in case $1 < \mu \leq \rho$, set $m = (r + \ell - 1)n + k$, $k = 0, \dots, r + \ell - 2$. Then as above

$$\overline{\lim_{n \to \infty}} \max_{|z|=\mu} |\Delta_{rn-1,\ell}(f;z)|^{\frac{1}{rn}} \geq \overline{\lim_{m \to \infty}} [\mu^{(r-1)n} |a_m| - O((K_{r,\ell+1}(\rho,\mu) + \varepsilon)^{rn})]^{\frac{1}{rn}}$$

$$\geq \mu^{1 - \frac{1}{r}} \rho^{-1 - \frac{\ell-1}{r}}.$$

When $\mu > \rho$, we take $m = (r + \ell - 1)n + k$, $k = n - r - \ell + 1, \dots n - 1$. Then from (2.5) we have

$$\overline{\lim_{n \to \infty}} \max_{|z|=\mu} |\Delta_{rn-1,\ell}(f;z)|^{\frac{1}{rn}} \geq \overline{\lim_{m \to \infty}} [\mu^{rn} |a_m| - O((K_{r,\ell+1}(\rho,\mu) + \varepsilon)^{rn})]^{\frac{1}{rn}}$$

$$\geq \mu \rho^{-1 - \frac{\ell}{r}}.$$

Theorem 6 is completely proved. □

4.3. The Distinguished Role of the Roots of Unity for the Circle

In the previous sections we described the exact order of magnitude of the overconvergence in case of the roots of unity as nodes. The question naturally arises whether these nodes are optimal among all possible systems of nodes. The conjecture is that the roots of unity play a distinguished role and, as we will see, this is indeed the situation. However, this is a highly nontrivial statement which requires a number of new ideas.

First of all we make the following natural restriction on the nodes of interpolation (which can be multiple):

$$|z_k| < \rho, \quad k = 1, \ldots, n. \tag{3.1}$$

This upper limit is the consequence of the fact that the function is analytic (and
defined) only in $|z| < \rho$.

The other restriction is the following: let $\omega_n(z) = \prod_{k=1}^{n} (z - z_k)$, and let

$$\gamma_n(\rho) = \text{ the modulus of the first nonzero term of } \begin{cases} \omega_n(\rho) & \text{if } \ell > 1 \\ \omega_n(\rho) - \rho^n & \text{if } \ell = 1 \end{cases}$$
$$\tag{3.2}$$

(as for ℓ, see (4.4) below). This quantity is always positive, except when $\omega_n(z) = z^n$ and $\ell = 1$. In the latter case let $\gamma_n(\rho) = 0$. Now our restriction is that

$$\mu(\rho) := \varlimsup_{n \to \infty} \gamma_n(\rho)^{1/n} \geq 1 \text{ for all } \rho > 1. \tag{3.3}$$

In case of roots of unity on the circle $|z| = r$ we have $\mu(\rho) = r$. Thus the
case $r < 1$ is excluded by (3.3). The reason for this exclusion is that in this
case (as is easily seen) the region of overconvergence is $|z| < \rho^{\ell+1}/r$, which is
larger than $|z| < \rho^{\ell+1}$, the region of overconvergence for the roots of unity on
$|z| = 1$. Hence by letting $r \to 0$, the region of overconvergence becomes the
whole complex plane. In order to avoid this situation, we make the restriction
(3.3).

Now the main result of this section is the following: Let $L_{n-1}(f, z, Z)$ be
the Lagrange (or Hermite, depending on the multiple nodes z_k) interpolation
polynomial of $f \in A_\rho$ based on the nodes Z, and let

$$\Delta_{n,\ell}(f, z, Z) := L_{n-1}(f, z, Z) - \sum_{j=0}^{\ell-1} p_{n-1,j}(f; z) \tag{3.4}$$

(see the analogous definition (1.2) in case of roots of unity on $|z| = 1$). Further
let

$$\Delta_\ell(r, \rho, Z) := \sup_{f \in A_\rho} \varlimsup_{n \to \infty} \left\{ \max_{|z|=r} |\Delta_{n,\ell}(f, z, Z)| \right\}^{1/n}, \tag{3.5}$$

and denote by E the matrix of nodes of the roots of unity on $|z| = 1$.

THEOREM 7. Let $Z = \{z_{kn}\}_{k=1}^{n}$ be a system of nodes of Lagrange interpo-
lation satisfying (3.1) and (3.3). Then for any $|\hat{z}| > \max \left(\frac{\rho^{\ell+1}}{\mu(\rho)}, \rho \right)$ there is an
$f_0 \in A_\rho$ such that

$$\varlimsup_{n \to \infty} |\Delta_{n,\ell}(f_0, \hat{z})| = \infty. \tag{3.6}$$

In addition we have

$$\Delta_\ell(r, \rho, Z) \geq \mu(\rho) r / \rho^{\ell+1} \geq \Delta_\ell(r, \rho, E) \quad \text{for all} \quad r > \rho > 1. \qquad (3.7)$$

(3.7) shows that indeed, the matrix E is the "best" among all matrices satisfying (3.1) and (3.3).

PROOF. Let $|u| = \rho$ and

$$f_u(z) := \frac{1}{u - z} \in A_\rho. \qquad (3.8)$$

A simple computation shows that

$$L_{n-1}(f_u, z, Z) = \frac{\omega_n(u) - \omega_n(z)}{\omega_n(u)(u - z)}, \quad \sum_{j=0}^{\ell-1} p_{n-1}(f_u; z) = \frac{(u^{\ell n} - 1)(u^n - z^n)}{(u^n - 1)(u - z)u^{\ell n}}. \qquad (3.9)$$

Thus for any $|z| = r > \rho$ we have

$$|\Delta_{n,\ell}(f_u, z, Z)| \geq \frac{|\Omega_n(u; z)|}{(r + \rho)\rho^{\ell n}|\omega_n(u)|} \qquad (3.10)$$

where

$$\Omega_n(u; z) := u^{\ell n}\big(\omega_n(u) - \omega_n(z)\big) - \frac{u^{\ell n} - 1}{u^n - 1}(u^n - z^n)\omega_n(u) \qquad (3.11)$$

is a polynomial in u.

Denote j_n, $0 \leq j_n \leq n$, the multiplicity of 0 among the z_k's. Then we can write

$$\omega_n(u) = u^{j_n} \widetilde{\omega}_n(u), \quad \text{where} \quad \widetilde{\omega}_n(0) \neq 0. \qquad (3.12)$$

Similarly,

$$\Omega_n(u; z) = u^{j_n} \widetilde{\Omega}_n(u; z) \qquad (3.13)$$

where

$$\widetilde{\Omega}_n(u; z) := \left\{ u^{\ell n} - \frac{u^{\ell n} - 1}{u^n - 1}(u^n - z^n) \right\} \widetilde{\omega}_n(u) - u^{\ell n - j_n} z^{j_n} \widetilde{\omega}_n(z). \qquad (3.14)$$

(3.12) and (3.13) imply

$$\max_{|u|=\rho} \left| \frac{\Omega_n(u; z)}{\omega_n(u)} \right| = \max_{|u|=\rho} \left| \frac{\widetilde{\Omega}_n(u; z)}{\widetilde{\omega}_n(u)} \right|. \qquad (3.15)$$

Let

$$\widetilde{\omega}_n(u) := \prod_{k=1}^{n-j_n}{}'(u - z'_k) \quad \text{with} \quad |z'_k| < \rho, \tag{3.16}$$

where the dash indicates that in case $j_n = n$, the product is identically 1. Now

$$\widetilde{R}_n(u; z) := \frac{\widetilde{\Omega}_n(u; z)}{\displaystyle\prod_{k=1}^{n-j_n}{}'\left(\rho - \frac{\overline{z}'_k u}{\rho}\right)} \tag{3.17}$$

as a function of u, is analytic in $|u| \le \rho \left(< \frac{\rho^2}{\max |z'_k|} \right)$. Thus by (3.15) and (3.16)

$$\max_{|u|=\rho} |\widetilde{R}_n(u; z)| = \max_{|u|=\rho} \frac{|\widetilde{\Omega}_n(u; z)|}{\displaystyle\prod_{k=1}^{n-j_n}{}'\left|\rho - \frac{\overline{z}'_k u}{\rho}\right|} \tag{3.18}$$

$$= \max_{|u|=\rho} \left| \frac{\widetilde{\Omega}_n(u; z)}{\widetilde{\omega}_n(u)} \right| = \max_{|u|=\rho} \left| \frac{\Omega_n(u; z)}{\omega_n(u)} \right|.$$

Now fix an arbitrary $\widehat{z}$ such that

$$\mu(\rho)|\widehat{z}| > \rho^{\ell+1}, \tag{3.19}$$

and let $u_n = u_n(\widehat{z})$ denote a point on $|u| = \rho$ where $|\widetilde{R}_n(u; \widehat{z})|$ attains its maximum. By the maximum principle, (3.17) - (3.18) yield

$$\max_{|u|=\rho} \left| \frac{\Omega_n(u; \widehat{z})}{\omega_n(u)} \right| = |\widetilde{R}_n(u_n; \widehat{z})| \ge |\widetilde{R}_n(0; \widehat{z})| = \frac{|\widetilde{\Omega}_n(0; \widehat{z})|}{\rho^{n-j_n}}. \tag{3.20}$$

(3.14) implies, with $|\widehat{z}| = r$,

$$\widetilde{\Omega}_n(0; \widehat{z}) = \begin{cases} r^n |\widetilde{\omega}_n(0)| & \text{if } \ell > 1, \text{ or } \ell = 1 \text{ and } j_n < n \\ 0 & \text{if } \ell = 1 \text{ and } j_n = n. \end{cases}$$

From the definition (3.2) it follows in all cases that

$$|\widetilde{\Omega}_n(0; \widehat{z})| = \frac{r^n \gamma_n(\rho)}{\rho^{j_n}}. \tag{3.21}$$

Thus combining (3.10), (3.20) and (3.21) yields

$$|\Delta_{n,\ell}(f_u, \widehat{z}, Z)| \ge \left(\frac{r}{\rho^{\ell+1}} \right)^n \frac{\gamma_n(\rho)}{r + \rho} \quad \text{for all} \quad n \ge 1. \tag{3.22}$$

Recalling the hypothesis (3.3), let $\varepsilon > 0$ be arbitrary such that

$$(\mu(\rho) - \varepsilon)r > \rho^{\ell+1}, \quad |\widehat{z}| = r \tag{3.23}$$

(cf. (3.19)), and let $n_1 < n_2 < \ldots$ be a sequence of integers such that

$$\gamma_{n_j}(\rho) \geq (\mu(\rho) - \varepsilon)^{n_j}, \quad \text{for all} \quad j \geq 1. \tag{3.24}$$

Then (3.22) yields

$$|\Delta_{n_j,\ell}(f_j, \widehat{z}, Z)| \geq \frac{1}{r + \rho} \left[\frac{(\mu(\rho) - \varepsilon)r}{\rho^{\ell+1}} \right]^{n_j} \quad \text{for all} \quad j \geq 1, \tag{3.25}$$

where, for convenience, we have set $f_j := f_{u_{n_j}}$. Further let

$$\rho_n := \max_{1 \leq k \leq n} |z_k| < \rho, \quad n \geq 1. \tag{3.26}$$

We may suppose that there exists a sequence of indices $\{n_j\}_{j=1}^{\infty}$ such that (3.24), as well as

$$n_{j+1} \geq \beta n_j \left(\frac{2\rho^{\ell+1}}{(\mu(\rho) - \varepsilon)(\rho - \rho_{n_j})} \right)^{n_j}, \quad j = 1, 2, \ldots \tag{3.27}$$

and

$$|\Delta_{n_j;\ell}(f_k, \widehat{z}, Z)| \leq \frac{\alpha}{\beta n_j} \left(\frac{(\mu(\rho) - \varepsilon)r}{\rho^{\ell+1}} \right)^{n_j} \quad \text{for all} \quad j > k, \quad k = 1, 2, \ldots \tag{3.28}$$

hold, where

$$\alpha = \frac{\beta - 1}{3(r + \rho)} > 0, \quad \beta = 1 + \frac{12(r + \rho)}{r - \rho}. \tag{3.29}$$

Namely, if this were not true, then for some k_0 we would have

$$|\Delta_{n,\ell}(f_{k_0}, \widehat{z}, Z)| \geq \frac{\alpha}{\beta n} \left(\frac{(\mu(\rho) - \varepsilon)r}{\rho^{\ell+1}} \right)^r \to \infty \quad \text{as} \quad n \to \infty$$

by (3.23), and this would prove (3.6) with $f_0 = f_{k_0}$.

So assuming (3.27) and (3.28), define

$$f_0(z) := \sum_{k=1}^{\infty} \frac{f_k(z)}{n_k} = \sum_{k=1}^{\infty} \frac{1}{n_k(u_{n_k} - z)} \tag{3.30}$$

where $|u_{n_k}| = \rho$ for all $k \geq 1$. Now (3.27) implies

$$n_k \geq \beta^{k-j} n_j \quad \text{for all} \quad k \geq j \geq 1, \tag{3.31}$$

i.e. the series (3.30) converges uniformly in $|z| < \rho$, and $f_0(z)$ is analytic in $|z| < \rho$. In order to show that $f_0(z)$ has a singularity on $|z| = \rho$, let $f_0(z) = \sum\limits_{j=0}^{\infty} c_j z^j$. Then by (3.30) - (3.31)

$$|c_j| = \left| \sum_{k=1}^{\infty} \frac{1}{n_k u_{n_k}^{j+1}} \right| \geq \frac{1}{\rho^{\ell+1}} \left(\frac{1}{n_1} - \sum_{k=2}^{\infty} \frac{1}{n_k} \right) \geq \frac{\beta - 2}{(\beta - 1) n_1 \rho^{j+1}}, \quad j = 0, 1, \ldots,$$

whence $\varlimsup\limits_{j \to \infty} |c_j|^{1/j} \geq 1/\rho$, and so $f_0 \in A_\rho$.

Now (3.30) implies

$$|\Delta_{n_j, \ell}(f_0, \hat{z}, Z)| \geq S_1 - S_2 - S_3 - S_4$$

where (see (3.4))

$$S_1 := \frac{1}{n_j} |\Delta_{n_j, \ell}(f_j, \hat{z}, Z)|, \qquad S_2 := \sum_{k=1}^{j-1} \frac{1}{n_k} |\Delta_{n_j, \ell}(f_k, \hat{z}, Z)|,$$

$$S_3 := \sum_{k=j+1}^{\infty} \frac{1}{n_k} |L_{n_j-1}(f_k, \hat{z}, Z)|, \quad S_4 := \sum_{k=j+1}^{\infty} \frac{1}{n_k} \left| \sum_{t=0}^{\ell-1} p_{n_j-1,t}(f_k; \hat{z}) \right|.$$

From (3.25) we have

$$s_1 \geq \frac{1}{(r+\rho)n_j} \left(\frac{(\mu(\rho) - \varepsilon)r}{\rho^{\ell+1}} \right)^{n_j}, \quad \text{for all} \quad j \geq 1,$$

while from (3.28) and (3.31) we get

$$S_2 \leq \frac{\alpha}{\beta n_j} \left(\frac{(\mu(\rho) - \varepsilon)r}{\rho^{\ell+1}} \right)^{n_j} \sum_{k=1}^{j-1} \frac{1}{n_k}$$

$$\leq \frac{\alpha}{(\beta - 1) n_j} \left(\frac{(\mu(\rho) - \varepsilon)r}{\rho^{\ell+1}} \right)^{n_j}, \quad \text{for} \quad j \geq 2.$$

Next, from the first equation in (3.9) and from (3.26) we obtain

$$|L_{n_j-1}(f_k, \hat{z}, Z)| \leq \frac{(\rho + \rho_{n_j})^{n_j} + (r + \rho_{n_j})^{n_j}}{(\rho - \rho_{n_j})^{n_j}(r - \rho)}$$

$$\leq \frac{2(2r)^{n_j}}{(\rho - \rho_{n_j})^{n_j}(r - \rho)}, \quad \text{for all} \quad k \geq j+1.$$

Similarly, from the second equation in (3.9), we deduce

$$\left| \sum_{t=0}^{\ell-1} p_{n_j-1,t}(f_k; \hat{z}) \right| \leq \frac{(1 + \rho^{-\ell n_n})(\rho^{n_j} + r^{n_j})}{(\rho^{n_j} - 1)(r - \rho)}$$

$$\leq \frac{4r^{n_j}}{(\rho^{n_j} - 1)(r - \rho)} \leq \frac{8r^{n_j}}{(\rho - \rho_{n_j})^{n_j}(\rho - r)}$$

for all $k \geq j + 1$ and j large enough.

Thus by (3.31) and (3.27)

$$S_3 + S_4 \leq \frac{4(2r)^{n_j}}{(\rho - \rho_{n_j})^{n_j}(r - \rho)} \sum_{k=j+1}^{\infty} \frac{1}{n_k}$$

$$\leq \frac{4\beta(2r)^{n_j}}{(\beta - 1)(\rho - \rho_{n_j})^{n_j}(r - \rho)n_{j+1}}$$

$$\leq \frac{4}{(r - \rho)(\beta - 1)n_j} \left(\frac{(\mu(\rho) - \varepsilon)r}{\rho^{\ell+1}} \right)^{n_j},$$

for all j sufficiently large. On using (3.29), these inequalities imply

$$S_1 - S_2 - S_3 - S_4 \geq \frac{1}{3(r + \rho)n_j} \left(\frac{(\mu(\rho) - \varepsilon)r}{\rho^{\ell+1}} \right)^{n_j},$$

for all j sufficiently large. Hence and by (3.23)

$$|\Delta_{n_j;\ell}(f_0, \widehat{z}, Z)| \geq \frac{1}{3(r + \rho)n_j} \left(\frac{(\mu(\rho) - \varepsilon)r}{\rho^{\ell+1}} \right)^{n_j} \to \infty \text{ as } j \to \infty \qquad (3.32)$$

which proves (3.6).

To conclude the proof of Theorem 6, we note that the above construction is valid for *any* choice of the complex number $\widehat{z}$ such that $|\widehat{z}| = r > \rho$ and any ε with $0 < \varepsilon < \mu|\rho|$. So from (3.32) we get

$$\overline{\lim_{n \to \infty}} \left\{ \max_{|z|=r} |\Delta_n(f_0, z, Z)| \right\}^{1/n} \geq \frac{(\mu(\rho) - \varepsilon)r}{\rho^{\ell+1}}, \quad r > \rho,$$

and since $f_0 \in A_\rho$, by (3.5) we have

$$\Delta_\ell(r, \rho, Z) \geq \frac{(\mu(\rho) - \varepsilon)r}{\rho^{\ell+1}}, \quad r > \rho.$$

Since $\varepsilon > 0$ is arbitrary and $\Delta_\ell(r, \rho, E) = \frac{r}{\rho^{\ell+1}}$ (cf. Theorem 1) we obtain (3.7). $\square$

4.4. Equiconvergence of Hermite Interpolation on Concentric Circles

So far we considered the difference of interpolation polynomials and "shifted" Taylor-type series, or superpositions of these, in order to demonstrate the phenomenon of overconvergence. In this section we will see that differences of Hermite interpolation polynomials based on roots of unity on concentric circles also show corresponding properties.

If $f(z) \in A_\rho$ $(\rho > 1)$, $p \geq 1$ is an integer and $0 < \alpha < \rho$, then let $h_{pn-1}(f; \alpha; z) \in \pi_{pn-1}$ denote the Hermite interpolation polynomial to $f(z)$ in the zeros of $(z^n - \alpha^n)^p$, i.e.

$$h_n^{(j)}(f, \alpha, \omega^k) = f^{(j)}(\alpha^k \omega^k), \quad j = 0, \ldots, p-1; \quad k = 0, \ldots, n-1$$

where ω is a primitive n^{th} root of unity. Let $f(z) = \sum_{k=0}^{\infty} a_k z^k$, then using (2.1) of Ch. 2 we obtain

$$h_{pn-1}(f, \alpha, z) = \sum_{k=0}^{pn-1} a_k z^k + \sum_{k=0}^{n-1} \left\{ \sum_{j=1}^{\infty} \alpha^{n(p+j-1)} \beta_{j,p}(z^n/\alpha^n) a_{k+n(p+j-1)} \right\} z^k$$

$$(4.1)$$

where the polynomials $\beta_{j,p} \in \pi_{p-1}$ are defined in (1.7) of Ch. 2.

After these preliminaries, we can state the equiconvergence result. If $0 < \alpha < \gamma < \rho$, set

$$\Delta_n(f; \alpha, \gamma, z) := h_{pn-1}(f, \alpha, z) - h_{pn-1}(f, \gamma, z).$$

THEOREM 8. Let $f(z) \in A_\rho$ $(\rho > 1)$ and $0 < \alpha < \gamma < \rho$. Then

$$\overline{\lim_{n \to \infty}} |\Delta_n(f, \alpha, \gamma, z)|^{1/n} = \begin{cases} \frac{\gamma |z|^p}{\rho^{p+1}} & \text{if } |z| > \rho, \\ \frac{\gamma |z|^{p-1}}{\rho^p} & \text{if } \gamma < |z| < \rho, \\ \frac{\gamma^p}{\rho^p} & \text{if } 0 < |z| < \gamma, \end{cases} \quad (4.2)$$

with the exception of at most p points in $|z| > \rho$ and $p-1$ points in $0 < |z| < \rho$, at which points the left hand side of (4.2) is strictly less than the right hand side. At $|z| = \gamma$ and at $|z| = \rho$, the " $\leq$ " holds in (4.2).

The result shows that $|z| < \frac{\rho^{1+1/p}}{\gamma^{1/p}}$ is the exact domain of overconvergence.

PROOF. (4.1) implies

$$\Delta_n(f, \alpha, \gamma, z) = \sum_{k=0}^{n-1} \left\{ \sum_{j=1}^{\infty} \left[\alpha^{n(p+j-1)} \beta_{j,p} \left(\frac{z}{\alpha} \right)^n - \right. \right.$$

$$\left. \left. - \gamma^{n(p+j-1)} \beta_{j,p} \left(\frac{z}{\gamma} \right)^n \right] a_{k+n(p+j-1)} \right\} z^k.$$

In all three cases to be considered, the term $j = 1$ will dominate. To estimate the terms corresponding to $j \geq 2$, we use $a_k = O((\rho-\varepsilon)^{-k})$ $(k = 0, 1, \ldots, \varepsilon > 0$ arbitrary) as well as Lemma 2 of Ch. 2 to get

$$\Delta_n(f, \alpha, \gamma, z) = \sum_{k=0}^{n-1} [\alpha^{pn} \beta_{1,p}(z^n/\alpha^n) - \beta^{pn} \beta_{1,p}(z^n/\gamma^n)] a_{k+pn} z^k + \varepsilon_n(z),$$

where

$$\varepsilon_n(z) = \begin{cases} O\left(\frac{\beta^{2n}|z|^{pn}}{(\rho-\varepsilon)^{(p+2)n}}\right) & \text{if } \beta < \rho - \varepsilon < |z|, \\ O\left(\frac{\gamma^{2n}|z|^{(p-1)n}}{(\rho-\varepsilon)^{(p+1)n}}\right) & \text{if } \gamma \le |z| < \rho - \varepsilon, \\ O\left(\frac{\gamma^{(p+1)n}}{(\rho-\varepsilon)^{(p+1)n}}\right) & \text{if } |z| \le \gamma. \end{cases}$$

Since $\beta_{1,p}(z) = z^p - (z-1)^p$ (cf. (1.8) of Ch. 2), we get

$$\Delta_n(f, \alpha, \gamma, z) = [(z^n - \alpha^n)^p - (z^n - \gamma^n)^p] \sum_{k=0}^{n-1} a_{k+pn} z^k + \varepsilon_n(z).$$

Here

$$(z^n - \alpha^n)^p - (z^n - \gamma^n)^p = \begin{cases} pz^{(p-1)n}\gamma^n + O(|z|^{(p-1)n}\alpha^n + |z|^{(p-2)n}\gamma^{2n}) \\ \qquad \text{if } |z| \ge \gamma \\ (-1)^{p+1}\gamma^{pn} + O(\alpha^{pn} + |z|^n \gamma^{(p-1)n}) \\ \qquad \text{if } |z| \le \gamma. \end{cases}$$

Thus we obtain

$$\Delta_n(f, \alpha, \gamma, z) = c_n(z) \sum_{k=0}^{n-1} a_{k+pn} z^k + \xi_n(z), \tag{4.3}$$

where

$$c_n(z) = \begin{cases} pz^{(p-1)n}\gamma^n & \text{if } |z| \ge \gamma, \\ (-1)^{p+1}\gamma^{pn} & \text{if } |z| \le \gamma \end{cases} \tag{4.4}$$

and

$$\xi_n(z) = \begin{cases} O\left(\frac{\gamma^{2n}|z|^{pn}}{(\rho-\varepsilon)^{(p+2)n}} + \frac{\alpha^n |z|^{pn}}{(\rho-\varepsilon)^{(p+1)n}}\right) & \text{if } |z| > \rho \\ O\left(\frac{\gamma^{2n}|z|^{(p-1)n}}{(\rho-\varepsilon)^{(p+1)n}} + \frac{\alpha^n |z|^{(p-1)n}}{(\rho-\varepsilon)^{pn}}\right) & \text{if } \gamma \le |z| < \rho \\ O\left(\frac{\gamma^{(p+1)n}}{(\rho-\varepsilon)^{(p+1)n}} + \frac{\alpha^{pn} + |z|^n \gamma^{(p-1)n}}{(\rho-\varepsilon)^{pn}}\right) & \text{if } |z| \le \gamma. \end{cases} \tag{4.5}$$

Therefore

$$|\Delta_n(f, \alpha, \gamma, z)| = \begin{cases} O\left(\frac{\gamma^n |z|^{pn}}{(\rho-\varepsilon)^{(p+1)n}}\right) & \text{if } |z| > \rho \\ O\left(\frac{\gamma^n |z|^{(p-1)n}}{(\rho-\varepsilon)^{pn}}\right) & \text{if } \gamma \le |z| < \rho \\ O\left(\frac{\gamma^{pn}}{(\rho-\varepsilon)^{pn}}\right) & \text{if } |z| \le \gamma, \end{cases}$$

whence the upper estimate (i.e. " $\leq$ ") in (4.2) follows.

To prove the lower estimate, we distinguish two cases.

CASE 1: $|z| > \gamma$. Then by (4.3) - (4.5) we have

$$\delta_n(z) := \gamma \Delta_n(f, \alpha, \gamma, z) - z \Delta_{n+1}(f, \alpha, \gamma, z) \tag{4.6}$$

$$= p\gamma^{n+1} z^{(p-1)n} \left(\sum_{k=0}^{n-1} a_{k+pn} z^k - \sum_{k=0}^{n} a_{k+(n+1)p} z^{k+p} \right) + O(\xi_n(z))$$

$$= p\gamma^{n+1} z^{(p-1)n} \left(\sum_{k=0}^{p-1} - \sum_{k=n}^{n+p} \right) a_{k+pn} z^k + O(\xi_n(z))$$

$$= \begin{cases} -p\gamma^{n+1} z^{pn} \sum_{k=0}^{p} a_{k+(p+1)n} z^k + O\left(\dfrac{\gamma^n |z|^{(p-1)n}}{(\rho - \varepsilon)^{pn}} + \xi_n(z) \right) \\ \qquad\qquad \text{if} \quad |z| > \rho, \\[2em] p\gamma^{n+1} z^{(p-1)n} \sum_{k=0}^{p-1} a_{k+pn} z^k + O\left(\dfrac{\gamma^n |z|^{pn}}{(\rho - \varepsilon)^{(p+1)n}} + \xi_n(z) \right) \\ \qquad\qquad \text{if} \quad \gamma < |z| < \rho. \end{cases}$$

Now suppose that for some pairwise different $z_0, \ldots, z_p$ such that $|z_j| > p$ $(j = 0, \ldots, p)$ the equality in (4.1) does not hold, i.e.

$$|\Delta_n(f, \alpha, \gamma, z_j)| = O\left(\left(\frac{\gamma |z|^p}{(\rho - \varepsilon)^{p+1}} - \eta \right)^n \right), \quad j = 0, \ldots, p$$

with some $\eta > 0$. Then by (4.6)

$$|\delta_n(z_j)| = O\left(\left(\frac{\gamma |z_j|^p}{(\rho - \varepsilon)^{p+1}} - \eta \right)^n \right), \quad j = 0, \ldots, p;$$

and

$$\sum_{k=0}^{p} a_{k+(p+1)n} z_j^k = O\left(\left(\frac{1}{(\rho - \varepsilon)^{p+1}} - \eta_1 \right)^n \right), \quad j = 0, \ldots, p,$$

provided $\eta_1 = \frac{\eta}{\gamma \max |z_j|^p} > 0$ is small enough. Solving this system of equations for the unknowns $a_{k+(p+1)n}$ we get

$$|a_{k+(p+1)n}| = O\left(\left(\frac{1}{(\rho - \varepsilon)^{p+1}} - \eta_1 \right)^n \right), \quad k = 0, \ldots, p.$$

Hence $\varliminf_{n \to \infty} |a_n|^{1/n} < \frac{1}{\rho - \varepsilon} - \frac{1}{\eta^{\frac{1}{p+1}}}$, for arbitrary $\varepsilon > 0$. Thus $\varliminf_{n \to \infty} |a_n|^{1/n} < \frac{1}{\rho}$ which contradicts $f(z) \in A_\rho$.

Similarly, suppose that for some pairwise different $z_0, \ldots, z_{p-1}$ such that $\beta < |z_j| < \rho$ the equality in (4.2) does not hold, i.e.

$$|\Delta_n(f, \alpha, \gamma, z_j)| = O\left(\left(\frac{\gamma|z_j|^{p-1}}{(\rho - \varepsilon)^p} - \eta\right)^n\right), \quad j = 0, \ldots, p-1 \qquad (4.7)$$

with some $\eta > 0$. Then by (4.6)

$$|\delta_n(z_j)| = O\left(\left(\frac{\gamma|z_j|^{p-1}}{(\rho - \varepsilon)^p} - \eta\right)'^n\right), \quad j = 0, \ldots, p-1 \qquad (4.8)$$

and

$$\sum_{k=0}^{p-1} a_{k+pn} z_j^k = O\left(\left(\frac{1}{(\rho - \varepsilon)^p} - \eta_1\right)^n\right), \quad j = 0, \ldots, p-1 \qquad (4.9)$$

provided $\eta_1 = \frac{\eta}{\gamma \max |z_j|^{p-1}} > 0$ is small enough. Solving this system we get

$$a_{k+pn} = O\left(\left(\frac{1}{(\rho - \varepsilon)^p} - \eta_1\right)^n\right), \quad k = 0, \ldots, p-1. \qquad (4.10)$$

Hence $\varlimsup_{n \to \infty} |a_n|^{1/n} < \frac{1}{\rho - \varepsilon} - \frac{1}{\eta_1^{1/p}}$ which leads again to contradiction.

CASE 2: $0 < |z| < \gamma$. Then by (4.3)-(4.5) we have

$$\delta_n(z) := (-1)^{p+1}(\gamma^p \Delta_n - z^p \Delta_{n+1})$$

$$= \gamma^{p(n+1)}\left(\sum_{k=0}^{n-1} a_{k+pn} z^k - \sum_{k=0}^{n} a_{k+(n+1)p} z^{k+p}\right) + O(\xi_n(z))$$

$$= \gamma^{p(n+1)}\left(\sum_{k=0}^{p-1} - \sum_{k=n}^{n+p}\right) a_{k+pn} z^k + O(\xi_n(z))$$

$$= \gamma^{p(n+1)} \sum_{k=0}^{p-1} a_{k+pn} z^k + O(\xi_n(z)).$$

Again, assuming the existence of $z_0, \ldots, z_{p-1}$ such that $0 < |z_j| < \gamma$, $j = 0, \ldots, p-1$ and (4.7) holds, then (4.8) will also be valid. Proceeding as in (4.9)-(4.10), we arrive at the same conclusion. $\qquad \square$

Note that this theorem does not settle the case of exceptional points on the circles $|z| = \gamma$ and $|z| = \rho$.

4.5. $(0, m)$-Pál type Interpolation

In this section we dwell upon the regularity of $(0; m)$ Pál type interpolation and obtain an equiconvergence result for it. It differs from $(0, m)$ interpolation where values and m^{th} derivatives are prescribed at the same nodes. The regularity and equiconvergence for $(0, m)$ interpolation on roots of unity is known. Here we show that the problem of $(0; m)$ Pál type interpolation on the zeros of $z^n - \alpha^n$ and $z^n + \alpha^n$ is regular and we obtain the precise region of overconvergence when $f \in A_\rho$.

Let α, $0 < \alpha < \rho$, be a real number. By Pál type $(0; m)$ interpolation on the zeros of $z^n - \alpha^n$ and of $z^n + \alpha^n$, we mean to find a polynomial $P_{2n,m}(z)$ of degree $\leq 2n - 1$ such that $P_{2n,m}(z)$ interpolates a given function $f(z)$ in the zeros of $z^n - \alpha^n$ and its mth derivative $P_{2n,m}^{(m)}(z)$ interpolates $f^{(m)}(z)$ in the zeros of $z^n + \alpha^n$. We first find an explicit formula for $P_{2n,m}(f, z)$, when

$$f(z) = \sum_{k=0}^{\infty} a_k z^k \in A_\rho.$$

Since interpolation is a linear process, it is enough to find the value of $P_{2n,m}(f_\lambda, z)$ when $f_\lambda(z) = z^{\lambda n + k}$, $0 \leq k \leq n - 1$ for any positive integer λ. We try to get our polynomial in the form

$$P_{2n,m}(f_\lambda; z) = az^k + bz^{n+k},$$

and require that

$$P_{2n,m}(f_\lambda; z_j) = z_j^{\lambda n + k}, \qquad\qquad z_j = \alpha w_j \quad \text{where} \quad w_j^n = 1,$$

$$P_{2n,m}^{(m)}(f_\lambda; z_j') = z_j'^{\lambda n + k - m} (\lambda n + k)_m, \quad z_j' = \alpha w_j e^{\frac{i\pi}{n}},$$

where $(k)_m = k(k-1)\ldots(k-m+1)$. We then have

$$\begin{cases} a + b\alpha^n = \alpha^{n\lambda}, \\ a(k)_m - (n+k)_m b\alpha^n = (\lambda n + k)_m \alpha^{\lambda n}(-1)^\lambda \end{cases}$$

Then

$$\begin{cases} b = \alpha^{(\lambda-1)n} \dfrac{(k)_m - (\lambda n + k)_m (-1)^\lambda}{(k)_m + (n+k)_m} =: \beta_{\lambda k} \\ a = \alpha^{\lambda n} \dfrac{(\lambda n + k)_m (-1)^\lambda + (n+k)_m}{(k)_m + (n+k)_m} =: \alpha_{\lambda k}. \end{cases} \tag{5.1}$$

If $\lambda = 0$, $\alpha_{0k} = 1$ and $\beta_{0k} = 0$. If $\lambda = 1$, $\alpha_{0k} = 0$ and $\beta_{1k} = 1$. This proves that Pál type $(0; m)$ interpolation is regular on the zeros of $z^n - \alpha^n$ and $z^n + \alpha^n$. Since

$$f(z) = \sum_{k=0}^{n-1} \sum_{\lambda=0}^{\infty} a_{\lambda n + k} z^{\lambda n + k}$$

we have

$$P_{2n,m}(f;z) = \sum_{k=0}^{n-1}\sum_{\lambda=0}^{\infty}(\alpha_{\lambda k}z^k + \beta_{\lambda k}z^{n+k})a_{\lambda n+k}.$$

For any integer $\ell \geq 1$, we set

$$S_{n,\ell}(f;z) := \sum_{k=0}^{n-1}\sum_{\lambda=0}^{\ell-1}a_{\lambda n+k}(\alpha_{\lambda k}z^k + \beta_{\lambda k}z^{n+k})$$

$$= P_{2n,m}\big(T_{\ell n-1}(f);z\big)$$

where $T_{\ell n-1}(f)$ is the Taylor expansion of f of degree $\ell n - 1$. (For $\ell = 1$ or 2, evidently $S_{n,\ell}(f;z) = T_{\ell n-1}(f;z)$.)

Consider the difference

$$\Delta_{n,\ell}(f;z) := P_{2n,m}(f;z) - S_{n,\ell}(f;z)$$

$$= \sum_{k=0}^{n-1}\sum_{\lambda=\ell}^{\infty}a_{\lambda n+k}(\alpha_{\lambda k}z^k + \beta_{\lambda k}z^{n+k}). \tag{5.2}$$

THEOREM 9. *If* $f(z) \in A_\rho$, $\rho > 1$, $\rho > \alpha > 0$ *then*

$$\overline{\lim_{n\to\infty}} |\Delta_n(f;z)|^{\frac{1}{2n}} \leq \begin{cases} \dfrac{\max\{\alpha^{\frac{1}{2}},|z|^{\frac{1}{2}}\}\alpha^{\frac{\ell-1}{2}}}{\rho^{\frac{\ell}{2}}} & \text{if } |z| < \rho \\[4mm] \dfrac{|z|\alpha^{\frac{\ell-1}{2}}}{\rho^{\frac{\ell+1}{2}}} & \text{if } |z| \geq \rho. \end{cases} \tag{5.3}$$

Hence we have overconvergence in the circle $|z| < \rho\big(\frac{\rho}{\alpha}\big)^{\frac{\ell-1}{2}}$.

PROOF. Let $f(z) \in A_\rho$, then $|a_n| = O\big(\frac{1}{(\rho-\varepsilon)^n}\big)$, for $\varepsilon > 0$, $n = 0,1,2\dots$ and by (5.1),

$$\alpha_{\lambda k} = O\big(\lambda^m\alpha^{\lambda n}\big), \qquad \beta_{\lambda k} = O\big(\lambda^m\alpha^{(\lambda-1)n}\big),$$

$$(n \text{ or } \lambda \to \infty, \ k = 0,1,\dots,n-1).$$

Therefore from (5.2), we get

$$|\Delta_{n,\ell}(f;z)| = O\Big(\sum_{k=0}^{n-1}\sum_{\lambda=\ell}^{\infty}\lambda^m\alpha^{(\lambda-1)n}(\rho-\varepsilon)^{-\lambda n-k}(|\alpha^n z|^k + |z|^{n+k})\Big)$$

$$= O\Big(\sum_{k=0}^{n-1}\alpha^{(\ell-1)n}(\rho-\varepsilon)^{-\ell n-k}(|\alpha^n z|^k + |z|^{n+k})\Big)$$

$$
= \begin{cases} O\left(\frac{\alpha^n + |z|^n}{(\rho-\varepsilon)^{\ell n}}\right)\alpha^{(\ell-1)n} & \text{if } |z| < \rho - \varepsilon \\[2mm] O\left(\frac{|z|^{2n}}{(\rho-\varepsilon)^{(\ell+1)n}}\right)\alpha^{(\ell-1)n} & \text{if } |z| > \rho - \varepsilon. \end{cases}
$$

$\square$

Since $\varepsilon > 0$ is arbitrary, we have (5.3).

THEOREM 10. *Under the same conditions as in Theorem 1, we have*

$$
\overline{\lim_{n\to\infty}} \max_{|z|=R} |\Delta_{n,\ell}(f,z)|^{\frac{1}{2n}} = \begin{cases} \dfrac{\max\{\alpha^{1/2}, R^{1/2}\}\alpha^{\frac{\ell-1}{2}}}{\rho^{\ell/2}} & \text{if } R < \rho, \\[4mm] \dfrac{R\alpha^{\frac{\ell-1}{2}}}{\rho^{\frac{\ell+1}{2}}} & \text{if } R \geq \rho. \end{cases}
\tag{5.4}
$$

PROOF. From (5.2), we have

$$
\Delta_{n,\ell}(f;z) = \sum_{k=0}^{n-1} a_{\ell n+k}(\alpha_{\ell k} z^k + \beta_{\ell k} z^{n+k})
$$

$$
+ O\left(\sum_{k=0}^{n-1}\sum_{\lambda=\ell+1}^{\infty} \lambda^m (\rho-\varepsilon)^{-\lambda n-k}|\alpha|^{(\lambda-1)n}(|\alpha^n z|^k + |z|^{n+k})\right)
$$

that is

$$
\Delta_{n,\ell}(f;z) = \sum_{k=0}^{n-1} a_{\ell n+k}(\alpha_{\ell k} z^k + \beta_{\ell k} z^{n+k})
\tag{5.5}
$$

$$
+ \begin{cases} O\left(\frac{\alpha^n + |z|^n}{(\rho-\varepsilon)^{(\ell+1)n}}|\alpha|^{\ell n}\right) & \text{if } |z| < \rho - \varepsilon \\[2mm] O\left(\frac{|z|^{2n}}{(\rho-\varepsilon)^{(\ell+2)n}}|\alpha|^{\ell n}\right) & \text{if } |z| \geq \rho - \varepsilon. \end{cases}
$$

Case 1: $R = |z| < \rho$. Then divide (2.5) by z^{k+1} $(k = 0, 1, \ldots, n-1)$ and integrate on the circle $|z| = R$. Then

$$
\frac{1}{2\pi i}\int_{|z|=R} \frac{\Delta_{n,\ell}(f;z)}{z^{k+1}}\, dz = \alpha_{\ell k} a_{\ell n+k} + O\left(\frac{|\alpha|^{(\ell+1)n}}{(\rho-\varepsilon)^{(\ell+1)n}}\right),
$$

$$
k = 0, 1, \ldots, \ell-1.
$$

Since $(\alpha_{\ell k}) \geq C_m \alpha^{\ell n} > 0$, we get

$$
\limsup_{n\to\infty}\left(\max_{|z|=R}|\Delta_{n,\ell}(f;z)|^{\frac{1}{2n}}\right) \geq \limsup_{n\to\infty}\left(\frac{\alpha^{\ell n}}{(\rho-\varepsilon)^{\ell n+k}} - \frac{\alpha^{(\ell+1)n}}{(\rho-\varepsilon)^{(\ell+1)n}}\right)^{\frac{1}{2n}}
$$

$$
= \frac{\alpha^{\frac{\ell}{2}}}{(\rho-\varepsilon)^{\frac{\ell}{2}}}.
$$

$$
\tag{5.6}
$$

Case 2: $\rho \leq |z| = R$. Then divide (5.5) by z^{n+k+1} $(0 \leq k \leq n-1)$ and integrate. We then have

$$\frac{1}{2\pi i} \int_{|z|=R} \frac{\Delta_{n,\ell}(f;z)}{z^{n+k+1}}\, dz = \beta_{\ell k} a_{\ell n+k} + \begin{cases} O\left(\frac{R^{-k-1}\alpha^{\ell n}}{(\rho-\varepsilon)^{(\ell+1)n}}\right) & \text{if } 1 < |z| < \rho-\varepsilon \\ & \text{and} \quad 0 \leq k \leq \ell-1 \\[2mm] O\left(\frac{R^{n-k-1}\alpha^{\ell n}}{(\rho-\varepsilon)^{(\ell+2)n}}\right) & \text{if } |z| \geq \rho-\varepsilon \\ & \text{and} \quad n-\ell+1 \leq k \leq n. \end{cases}$$

Since $|\beta_{\ell k}| \geq C_m \alpha^{(\ell-1)n}$, and $n-\ell+1 \leq k \leq n$ we have

$$\max_{|z|=R} |\Delta_{n,\ell}(f;z)| \geq C_m R^{n+k}\alpha^{(\ell-1)n}|a_{\ell n+k}|+$$

$$+ \begin{cases} O\left(\frac{R^n \alpha^{\ell n}}{(\rho-\varepsilon)^{(\ell+1)n}}\right) & \text{if} \quad \begin{array}{l} 1 < |z| < \rho-\varepsilon\,, \\ 0 \leq k < \ell \end{array} \\[4mm] O\left(\frac{R^{2n}\alpha^{\ell n}}{(\rho-\varepsilon)^{(\ell+2)n}}\right) & \text{if} \quad \begin{array}{l} |z| \geq \rho-\varepsilon\,, \\ n-\ell+1 \leq k \leq n. \end{array} \end{cases}$$

Hence

$$\limsup_{n \to \infty} \left(\max_{|z|=R} |\Delta_{n,\ell}(f;z)| \right)^{\frac{1}{2n}}$$

$$\geq \begin{cases} \limsup_{n \to \infty} \left(R^{n+k}(\rho-\varepsilon)^{-\ell n-k}\alpha^{(\ell-1)n} - O\left(\frac{R^n \alpha^{\ell n}}{(\rho-\varepsilon)^{(\ell+1)n}}\right) \right)^{\frac{1}{2n}}, \\ \qquad \text{if } 1 \leq |z| < \rho-\varepsilon \text{ and } 0 \leq k < \ell \\[4mm] \limsup_{n \to \infty} \left(R^{n+k}(\rho-\varepsilon)^{-\ell n-k}\alpha^{(\ell-1)n} - O\left(\frac{R^{2n}\alpha^{\ell n}}{(\rho-\varepsilon)^{(\ell+2)n}}\right) \right)^{\frac{1}{2n}}, \\ \qquad \text{if } |z| \geq \rho-\varepsilon \text{ and } n-\ell+1 \leq k \leq n \end{cases}$$

$$\geq \begin{cases} \dfrac{R^{\frac{1}{2}}\alpha^{\frac{\ell-1}{2}}}{(\rho-\varepsilon)^{\frac{\ell}{2}}}\,, & \text{if } 1 \leq |z| < \rho-\varepsilon\,, \\[4mm] \dfrac{R\,\alpha^{\frac{\ell-1}{2}}}{(\rho-\varepsilon)^{\frac{\ell+1}{2}}}\,, & \text{if } |z| \geq \rho-\varepsilon. \end{cases}$$

Combining these lower estimates in Cases 1 and 2 with the upper estimates of Theorem 1, we get the statements (5.4) of Theorem 11. $\qquad\square$

4.6. Historical Remarks

Theorems 1 and 2, as well as the special case $\lambda = \ell-1$ of Theorem 3 are due to Totik [111]. Theorem 4 can be found in Ivanov and Sharma [51], where interesting properties of the distinguished points (i.e., point systems when the general error term is not attained) can be found. We mention here some of them.

(a) When $\ell = 1$, then for any given function $f \in A_\rho$ there are no $(\rho, 1)$-distinguished points in $|z| < \rho$ and there is at most one $(\rho, 1)$-distinguished point in $|z| > \rho$ (by Corollary 1).

Similarly, for $\ell - 2$, there is at most one point in $|z| < \rho$ and at most two points in $|z| > \rho$ which are $(\rho, 2)$-distinguished.

(b) Let $\ell = 2$, $Z = \{z_1, z_2, z_3\}$, $|z_1| < \rho < |z_2|, |z_3|$. Then the set Z is $(\rho, 2)$-distinguished if and only if $z_1 = 0$ and $z_2 + z_3 = 0$.

Namely, $M = M(X, Y)$ is a 6×7 matrix with

$$X = [1 \, z_1] \quad \text{and} \quad Y = \begin{pmatrix} 1 & z_2 & z_2^2 \\ 1 & z_3 & z_3^2 \end{pmatrix}.$$

Let us denote by M_j the matrix obtained from from M when the jth row is deleted. From Theorem 5, Z is a $(\rho, 2)$-distinguished set if and only if $|M_j| = 0$ for $j = 1, \ldots, 7$. A direct computation shows that (to within a factor ± 1) we have

$$|M_1| = (z_2 - z_3)^2(z_2 + z_3 - z_1), \quad |M_3| = z_1 z_2 z_3 (z_2 - z_3)^2 (z_2 z_3 - z_1 z_2 - z_1 z_3).$$

Here $|M_j|$ denotes the determinant of the matrix M_j. If $z_1 \neq 0$, then from $|M_1| = 0$ and $|M_3| = 0$, it follows that $z_=z_2 + z_3$ and $z_1(z_2 + z_3) = z_2 z_3$. Hence $z_1^2 = z_2 z_3$, i.e., $\rho^2 < |z_2 z_3| = |z_1|^2 < \rho^2$, a contradiction. So for Z to be a $(\rho, 2)$-distinguished set, we necessarily have $z_1 = 0$. In this case the determinants of the M_j's are easily seen to be the following:

$$|M_1| = (z_2 - z_3)^2(z_2 + z_3), \quad |M_2| = z_2 z_3 (z_2 - z_3)^2(z_2 + z_3),$$

$$|M_4| = z_3(z_3 - z_2)(z_3 + z_2), \quad |M_5| = z_2(z_3 - z_2)(z_3 + z_2),$$

$$|M_3| = |M_6| = |M_7| = 0.$$

Hence $|M_j| = 0$ for $j = 1, \ldots, 7$ implies $z_2 + z_3 = 0$. Conversely, if $z_1 = 0$ and $z_2 + z_3 = 0$, we have rank $M < 6$. This proves the claim.

(c) Let $\ell = 3$. In this case, by Corollary 1, we can have at most two points in $|z| < \rho$ and at most three points in $|z| > \rho$ which are $(\rho, 3)$-distinguished. The cases which are not covered by Corollary 2 are:

(I) one point inside and three points outside the circle $|z| = \rho$;

(II) two points inside and two points outside;

(III) two points inside and three points outside.

Let $Z = \{z_1, z_2, z_3, z_4\}$. In (I), let $|z_1| < \rho$ and $|z_j| > \rho$, $j = 2, 3, 4$. We claim that Z is a $(\rho, 3)$-distinguished set if and only if one of the following cases occurs:

(a) $z_1 = 0$ and $z_2 + z_3 + z_4 = 0$,

(b) $z_1 = 0$ and $z_2^{-1} + z_3^{-1} + z_4^{-1} = 0$,

(c) $z_1 - (z_2 + z_3 + z_4) = 0$ and

$$z_1^2(z_2 + z_3 + z_4) - z_1(z_2 z_3 + z_2 z_4 + z_3 z_4) + z_2 z_3 z_4 = 0.$$

Observe in (c) that we necessarily have $z_1 \neq 0$, since otherwise we have $z_2 z_3 z_4 = 0$, which is impossible. Also remark that the solution of (c) is not vacuous. In fact, a solution with $|z_1| < \rho$ and $|z_2|, |z_3|, |z_4| > \rho$ is easily seen to be:

$$z_1 = a, \quad z_2 = -2a, \quad z_3, z_4 = a\left(\frac{3}{2} + \frac{i}{2\sqrt{3}}\right) \qquad \text{with} \quad \rho\sqrt{\frac{3}{7}} < a < \rho.$$

The proof of this assertion can be found in [51, Sec. 5].

The sharpness of overconvergence results for Hermite interpolation on roots of unity (Theorem 5) is due to Ivanov and Sharma [50].

The distinguished role of the roots of unity (Theorem 7) was settled by Szabados and Varga [107] when the roots are in $|z| < \rho$ and satisfy (4.1) and (4.3).

Results for roots which are sufficiently close to the roots of unity were proved by Baishanski [4] and Bokhari [7]. The results of Section 4.4 for concentric circles are new.

Pál-type interpolation was introduced in [82].

CHAPTER 5

CONVERSE RESULTS

In this chapter we will be concerned with so-called converse theorems of the theory of equiconvergence. This means that from some convergence properties of the operator in question we will deduce structural properties of the function approximated. The results are strongly connected with the sharpness theorems of Chapter 4, but here our assumptions on the function will be kept minimal so that the operators can be defined for it: analyticity in $|z| < 1$ (in order to have a convergent Taylor series), and continuity in $|z| \leq 1$ (in order to make sense of interpolations).

5.1. Lagrange Interpolation

Theorem 1 of Chapter 4 shows that for Lagrange interpolation and shifted Taylor series the equiconvergence takes place in $|z| < \rho^{\ell+1}$ for functions in the class A_ρ. To state the corresponding converse result, let $\overline{A}_\rho$, $\overline{A}_\rho C$ denote, respectively, the set of functions analytic, at least, in $|z| < \rho$ and analytic, at least, in $|z| < \rho$ and continuous on $|z| \leq \rho$. (In other words, in contrast to A_ρ, here we do not require the function to have singularities on $|z| = \rho$.)

THEOREM 1. *Let $f \in \overline{A}_1 C$. If the sequence*

$$\{\Delta_{n,\ell}(f;z)\}_{n=1}^\infty = \left\{ L_{n-1}(f;z) - \sum_{j=0}^{\ell-1} p_{n-1,j}(f;z) \right\}_{n=1}^\infty \tag{1.1}$$

is uniformly bounded on every closed subset of $|z| < \rho^{\ell+1}$, then $f(z) \in \overline{A}_\rho$ $(\rho > 1)$.

REMARK. The example $f(z) = (\rho - z)^{-1}$ shows that the conclusion of Theorem 1 is the best possible. Also, notice that we do not require the convergence of the sequence (1.1) to zero; only the boundednes.

The proof is based on Theorem 1.1 of Chapter 4, as well as the following lemma.

LEMMA 1. *If a_k $(k = 0, 1, \ldots)$ are arbitrary complex numbers then*

$$a_{(3\ell^2+1)m+p} - a_{(6\ell^2+1)m+p} = \sum_{j=\ell}^{2\ell-1} a_{(3j\ell+1)m+p} + \sum_{s=\ell+1}^{2\ell-1} \sum_{j=\ell}^{2\ell-1} a_{(3js+2\ell+s+1)m+p} \tag{1.2}$$

$$- \sum_{s=\ell}^{2\ell-1} \sum_{j=\ell}^{2\ell-1} a_{(3js+2\ell+j+1)m+p}$$

$$(\ell \geq 2,\ p \geq 0,\ m \geq 1).$$

PROOF. Changing j to $j + 1$, the first sum on the right hand side can be written as

$$a_{(3\ell^2+1)m+p} + \sum_{j=\ell}^{2\ell-1} a_{(3j\ell+3\ell+1)m+p} - a_{(6\ell^2+1)m+p} \cdot$$

But here the sum is nothing else but the missing term $s = \ell$ of the first double sum in (1.2). Thus the right hand side of (1.2) will be

$$a_{(3\ell^2+1)m+p} - a_{(6\ell^2+1)m+p} + \sum_{s=\ell}^{2\ell-1} \sum_{j=\ell}^{2\ell-1} a_{(3js+2\ell+s+1)m+p}$$

$$- \sum_{s=\ell}^{2\ell-1} \sum_{j=\ell}^{2\ell-1} a_{(3js+2\ell+j+1)m+p} \cdot$$

Here the double sums are the same, which can be seen by switching the roles of j and s.   □

PROOF OF THEOREM 1. By the remark made at the beginning of Section 4 in Chapter 1, we have

$$\Delta_{n,\ell}(f; z) = L_{n-1}\left(\sum_{k=\ell n}^{\infty} a_k z^k; z \right)$$

provided

$$f(z) = \sum_{k=0}^{\infty} a_k z^k.$$

Thus by assumption

$$\sup_{n} \max_{|z|=r} \left| L_{n-1}\left(\sum_{k=\ell n}^{\infty} a_k z^k; z \right) \right| := M(r) < \infty, \quad r < p^{\ell+1}. \tag{1.3}$$

Therefore

$$\max_{|z|=r} \left| L_{2n-1}\left(\sum_{k=2\ell n}^{\infty} a_k z^k; z \right) \right| \leq M(r).$$

If we denote

$$L_{2n-1}\left(\sum_{k=2\ell n}^{\infty} a_k z^k; z \right) = \sum_{k=0}^{2n-1} b_k z^k,$$

then

$$|b_k| = \frac{1}{2\pi} \left| \int_{|z|=r} z^{-k-1} L_n\left(\sum_{k=2\ell n}^{\infty} a_k z^k; z \right) dz \right|$$

$$\leq \frac{M(r)}{r^k}, \quad k = 0, 1, \dots, 2n - 1.$$

Thus using that the nth roots of unity are also $(2n)^{\text{th}}$ roots of unity we obtain

$$\left| L_{n-1}\left(\sum_{k=2\ell n}^{\infty} a_k z^k; z \right) \right| = \left| L_{n-1}\left(L_{2n-1}\left(\sum_{k=2\ell n}^{\infty} a_k z^k; z \right); z \right) \right|$$

$$= \left| \sum_{k=0}^{n-1} (b_k + b_{k+n}) z^k \right|$$

$$\leq M(r) \sum_{k=0}^{n-1} (r^{-k} + r^{-k-n}) r^k < 2nM(r), \quad |z| = r.$$

This together with (1.3) yields

$$\max_{|z|=r} \left| L_{n-1}\left(\sum_{k=\ell n}^{2\ell n-1} a_k z^k; z \right) \right| < (2n+1)M(r). \tag{1.4}$$

On the other hand, using the obvious property

$$L_{n-1}\left(z^{mn} g(z); z \right) \equiv L_{n-1}\left(g(z); z \right)$$

several times,

$$L_{n-1}\left(\sum_{k=\ell n}^{2\ell n-1} a_k z^k; z \right) = \sum_{k=0}^{n-1} \left(\sum_{j=\ell}^{2\ell-1} a_{k+jn} \right) z^k,$$

hence and by (1.4)

$$\left| \sum_{j=\ell}^{2\ell-1} a_{k+jn} \right| = \frac{1}{2\pi} \left| \int_{|z|=r} z^{-k-1} L_{n-1}\left(\sum_{j=\ell n}^{2\ell n-1} a_j z^j; z \right) dz \right| \tag{1.5}$$

$$\leq r^{-k} (2n+1) M(r), \quad k = 0, \dots, n-1.$$

Now if $\ell = 1$ then this yields for $k = n - 2$ and $k = n - 1$

$$|a_{2n-2}| < r^{2-n}(2n+1)M(r), \quad |a_{2n-1}| < r^{1-n}(2n+1)M(r),$$

respectively, i.e. $\varlimsup\limits_{n\to\infty} |a_n|^{1/n} \le r^{-1/2}$. But $r < \rho^{\ell+1} = \rho^2$ was arbitrary, thus

$$\limsup_{n\to\infty} |a_n|^{1/n} \le 1/\rho.$$

This shows that $f \in \overline{A}_\rho$.

So from now on we may assume that $\ell \ge 2$. Using (1.5) with

$$k = m + p, \ (2\ell + s + 1)m + p, \ (2\ell + 1)m + p \quad \text{and} \quad n = 3\ell m, \ 3sm, \ (3s + 1)m,$$

respectively, we obtain

$$\left| \sum_{j=\ell}^{2\ell-1} a_{(3j\ell+1)m+p} \right| \le r^{-m-p}(6\ell m + 1)M(r) < 7\ell m r^{-m} M(r) , \quad 0 \le p < m,$$

$$\left| \sum_{j=\ell}^{2\ell-1} a_{(3js+2\ell+s+1)m+p} \right| \le r^{-(2\ell+s+1)m-p}(6sm + 1)M(r)$$

$$< 12\ell m r^{-m} M(r) , \quad 0 \le p < m, \ \ell + 1 \le s \le 2\ell - 1,$$

and

$$\left| \sum_{j=\ell}^{2\ell-1} a_{(3js+2\ell+j+1)m+p} \right| \le r^{-(2\ell+1)m-p}[(6s + 2)m + 1]M(r)$$

$$< 12\ell m r^{-m} M(r) , \quad 0 \le p < m, \ \ell \le s \le 2\ell - 1.$$

Hence Lemma 1 yields

$$|a_{(3\ell^2+1)m+p} - a_{(6\ell^2+1)m+p}| \le 7\ell m r^{-m} M(r) + 12\ell^2 m r^{-m} M(r)$$
$$+ 12\ell^2 m r^{-m} M(r) \tag{1.6}$$
$$\le 28\ell^2 m r^{-m} M(r) , \quad 0 \le p < m, \ \ell \ge 2.$$

Now let

$$n \ge (3\ell^2 + 1) \max\left(6\ell^2 + 1, \frac{\log 9}{\log r} + 1 \right) \tag{1.7}$$

represented in the form

$$n = (3\ell^2 + 1)m + p , \quad 0 \le p \le 3\ell^2, \tag{1.8}$$

and define the sequences of integers $\{r_k\}_{k=0}^\infty$ and $\{s_k\}_{k=0}^\infty$ by

$$r_0 = m, \quad s_0 = p, \tag{1.9}$$

$$r_{k+1} = \left[\frac{6\ell^2+1}{3\ell^2+1}\,r_k\right], \quad \text{and} \quad s_{k+1} = s_k + (6\ell^2+1)r_k - (3\ell^2+1)\left[\frac{6\ell^2+1}{3\ell^2+1}\,r_k\right],$$

$$\text{for} \quad k = 1, 2, \ldots. \tag{1.10}$$

Then evidently

$$(6\ell^2+1)r_k + s_k = (3\ell^2+!)r_{k+1} + s_{k+1}, \quad k = 0, 1, \ldots. \tag{1.11}$$

We shall prove that

$$a_n = \sum_{k=0}^{\infty}(a_{(3\ell^2+1)r_k+s_k} - a_{(6\ell^2+1)r_k+s_k}). \tag{1.12}$$

Namely, (1.8), (1.9) and (1.11) yield that the Nth partial sum of this series can be written as

$$\sum_{k=0}^{N}(a_{(3\ell^2+1)r_k+s_k} - a_{(6\ell^2+1)r_k+s_k}) = a_n - a_{(6\ell^2+1)r_N+s_N}.$$

The assumption $f \in \overline{A}_1 C$ implies $\lim_{k\to\infty} a_k = 0$, namely if $p_{k-1}(z)$ denotes the best approximation of $f(z)$ on $|z| \leq 1$ by polynomials of degree at most $k-1$, i.e.

$$\max_{|z|=1} |f(z) - p_{k-1}(z)| = E_{k-1}(f),$$

then

$$|a_k| = \frac{1}{2\pi}\left|\int_{|z|=1} z^{-k-1}f(z)dz\right|$$

$$= \frac{1}{2\pi}\left|\int_{|z|=1} z^{-k-1}[f(z) - p_{k-1}(z)]dz\right|$$

$$\leq E_{k-1}(f) \to 0 \quad \text{as} \quad k \to 0.$$

From (1.9) - (1.10) it is clear that

$$\left(\frac{3}{2}\right)^k m \leq r_k \leq 2^k m, \quad p \leq s_k \leq 3(k+1)\ell^2, \quad k = 0, 1, \ldots, \tag{1.13}$$

thus $\lim_{N\to\infty} a_{(6\ell^2+1)r_N+s_N} = 0$, i.e. (1.12) holds. Since by (1.7) and (1.8)

$$m \geq \max\left(6\ell^2, \frac{\log 9}{\log r}\right),$$

we have from (1.13) $s_k < r_k$ $(k = 0, 1, \ldots)$, and applying (1.6) we get the result.

A more general formulation is the following.

THEOREM 2. *Suppose $f(z)$ is defined on the set*

$$U = \bigcup_{n=1}^{\infty} \{z \mid z = e^{2\pi i k/n}, \ k = 0, \ldots, n-1\}. \tag{1.14}$$

and $\{a_n\}_{n=0}^{\infty}$ is a sequence of complex numbers such that

$$\left\{ \begin{array}{l} \Delta_{n,\ell}(z) = \Delta_{n,\ell}\big(f, \{a_n\}_{n=0}^{\infty}; z\big) := L_{n-1}(f; z) - \displaystyle\sum_{j=0}^{\ell-1} \sum_{k=0}^{n-1} a_{k+jn} z^k, \quad \ell \geq 1 \\[2mm] \text{is uniformly bounded in every compact subset of } |z| < \rho^{\ell+1}. \end{array} \right. \tag{1.15}$$

Then

(i) $g(z) := \displaystyle\sum_{k=0}^{\infty} a_k z^k$ *is analytic in $|z| < \rho$,*

and

(ii) $f(z) - g(z)$ *can be extended analytically to $|z| < \rho^{\ell+1}$.*

An obvious consequence of this theorem (with $f = g$) is the following

COROLLARY 1. *If $f(z) = \displaystyle\sum_{n=0}^{\infty} a_n z^n$ is analytic in $|z| < 1$ and continuous on $|z| \leq 1$; moreover (1.15) is uniformly bounded in every compact subset of $|z| < \rho^{\ell+1}$, then $f(z)$ is analytic in $|z| < \rho$.*

PROOF OF THEOREM 2. Denote

$$S_m(z) = \sum_{k=0}^{m} a_k z^k,$$

then evidently

$$L_{n-1}(S_{\ell n-1}; z) = \sum_{k=0}^{n-1} \sum_{j=0}^{\ell-1} a_{k+jn} z^k. \tag{1.16}$$

Thus by (1.15) we have for $1 < r < \rho^{\ell+1}$,

$$M(r) := \sup_n \max_{|z|=r} |\Delta_{n,\ell}(z)| = \sup_n \max_{|z|=r} |f(z) - S_{\ell n-1}(z)| < \infty. \tag{1.17}$$

Applying (1.16) with ℓN instead of ℓ,

$$L_{n-1}\big(S_{\ell n N-1} - S_{\ell n-1}; z\big) = \sum_{k=0}^{n-1} \sum_{j=\ell}^{\ell N-1} a_{k+jn} z^k,$$

and thus by (1.17) for $0 \leq k \leq n-1$,

$$\left| \sum_{j=\ell}^{\ell N-1} a_{k+jn} \right| = \frac{1}{2\pi} \left| \int_{|z|=r} z^{-k-1} L_{n-1}\big(S_{\ell n N-1} - S_{\ell n-1}; z\big) dz \right| \tag{1.18}$$

$$\leq M(r) r^{-k} + \frac{1}{2\pi} \left| \int_{|z|=r} z^{-k-1} L_{n-1}\big(f - S_{\ell n N-1}; z\big) dz \right|.$$

Let

$$L_{nN-1}\big(f - S_{\ell nN-1}; z\big) = \sum_{t=0}^{nN-1} b_t z^t, \tag{1.19}$$

then

$$\frac{1}{2\pi i} \int_{|z|=r} \Big(\sum_{\nu=0}^{N-1} z^{-\nu n}\Big) z^{-k-1} L_{nN-1}\big(f - S_{\ell nN-1}; z\big) dz = \sum_{\nu=0}^{N-1} b_{k+\nu n},$$

while by (1.19)

$$\frac{1}{2\pi i} \int_{|z|=r} z^{-k-1} L_{n-1}\big(f - S_{\ell nN-1}; z\big) dz$$

$$= \frac{1}{2\pi i} \int_{|z|=r} z^{-k-1} L_{n-1}\big(L_{nN-1}(f - S_{\ell nN-1}; z); t\big) dt$$

$$= \frac{1}{2\pi i} \int_{|z|=r} z^{-k-1} L_{n-1}\Big(\sum_{t=0}^{nN-1} b_t z^t; z\Big) dt$$

$$= \frac{1}{2\pi i} \int_{|z|=r} z^{-k-1} \sum_{t=0}^{n-1} \sum_{\nu=0}^{N-1} b_{t+\nu n} z^t dt$$

$$= \sum_{\nu=0}^{N-1} b_{k+\nu n}.$$

Thus (1.18) can be written in the form (using (1.17) again, with nN instead of n).

$$\Big|\sum_{j=\ell}^{\ell N-1} a_{k+jn}\Big| \le M(r) r^{-k} + \frac{1}{2\pi}\Big|\int_{|z|=r} \sum_{\nu=0}^{N-1} z^{-\nu n-k-1} L_{nN-1}\big(f - S_{\ell nN-1}; z\big) dt\Big| \tag{1.20}$$

$$\le M(r) r^{-k}\Big(1 + \sum_{\nu=0}^{\infty} r^{-\nu n}\Big) \le \frac{2r^n - 1}{r^n - 1} M(r) r^{-k}$$

$$\le \frac{2r - 1}{r - 1} M(r) r^{-k}, \quad k = 0, \ldots, n - 1.$$

Now we make use of the identity (cf. also Lemma 1)

$$a_{(3\ell+1)\ell m+p} = a_{(3\ell+1)\ell mN+p} + \sum_{j=\ell}^{\ell N-1} a_{(3\ell+1)jm+p} \tag{1.21}$$

$$+ \sum_{s=\ell+1}^{\ell N-1} \sum_{j=\ell}^{\ell N-1} a_{(3sj+2s+j+\ell+1)m+p}$$

$$- \sum_{j=\ell}^{\ell N-1} \sum_{s=\ell}^{\ell N-1} a_{(3sj+2s+j+\ell+1)m+p}$$

which is easily seen if we add the term $s = \ell$ so the first double sum on the right hand side, and subtract it from the first (single) sum. Apply first (1.20) successively with $n = (3\ell + 1)m$, $k = p$; then with $n = (3s + 1)m$, $k = (2s + \ell + 1)m + p$; and finally with $n = (3j + 2)m$, $k = (j + \ell + 1)m + p$, where $0 \leq p < m$. We obtain

$$
\begin{cases}
\left| \sum_{j=\ell}^{\ell N - 1} a_{j(3\ell+1)m+p} \right| & \leq \frac{2r-1}{r-1} M(r) r^{-p}, \\[2em]
\left| \sum_{j=\ell}^{\ell N - 1} a_{j(3s+1)m+(2s+\ell+1)m+p} \right| & \leq \frac{2r-1}{r-1} M(r) r^{-(2s+\ell+1)m-p} \\[0.5em]
& \quad \text{for} \quad \ell+1 \leq s \leq \ell N - 1, \\[2em]
\left| \sum_{s=\ell}^{\ell N - 1} a_{(3j+2)sm+(j+\ell+1)m+p} \right| & \leq \frac{2r-1}{r-1} M(r) r^{-(j+\ell+1)m-p} \\[0.5em]
& \quad \text{for} \quad \ell \leq j \leq \ell N - 1.
\end{cases}
\tag{1.22}
$$

Hence

$$
\left| \sum_{s=\ell+1}^{\ell N - 1} \sum_{j=\ell}^{\ell N - 1} a_{(3js+2s+j+\ell+1)m+p} \right| \leq \frac{2r-1}{r-1} M(r) r^{-p} \sum_{s=\ell+1}^{\ell N - 1} r^{-2sm}
$$

$$
\left| \sum_{j=\ell}^{\ell N - 1} \sum_{s=\ell}^{\ell N - 1} a_{(3js+2s+j+\ell+1)m+p} \right| \leq \frac{2r-1}{r-1} M(r) r^{-p} \left(r^{-(\ell+1)m} \sum_{j=\ell}^{\ell N - 1} r^{-jm} \right).
$$

Thus using (1.22) in (1.21), we have

$$
|a_{(3\ell+1)\ell m+p}| \leq |a_{(3\ell+1)\ell m N+p}| + C(r) r^{-p}.
$$

Letting $N \to \infty$, $a_N \to 0$, we obtain

$$
|a_{(3\ell+1)\ell m+p}| \leq C(r) r^{-p}, \quad 0 \leq p < m.
$$

Hence with $n = (3\ell + 1)\ell m + p$, $m - (3\ell + 1)\ell \leq p \leq m - 1$ we obtain

$$
\rho^{-1} = \varlimsup_{n \to \infty} |a_n|^{1/n} \leq r^{-\frac{1}{(3\ell+1)\ell+1}} < 1.
$$

We now show that $\rho_1 \geq \rho$. If we had $\rho_1 < \rho$, then choose r such that $\rho_1 < r^{\frac{1}{\ell+1}} < \rho$. Then from (1.20) we have

$$
|a_{\ell n+k}| \leq \left| \sum_{j=\ell}^{\ell N - 1} a_{jn+k} \right| + \left| \sum_{j=\ell+1}^{\ell N - 1} a_{jn+k} \right|
$$

$$
\leq c(r) r^{-k} + O\left(\sum_{j=\ell+1}^{\infty} (\rho_1 - \varepsilon)^{-jn-k} \right)
$$

$$
\leq c(r) r^{-k} + O\left((\rho_1 - \varepsilon)^{-(\ell+1)n-k} \right), \quad 0 \leq k \leq n - 1.
$$

Using this with $n - \ell \le k \le n - 1$, we see from the definition of ρ_1^{-1} and the above inequality that

$$\rho_1^{-1} = \varlimsup_{\nu \to \infty} |a_\nu|^{1/\nu} \le \max\left\{ r^{-\frac{1}{\ell+1}}, (\rho_1 - \varepsilon)^{-\frac{\ell+z}{\ell+1}} \right\} < \frac{1}{\rho_1}$$

for sufficiently small $\varepsilon > 0$, a contradiction. This proves that $g(z)$ is analytic in $|z| < \rho$.

In order to prove (ii) we observe that by Theorem 1 of Chapter 4, $\Delta_{n,\ell}(g; z)$ is uniformly bounded on compact subsets of $|z| < \rho^{\ell+1}$. Thus by (1.15),

$$|L_{n-1}(f - g; z)| \le |\Delta_{n,\ell}(z)| + |\Delta_{n,\ell}(g; z)| \le M < \infty$$

on compact subsets of $|z| < \rho^{\ell+1}$. If we apply this to $n!$ instead of n, then by Vitali's selection principle there exists a subsequence $n_1 < n_2 < \dots$ such that $L_{n_k!-1}(f - g; z)$ converges uniformly to some function $F(z)$ analytic in $|z| < \rho^{\ell+1}$. Since the (1.14) is dense in the unit circle, it follows that $F(z) = f(z) - g(z)$, $z \in U$. $\qquad\square$

Next we investigate the question whether assuming the condition

$$\lim_{n \to \infty} \Delta_{n,\ell}(f; z) = 0 \quad \text{uniformly in} \quad |z| \le \rho^{\ell+1} \tag{1.23}$$

is stronger than (1.15), i.e. can we conclude that $f(z)$ is continuous in $|z| \le \rho$. The negative answer will follow from the following

THEOREM 3. *If* $f \in A_\rho$ $(\rho > 1)$ *and* $\{\varphi_n\}_{n=1}^\infty$ *is a positive monotone (increasing or decreasing) sequence with* $\varphi_{2n} \sim \varphi_n$, *then*

$$\max_{|z|=\rho^{\ell+1}} |\Delta_{n,\ell}(f; z)| = O(\varphi_n), \quad n \to \infty \tag{1.24}$$

and

$$a_n = O(\rho^{-n}\varphi_n), \quad n \to \infty \tag{1.25}$$

are equivalent. (Here a_n *are the Taylor coefficients of* f.)

PROOF. By the assumption on φ_n, there exists a constant $c > 0$ such that $(cn^c)^{-1} < \varphi_n < cn^c$. Thus, using the representation (1.2) from Chapter 4 we get by (1.25)

$$|\Delta_{n,\ell}(f; z)| \le \sum_{j=\ell}^{\infty} \sum_{k=0}^{n-1} |a_{k+jn}| \cdot |z|^k$$

$$= O\left(\sum_{j=\ell}^{\infty} \sum_{k=0}^{n-1} \varphi_{k+jn} \rho^{-k-jn+k(\ell+1)} \right)$$

$$= O\left(\sum_{j=\ell}^{\infty} j^c \varphi_n \rho^{(\ell-j)n} \right) = o(\varphi_n), \quad |z| = \rho^{\ell+1}.$$

Conversely, for $f \in A_\rho$ we get

$$
\begin{aligned}
\Delta_{n,\ell}(f;z) &= \sum_{k=0}^{n-1} a_{k+\ell n} z^k + O\Big(\sum_{j=\ell+1}^{\infty} \sum_{k=0}^{n-1} a_{k+jn} z^k \Big) \\
&= \sum_{k=0}^{n-1} a_{k+\ell n} z^k + O\Big(\sum_{j=\ell+1}^{\infty} \sum_{k=0}^{n-1} (\rho - \varepsilon)^{-k-jn} \rho^{k(\ell+1)} \Big) \\
&= \sum_{k=0}^{n-1} a_{k+\ell n} z^k + O\big((\rho - \varepsilon)^{-(\ell+2)n} \rho^{n(\ell+1)} \big), \quad |z| = \rho^{\ell+1}
\end{aligned}
$$

whence by (1.24), for $n = \ell n + k$, $n - \ell - 1 \le k \le n - 1$ we obtain

$$
\begin{aligned}
|a_m| &= \frac{1}{2\pi} \left| \int_{|z|=\rho^{\ell+1}} \frac{\Delta_{n,\ell}(f;z)}{z^{k+1}} \, dz \right| + O\big((\rho - \varepsilon)^{-(\ell+2)n} \rho^{(n-k)(\ell+1)} \big) \\
&= O\big(\rho^{-k(\ell+1)} \varphi_n + (\rho - \varepsilon)^{-(\ell+2)n} \big) = O(\rho^{-m} \varphi_m),
\end{aligned}
$$

since $k(\ell+1) \ge m - \ell(\ell+1)$, $n(\ell+1) \ge m$, and φ_n is of polynomial growth. $\square$

Now we can answer the question raised after the proof of Theorem 1. Let

$$
f(z) = \sum_{k=1}^{\infty} \frac{z^k}{k\rho^k} \in A_\rho.
$$

This function is not in $\overline{A}_\rho C$, since at $z = \rho$ it is not even defined. Nevertheless,

$$
a_n = \frac{1}{n\rho^n} = \rho^{-n} \varphi_n, \quad \varphi_n = \frac{1}{n}
$$

and thus by Theorem 3, (1.24) holds which implies (1.23).

Thus uniform convergence of $\Delta_{n,\ell}(f;z)$ to zero on $|z| \le \rho^{\ell+1}$ does not imply even the boundedness of $f(z)$ on $|z| \le \rho$.

5.2. Hermite Interpolation

For any $f(z) \in A_\rho$ ($\rho > 1$), $r-1$ times differentiable on $|z| \le \rho$, let $h_{rn-1}(f;z)$ denote the Hermite interpolant to f in the zeros of $(z^n - 1)^r$:

$$
h_{rn-1}^{(j)}(f;\omega^k) = f^{(j)}(\omega^k), \quad j = 0, \ldots, r-1; k = 0, \ldots, n-1
$$

where $\omega^n = 1$ (see (1.1) of Chapter 2). We shall also use the definition of $\beta_{s,r}(z)$ in (1.7) and $\Delta_{rn-1,\ell}(f;z)$ in (1.14) of Chapter 2. Now the analogue of Corollary 1 can be stated as follows:

THEOREM 4. *Assume that $f(z) \in A_1$ is $r-1$ times differentiable on $|z| = 1$, and*

$$\Delta_{rn-1,\ell}(f;z) \quad \text{is uniformly bounded on every closed subset of} \quad |z| < \rho^{1+\ell/r}. \tag{2.1}$$

Then $f(z) \in A_\rho$.

PROOF. As usual, let $f(z) = \sum_{k=0}^{\infty} a_k z^k$ be the Taylor expansion of f in $|z| < 1$. First we prove the following identity:

$$\Delta_{rn-1,\ell}(f;z) = h_{rn-1}\left(\sum_{k=(r+\ell-1)n}^{\infty} a_k z^k; z \right). \tag{2.2}$$

Namely, since h_{rn-1} is a linear operator which reproduces polynomials of degree at most $rn - 1$, we have

$$h_{rn-1}(f;z) - h_{rn-1}\left(\sum_{k=(r+\ell-1)n}^{\infty} a_k z^k; z \right) \tag{2.3}$$

$$= h_{rn-1}\left(\sum_{k=0}^{(r+\ell-1)n-1} a_k z^k; z \right)$$

$$= h_{rn-1}\left(\sum_{k=0}^{rn-1} a_k z^k; z \right) + h_{rn-1}\left(\sum_{k=rn}^{(r+\ell-1)n-1} a_k z^k; z \right)$$

$$= \sum_{k=0}^{rn-1} a_k z^k + \sum_{k=rn}^{(r+\ell-1)n-1} a_k h_{rn-1}(z^k; z)$$

$$= \sum_{k=0}^{rn-1} a_k z^k + \sum_{j=1}^{\ell-1} \sum_{k=0}^{n-1} a_{k+(r+j-1)n} h_{rn-1}\left(z^{k+(r+j-1)n}; z \right).$$

Here, according to (2.1) in Chapter 2 (applied with $j = k$, $s = j-1$, and $\alpha = 1$), we have

$$h_{rn-1}\left(z^{k+(r+j-1)n}; z \right) = z^k \beta_{j,r}(z^n), \quad j = 1, 2, \ldots, \quad 0 \le k \le n - 1. \tag{2.4}$$

Substituting this into (2.3), and using also (1.14) from Chapter 2 we get (2.2).

Now let $1 < R < \rho^{1+\ell/r}$ be arbitrary. Then by (2.1) and (2.2)

$$\sup_n \max_{|z|=R} \left| b_{rn-1}\left(\sum_{k=(r+\ell-1)n}^{\infty} a_k z^k; z \right) \right| \le M(R) < \infty. \tag{2.5}$$

The last inequality will be used with Nn instead n as well:

$$\max_{|z|=R} \left| h_{Nrn-1}\left(\sum_{k=N(r+\ell-1)n}^{\infty} a_k z^k; z \right) \right| \le M(R). \tag{2.6}$$

Thus if we set

$$h_{Nrn-1}\Big(\sum_{k=N(r+\ell-1)n}^{\infty} a_k z^k; z\Big) := \sum_{k=0}^{Nrn-1} b_k z^k \tag{2.7}$$

we get from Cauchy's formula

$$|b_k| \le M(R)R^{-k}, \quad k = 0, 1, \ldots, Nrn - 1. \tag{2.8}$$

Since the set of $(Nn)^{\text{th}}$ roots of unity includes all n^{th} roots of unity, evidently we have:

$$h_{rn-1}(g; z) = h_{rn-1}\big(h_{Nrn-1}(g; z); z\big).$$

Using this with $g(z) = \sum_{k=N(r+\ell-1)n}^{\infty} a_k z^k$ successively, as well as (2.4) and (2.7) to obtain

$$h_{rn-1}\Big(\sum_{k=N(r+\ell-1)n}^{\infty} a_k z^k; z\Big) = h_{rn-1}\Big(h_{Nrn-1}\Big(\sum_{k=N(r+\ell-1)n}^{\infty} a_k z^k; z\Big); z\Big)$$

$$= h_{rn-1}\Big(\sum_{k=0}^{Nrn-1} b_k z^k; z\Big)$$

$$= \sum_{k=0}^{rn-1} b_k z^k + \sum_{k=0}^{(N-1)r-1} b_{k+rn} h_{rn-1}\big(z^{k+rn}; z\big)$$

$$= \sum_{k=0}^{rn-1} b_k z^k + \sum_{k=0}^{n-1} \sum_{\lambda=0}^{(N-1)r-1} b_{k+(r+\lambda)n} h_{rn-1}\big(z^{k+(r+\lambda)n}; z\big)$$

$$= \sum_{k=0}^{rn-1} b_k z^k + \sum_{\lambda=0}^{(N-1)r-1} \beta_{\lambda+1,r}(z^n) \sum_{k=0}^{n-1} b_{k+(r+\lambda)n} z^k.$$

Here, by Lemma 2 of Chapter 2,

$$\max_{|z|=R} |\beta_{\lambda+1,r}(z^n)| \le c_r \lambda^r R^{(r-1)n}$$

where $c_r > 0$ depends only on r. Thus we get, using also (2.8)

$$\max_{|z|=R}\Big| h_{rn-1}\Big(\sum_{k=N(r+\ell-1)n}^{\infty} a_k z^k; z\Big)\Big| \tag{2.9}$$

$$\le rnM(R) + c_r \sum_{\lambda=0}^{(N-1)r-1} \lambda^r R^{(r-1)n} \sum_{k=0}^{n-1} M(R) R^{-k-(r+\lambda)n+k}$$

$$\le rnM(R) + c_r M(R) R^{-n} \sum_{\lambda=0}^{(N-1)r-1} \lambda^r R^{-\lambda n} \le c_r' n M(R)$$

with some other constant $c'_r > 0$.

This last estimate can be used as follows. Since

$$h_{rn-1}\left(\sum_{k=(r+\ell-1)n}^{N(r+\ell-1)n-1} a_k z^k; z\right) = h_{rn-1}\left(\sum_{k=(r+\ell-1)n}^{\infty} a_k z^k; z\right) -$$

$$- h_{rn-1}\left(\sum_{k=N(n+\ell-1)n}^{\infty} a_, z^k; z\right),$$

(2.5) and (2.9) yield

$$\max_{|z|=R}\left| h_{rn-1}\left(\sum_{k=(r+\ell-1)n}^{N(r+\ell-1)n-1} a_k z^k; z\right)\right| \le (c'_r n + 1)M(R). \qquad (2.10)$$

Using again (2.5) we get

$$h_{rn-1}\left(\sum_{k=(r+\ell-1)n}^{N(r+\ell-1)n-1} a_k z^k; z\right) =$$

$$= \sum_{k=0}^{n-1}\sum_{\lambda=0}^{(r+\ell-1)(N-1)-1} a_{k+(r+\lambda+\ell-1)n} h_{rn-1}\left(z^{k+(r+\lambda+\ell-1)n}; z\right)$$

$$= \sum_{k=0}^{n-1}\sum_{\lambda=0}^{(r+\ell-1)(N-1)-1} a_{k+(r+\lambda+\ell-1)n} z^k \beta_{\ell+\lambda,r}(z^n).$$

Here, by (1.7) of Chapter 2,

$$\beta_{\ell+\lambda,r}(z^n) = \sum_{\mu=0}^{r-1}\binom{r+\ell+\lambda-1}{\mu}(z^n-1)^\mu$$

$$= \sum_{\mu=0}^{r-1}\binom{r+\ell+\lambda-1}{\mu}\sum_{\nu=0}^{\mu}\binom{\mu}{\nu}(-1)^{\mu-\nu}z^{\nu n}$$

$$= \sum_{\nu=0}^{r-1} z^{\nu n}\sum_{\mu=0}^{r-1}(-1)^{\mu-\nu}\binom{r+\ell+\lambda-1}{\mu}\binom{\mu}{\nu}$$

and thus

$$h_{rn-1}\left(\sum_{k=(n+\ell-1)n}^{N(r+\ell-1)n-1} a_k z^k; z\right) = \sum_{k=0}^{n-1}\sum_{\nu=0}^{r-1} z^{k+\nu n} \times$$

$$\times \sum_{\lambda=0}^{r+\ell-2}\sum_{\mu=0}^{r-1}(-1)^{\mu-\nu}\binom{n+\ell+\lambda-1}{\mu}\binom{\mu}{\nu} a_{k+(n+\lambda+\ell-1)n} \cdot$$

Applying Cauchy's formula and (2.10) we get

$$
\left| \sum_{\lambda=0}^{(r+\ell-1)(N-1)-1} \sum_{\mu=\nu}^{r-1} (-1)^{\mu-\nu} \binom{r+\ell+\lambda-1}{\mu} \binom{\mu}{\nu} a_{k+(r+\ell+\lambda-1)n} \right| \leq
$$

$$
\leq \frac{(c_r'n+1)M(R)}{R^{k+\nu n}}, \quad k = 0, \ldots, n-1; \nu = 0, \ldots, r-1. \tag{2.11}
$$

Denoting

$$
\mu_{k,\nu,n} := \sum_{\lambda=0}^{(r+\ell-1)(N-1)-1} \sum_{\mu=0}^{r-1} (-1)^{\mu-\nu} \binom{r+\ell+\lambda-1}{\mu} \binom{\mu}{\nu} a_{k+(r+\ell+\lambda-1)n}
$$

and summing up for ν we obtain

$$
\sum_{\nu=0}^{r-1} \mu_{k,\nu,n} = \sum_{\lambda=0}^{(r+\ell-1)(N-1)-1} \sum_{\mu=0}^{r-1} (-1)^{\mu} \binom{r+\ell+\lambda-1}{\mu} a_{k+(r+\ell+\lambda-1)n} \times
$$

$$
\times \sum_{\nu=0}^{r-1} (-1)^{\nu} \binom{\mu}{\nu}
$$

$$
= \sum_{\lambda=0}^{(r+\ell-1)(N-1)-1} a_{k+(r+\ell+\lambda-1)n}
$$

$$
= \sum_{j=r+\ell-1}^{N(r+\ell-1)-1} a_{k+jn}.
$$

Thus by (2.10) we get

$$
\left| \sum_{j=r+\ell-1}^{N(r+\ell-1)-1} a_{k+jn} \right| \leq \sum_{\nu=0}^{r-1} |\mu_{k,\nu,n}| \leq \frac{r(c_r'n+1)(M(R))}{R^k}, \quad k = 0, 1, \ldots, n-1.
$$

From these inequalities we can conclude (just like in the proof of Theorem 1) that $f(z)$ can be analytically continued from $|z| \leq 1$ into a larger circle. Let $\overline{\rho} > 1$ be the maximal radius for which $f(z)$ is analytic in $|z| < \overline{\rho}$. By Theorem 3(ii) of Chapter 4, the sequence (2.2) can be bounded in at most $r + \ell - 1$ distinct points in $|z| > \overline{\rho}^{1+\ell/r}$. From the hypothesis of the theorem, this means that $\rho \leq \overline{\rho}$, i.e. $f(z)$ is analytic in $|z| < \rho$. □

5.3. Historical Remarks

Theorem 1 is due to Szabados [106], and later it was extended (in the weaker form) to Hermite interpolation by Cavaretta, Sharma and Varga [32] (Theorem 4). The stronger form Theorem 2 (and Corollary 1) was proved by Ivanov and Sharma [52]. The negative answer (Theorem 3) to the question if continuity property of the function is inherited in the order of overconvergence can be found in Totik [111].

CHAPTER 6

PADÉ APPROXIMATION AND
WALSH EQUICONVERGENCE FOR
MEROMORPHIC FUNCTIONS WITH ν POLES

6.1. Introduction

In the previous chapters extensions of the Walsh equiconvergence theorem for functions in A_ρ, $\rho > 1$, were estimated. In all these extensions of the Walsh equiconvergence theorem the differences of two polynomial approximation operators played a role. In this chapter we bring extensions of the Walsh theorem to differences of rational approximation operators of a certain type. It seems that most equiconvergence theorems for polynomial approximation operators can be extended to equiconvergence theorems for rational interpolants.

For each non-negative integer ν and for each ρ with $1 < \rho < \infty$, let $M_\rho(\nu)$ denote the set of functions $F(z)$ which are meromorphic with precisely ν poles (counting multiplicities) in the disc D_ρ but not in $\overline{D}_\rho$ and which are regular at $z = 0$ and on $|z| = 1$. A rational function

$$\frac{U_{n,\nu}(z)}{V_{n,\nu}(z)}, \quad U_{n,\nu}(z) \in \pi_n \quad \text{and} \quad V_{n,\nu} \in \pi_\nu \quad \text{is monic} \qquad (1.1)$$

is called of type (n, ν).

For a given $F \in M_\rho(\nu)$, denote by $S_{n,\nu}(z) = S_{n,\nu}(F, z)$ the rational interpolant of type (n, ν) to $F(z)$ in the $(n + \nu + 1)^{\text{th}}$ roots of unity. This means that if ω is a primitive $(n + \nu + 1)^{\text{th}}$ root of unity, we require that

$$S_{n,\nu}(\omega^k) = F(\omega^k), \quad 0 \le k \le n + \nu. \qquad (1.1a)$$

When $\nu = 0$ we have $S_{n,0} = L_n(f, z)$. Similarly, consider the Hermite-Padé rational interpolant $R_{n,\nu}(F; z)$ of type (n, ν) defined by

$$R_{n,\nu}(F, z) := R_{n,\nu}(z) := \frac{P_{n,\nu}(z)}{Q_{n,\nu}(z)}, \quad P_{n,\nu}(z) \in \pi_n, \quad Q_{n,\nu}(z) \in \pi_\nu \text{ is monic} \qquad (1.2)$$

and

$$R_{n,\nu}(z) - F(z) = O(z^{n+\nu+1}) \quad \text{as} \quad z \to 0. \qquad (1.2a)$$

It is easy to see that for $\nu = 0$, $R_{n,0} = S_n(F, z)$.

It is known from a theorem of de Montessus de Ballore [39] that the rational interpolants $R_{n,\nu}(z)$ and $S_{n,\nu}(z)$ exist and are unique for large n.

The first extension of Walsh theorem of equiconvergence to functions in the class $M_\rho(\nu)$ is the following:

THEOREM 1. *If $F \in M_\rho(\nu)$ ($\rho > 1$) and if $\{\alpha_j\}_{j=1}^\nu$ are the ν poles of F in D_ρ (listed according to multiplicities), then for the rational interpolants $S_{n,\nu}(z)$ of (1.1) and $R_{n,\nu}(z)$ of (1.2), we have*

$$\lim_{n\to\infty} [S_{n,\nu}(F, z) - R_{n,\nu}(F, z)] = 0, \quad \forall z \in D_{\rho^2}\backslash \bigcup_{j=1}^\nu \{\alpha_j\}, \tag{1.3}$$

the convergence being uniform and geometric on any closed subset of $D_{\rho^2}\backslash \overset{\nu}{\underset{j=1}{\cup}} \{\alpha_j\}$.

More precisely, on any closed subset $\mathcal{H}$ of $D_\tau\backslash \overset{\nu}{\underset{j=1}{\cup}} \{\alpha_j\}$ with $\rho \le \tau < \infty$, there holds

$$\limsup_{n\to\infty} \{\max_{z\in\mathcal{H}}|S_{n,\nu}(F, z) - R_{n,\nu}(F, z)|\}^{1/n} \le \frac{\tau}{\rho^2}. \tag{1.4}$$

The result (1.3) is best possible in the sense that for any $\nu \ge 0$ and for any ρ, $1 < \rho < \infty$, there exists an $\widetilde{F}_\nu \in M_\rho(\nu)$ such that

$$\limsup_{n\to\infty} \{ \min_{|z|=\rho^2} |S_{n,\nu}(\widetilde{F}, z) - R_{n,\nu}(\widetilde{F}, z)|\} > 0. \tag{1.5}$$

Note that in the special case $\nu = 0$, this theorem reduces to Walsh's theorem in Chapter 1 for functions in the class A_ρ.

We illustrate this theorem with an example before stating and proving its generalization.

EXAMPLE. For any given ρ with $1 < \rho < \infty$ and any fixed α with $0 < |\alpha| < \rho$, $|\alpha| \ne 1$, we consider the function $F(z) = \frac{1}{z-\alpha} + \frac{1}{z-\rho}$ which belongs to $M_\rho(1)$. Because there is only one pole inside the disc D_ρ, it is easy to check that

$$S_{n,1}(F; z) = \frac{U_{n,1}(z)}{V_{n,1}(z)} \quad \text{and} \quad R_{n,1}(F, z) = \frac{P_{n,1}(z)}{Q_{n,1}(z)}$$

where we write $V_{n,1}(z) = z + \lambda_n$ and $Q_{n,1}(z) = z + \gamma_n$. Observe that $U_{n,1}(z)$ interpolates $F(z)V_{n,1}(z)$ in the zeros of $z^{n+2} - 1$. Since $F(z)V_{n,1}(z) = 2 + \frac{\alpha+\lambda_n}{z-\alpha} + \frac{\rho+\lambda_n}{z-\rho}$, the interpolant to it at the zeros of $z^{n+2} - 1$ is clearly

$$2 + \frac{(\alpha + \lambda_n)(z^{n+2} - \alpha^{n+2})}{(z - \alpha)(1 - \alpha^{n+2})} + \frac{(\rho + \lambda_n)(z^{n+2} - \rho^{n+2})}{(z - \rho)(1 - \rho^{n+2})}.$$

In order that it is equal to $U_{n,1}(z)$ which $\in \pi_n$, we choose λ_n so that the coefficient of z^{n+1} in the above vanishes. This gives

$$\lambda_n = \frac{\alpha \rho^{n+2} + \alpha^{n+2}\rho - \rho - \alpha}{2 - \rho^{n+2} - \alpha^{n+2}} .$$

(For sufficiently large n, the denominator is different from zero.)

Similarly, we have

$$\gamma_n = -\rho\alpha \frac{\rho^{n+1} + \alpha^{n+1}}{\rho^{n+2} + \alpha^{n+2}} .$$

An easy calculation now gives

$$U_{n,1}(z) = 2 - \frac{\rho V_{n,1}(\rho)(z^{n+1} - \rho^{n+1})}{(\rho^{n+2} - 1)(z - \rho)} - \frac{\alpha V_{n,1}(\alpha)(z^{n+1} - \alpha^{n+1})}{(\alpha^{n+2} - 1)(z - \alpha)} ,$$

$$P_{n,1}(z) = 2 - \frac{Q_{n,1}(\rho)(z^{n+1} - \rho^{n+1})}{\rho^{n+1}(z - \rho)} - \frac{Q_{n,1}(\alpha)(z^{n+1} - \alpha^{n+1})}{\alpha^{n+1}(z - \alpha)} .$$

Since $\rho > |\alpha|$, both λ_n and γ_n tend to $-\alpha$ as $n \to \infty$. Thus the poles of $S_{n,1}(F, z)$ and $R_{n,1}(F, z)$ both tend to the inner pole of $F(z)$ as $n \to \infty$.

A straightforward but lengthy calculation yields

$$S_{n,1}(F, z) - R_{n,1}(F, z) = \frac{z^{n+2}(\rho - \alpha)^2(\rho + \alpha - 2z)}{\rho^{2n+4}(z - \alpha)^3(z - \rho)} + O(1)\frac{1}{\rho^n} \qquad (1.6)$$

as $n \to \infty$, the last term holding uniformly on any bounded set in $C\backslash(\{\alpha\}\cup\{\rho\})$. This shows overconvergence in $D_{\rho^2}\backslash\{\alpha\}$.

Moreover we have

$$\lim_{n \to \infty} \left\{ \min_{|z|=\rho^2} |S_{n,1}(F, z) - R_{n,1}(F; z)| \right\} \geq \frac{|\rho - \alpha|^2(2\rho^2 - \rho - |\alpha|)}{(\rho^2 + |\alpha|)^3(\rho^2 + \rho)} > 0. \qquad (1.7)$$

Thus we have verified the theorem in this case for $\nu = 1$.

The above example can also be modified to yield an example for the case when $\nu \geq 2$. In order to do this for any $\nu \geq 2$, we set

$$F_\nu(z) := F(z^\nu; \alpha^\nu, \rho^\nu) := \frac{1}{z^\nu - \alpha^\nu} + \frac{1}{z^\nu - \rho^\nu} \in M_\rho(\nu)$$

where $0 < |\alpha| < \rho$ and $|\alpha| \neq 1$. Then the rational interpolants $S_{n,\nu}(F_\nu; z)$ and $R_{n,\nu}(F_\nu, z)$ are easily seen to be related to the previous case of $S_{n,1}(F_1, z)$ and $R_{n,1}(F_1, z)$ as below:

$$S_{(m+1)\nu-1,\nu}(F_\nu, z) = S_{m,1}\big(F(\cdot, \alpha^\nu, \beta^\nu), z^\nu\big) ,$$

$$R_{(m+1)\nu-1,\nu}(F_\nu, z) = R_{m,1}\big(F(\cdot, \alpha^\nu, \beta^\nu), z^\nu\big), \quad m = 1, 2, \ldots .$$

It therefore follows from (1.6) that

$$S_{(m+1)\nu-1,\nu}(F_\nu, z) - R_{(m+1)\nu-1,\nu}(F_\nu, z) =$$
$$= \frac{z^{(m+1)\nu+1}(\rho^\nu - \alpha^\nu)^2(\rho^\nu + \alpha^\nu - 2z^\nu)}{\rho^{2(m+1)\nu+2}(z^\nu - \alpha^\nu)^3(z^\nu - \rho^\nu)} + O\left(\frac{1}{\rho^{(m+1)\nu}}\right), \quad \text{as} \quad m \to \infty$$

and from (1.7), we get

$$\lim_{m \to \infty} \left\{ \min_{|z|=\rho^2} |S_{(m+1)\nu-1}(F_\nu, z) - R_{(m+1)\nu-1}(F_\nu, z)| \right\}$$
$$\geq \frac{|\rho^\nu - \alpha^\nu|^2(2\rho^{2\nu} - \rho^\nu - |\alpha|^\nu)}{(\rho^{2\nu} + |\alpha|^\nu)^3(\rho^{2\nu} + \rho^\nu)} > 0.$$

Hence

$$\limsup_{n \to \infty} \left\{ \min_{|z|=\rho^2} |S_{n,\nu}(F_\nu, z) - R_{n,\nu}(F_\nu, z)| \right\} > 0$$

for each positive integer ν and each ρ with $1 < \rho < \infty$. This shows the sharpness of Theorem 1 in this case.

6.2 A Generalization of Theorem 1

We start with the representation of any function $F(z) \in M_\rho(\nu)$ in the form $\frac{f(z)}{B_\nu(z)}$, where $f(z) \in A_\rho$ and $B_\nu(z)$ is a monic polynomial of degree ν which may have multiple zeros. More precisely, let

$$f(z) := \sum_{k=0}^\infty a_k z^k, \quad \limsup_{n \to \infty} |a_n|^{1/n} = \frac{1}{\rho}, \quad f(z_j) \neq 0,$$

where

$$B_\nu(z) = \prod_{j=1}^\mu (z - z_j)^{\lambda_j} = \sum_{k=0}^\nu \alpha_k z^k, \quad \alpha_\nu = 1 \quad \text{and} \quad \sum_{j=1}^\mu \lambda_j = \nu.$$

For some $\alpha \in \mathbb{C}$, $|\alpha| < \rho$, we first find the (n, ν) Hermite-Padé interpolant to $F(z) \in M_\rho(\nu)$ on the zeros of $z^{n+r+1} - \alpha^{n+\nu+1}$. We suppose that ω is an arbitrary $(n + \nu + 1)^{\text{th}}$ primitive root of unity ($\omega^{n+\nu+1} = 1$). We shall denote this Hermite-Padé interpolant by

$$P_{n,\nu}^\infty(z) := \frac{U_n^\infty(z)}{B_{n,\nu}^\infty(z)} = \frac{\sum_{s=0}^n P_{s,n}^\infty z^s}{\sum_{s=0}^\nu \gamma_{s,n}^\infty z^s}, \quad \gamma_{\nu,n}^\infty = 1 \tag{2.1}$$

where $B_{n,\nu}^\infty(z)$ is monic of degree ν and $U_n^\infty(z) \in \pi_n$. We suppose that $\{\alpha\omega^k\}_{k=0}^{n+\nu}$ are different from the zeros of $B_\nu(z)$. For any other $\beta \in \mathbb{C}$ with $|\beta| < \rho$, $|\beta| \neq |\alpha|$ let $\{\beta\omega^k\}_{k=1}^{n+\nu}$ be different from the zeros of $B_\nu(z)$. For any integer $\ell \geq 1$ we denote the Lagrange interpolant to $f(z)$ on the zeros of $z^{\ell(n+\nu+1)} - \beta^{\ell(n+\nu+1)}$ by $L_{\ell(n+\nu+1)}(f;\beta;z)$ and for brevity we denote it by $L^{\beta}_{\ell(n+\nu+1)}(z)$. Now write

$$
\begin{cases}
L^{\beta}_{\ell(n+\nu+1)}(z) := \sum_{k=0}^{\ell(n+\nu+1)-1} B^{\ell}_{k,n} z^k, & \text{where} \\
B^{\ell}_{k,n} = \sum_{m=0}^{\infty} a_{m\ell(n+\nu+1)+k} \beta^{m\ell(n+\nu+1)}, & k = 0,1,\ldots,\ell(n+\nu+1)-1.
\end{cases}
\tag{2.2}
$$

The Lagrange interpolant to $L^{\beta}_{\ell(n+\nu+1)}$ on the zeros of $z^{n+\nu+1} - \alpha^{n+\nu+1}$ will be a polynomial of degree $n+\nu$ and we denote it by

$$
L_{n+\nu+1}(L^{\beta}_{\ell(n+\nu+1)};\alpha,z) = \sum_{j=0}^{n+\nu} A^{\ell}_{j,n} z^j,
\tag{2.3}
$$

where on using the value of $B^{\ell}_{k,n}$ in (2.2),

$$
\begin{aligned}
A^{\ell}_{j,n} &= \sum_{r=0}^{\ell-1} B^{\ell}_{r(n+\nu+1)+j,n} \alpha^{r(n+\nu+1)} \\
&= \sum_{r=0}^{\ell-1} \sum_{m=0}^{\infty} a_{(m\ell+r)(n+\nu+1)+j} \beta^{m\ell(n+\nu+1)} \alpha^{r(n+\nu+1)}, \quad j = 0,1,\ldots,n+\nu,
\end{aligned}
\tag{2.4}
$$

Set

$$
G^{\alpha,\beta}_{\ell,\nu}(F;z) := \frac{L_{n+\nu+1}(L^{\beta}_{\ell(n+\nu+1)};\alpha;z)}{B_{\nu}(z)}, \quad \ell \geq 1
$$

and consider the (n,ν) Hermite-Padé interpolant to $G^{\alpha,\beta}_{\ell,\nu}(F;z)$ on the zeros of $z^{n+\nu+1} - \alpha^{n+\nu+1}$. We denote it by

$$
P^{\ell}_{n,\nu}(z) := \frac{U^{\ell}_n(z)}{B^{\ell}_{n,\nu}(z)} = \frac{\sum_{s=0}^{n} p^{\ell}_{s,n} z^s}{\sum_{s=0}^{\nu} \gamma^{\ell}_{s,n} z^s}, \quad \gamma^{\ell}_{\nu,n} = 1.
$$

Using (2.3), this interpolation property of $P^{\ell}_{n,\nu}(z)$ to $G^{\alpha,\beta}_{\ell,\nu}(F;z)$ can be written in the form

$$
B_{\nu}(\alpha\omega^k) U^{\ell}_n(\alpha\omega^k) = B^{\ell}_{n,\nu}(\alpha\omega^k) \sum_{j=0}^{n+\nu} A^{\ell}_{j,n}(\alpha\omega^k)^j, \quad k = 0,1,\ldots,n+\nu.
\tag{2.5}
$$

Using the estimates $a_n = O((\rho-\varepsilon)^{-n})$ as $n \to \infty$, $\max(|\alpha|,|\beta|) < \rho - \varepsilon$, we easily obtain from (2.4)

$$
A^{\infty}_{j,n} := \lim_{\ell \to \infty} A^{\ell}_{jn} = \sum_{r=0}^{\infty} a_{r(n+\nu+1)+j} \alpha^{r(n+\nu+1)}, \quad n,j = 0,1,\ldots.
\tag{2.6}
$$

Thus with this definition of $A^{\infty}_{j,n}$ we can say that the interpolation property (2.5) remains valid for $\ell = \infty$. We now have the following

THEOREM 2. *Let $F \in M_\rho(\nu)$ $(\rho > 1,\ \nu \geq 0)$, and let $\alpha, \beta \in \mathbb{C}$ such that $|\alpha| \neq |\beta| < \rho$. Suppose that the poles of F are not on the circles $|z| = |\alpha|$ and $|z| = |\beta|$. Then for z such that $B_\nu(z) \neq 0$, we have*

$$\limsup_{n \to \infty} \left| P_{n,\nu}^\infty(z) - P_{n,\nu}^\ell(z) \right|^{1/n} \leq \begin{cases} \left(\dfrac{\gamma}{\rho} \right)^\ell & \text{if } |z| < \rho \\ \left(\dfrac{\gamma}{\rho} \right)^\ell \dfrac{|z|}{\rho} & \text{if } |z| \geq \rho, \end{cases}$$

where $\gamma := \max(|\alpha|, |\beta|, \max_{1 \leq j \leq \mu} |z_j|)$.

Thus we have overconvergence on compact subsets in $|z| < \frac{\rho^{\ell+1}}{\gamma^\ell}$, with zeros of $B_\nu(z)$ deleted.

The proof of this theorem will depend on several lemmas, but we first prove the unique existence of the operators $P_{n,\nu}^\ell$ and $P_{n,\nu}^\infty$ for n sufficiently large. Using the notation for coefficients introduced in (2.5) means that the polynomials

$$B_\nu(z) U_n^\ell(z) = \sum_{k=0}^\nu \alpha_k z^k \sum_{s=0}^n p_{s,n}^\ell z^s = \sum_{s=0}^{n+\nu} \left(\sum_{k+r=s} \alpha_k p_{r,n}^\ell \right) z^s$$

$$= \sum_{s=0}^{n+\nu} \left(\sum_{k=\max(0,s-n)}^{\min(\nu,s)} \alpha_k p_{s-k,n}^\ell \right) z^s \qquad (2.7)$$

and

$$B_{n,\nu}^\ell(z) \sum_{j=0}^{n+\nu} A_{j,n}^\ell z^j = \sum_{k=0}^\nu \gamma_{k,n}^\ell z^k \sum_{r=0}^{n+\nu} A_{r,n}^\ell z^r =$$

$$= \sum_{s=0}^{n+2\nu} \left(\sum_{k=\max(0,s-n-\nu)}^{\min(\nu,s)} \gamma_{k,n}^\ell A_{s-k}^\ell \right) z^s$$

$$= \sum_{s=0}^{n+\nu} \left(\sum_{k=0}^{\min(\nu,s)} \gamma_{k,n}^\ell A_{s-k,n}^\ell \right) z^s + z^{n+\nu+1} \sum_{s=0}^{\nu-1} \left(\sum_{k=s+1}^\nu \gamma_{k,n}^\ell A_{s+n+\nu+1-k,n}^\ell \right) z^s$$

coincide at the roots of the polynomial $z^{n+\nu+1} - \alpha^{n+\nu+1}$. Evidently this is also true of the polynomial

$$\sum_{s=0}^{n+\nu} \left(\sum_{k=0}^{\min(\nu,s)} \gamma_{k,n}^\ell A_{s-k,n}^\ell \right) z^s + \alpha^{n+\nu+1} \sum_{s=0}^{\nu-1} \left(\sum_{k=s+1}^\nu \gamma_{k,n}^\ell A_{s+n+\nu+1-k,n}^\ell \right) z^s.$$

$$(2.8)$$

Thus the polynomial (2.8) (being of degree at most $n+\nu$) is identically equal to the polynomial (2.7). Comparing coefficients we have

$$\sum_{k=\max(0,s-n)}^{\min(\nu,s)} \alpha_k p_{s-k,n}^\ell = \begin{cases} \displaystyle\sum_{k=0}^s \gamma_{k,n}^\ell A_{s-k,n}^\ell + \alpha^{n+\nu+1} \sum_{k=s+1}^\nu \gamma_{k,n}^\ell A_{s+n+\nu+1-k,n}^\ell \\ \qquad\qquad \text{if } 0 \leq s \leq \nu - 1 \\ \displaystyle\sum_{k=0}^\nu \gamma_{k,n}^\ell A_{s-k,n}^\ell, \quad \text{if } \nu \leq s \leq n+\nu. \end{cases}$$

$$(2.9)$$

REMARK. For $s < 0$, we make, for convenience, the convention that

$$A_{s,n}^\ell := \alpha^{n+\nu+1} A_{n+\nu+1+s,n}^\ell . \tag{2.10}$$

First we consider the case when $B_\nu(z)$ has simple zeros $z_1, z_2, \ldots, z_\nu$. Since (2.7) and (2.8) are identical this means that (2.8) is zero for the zeros of $B_\nu(z)$:

$$\sum_{s=0}^{n+\nu} \left(\sum_{k=0}^{\min(\nu,s)} \gamma_{k,n}^\ell A_{s-k,n}^\ell \right) z_j^s + \alpha^{n+\nu+1} \sum_{s=0}^{\nu-1} \left(\sum_{k=s+1}^{\nu} \gamma_{k,n}^\ell A_{s+n+\nu+1-k,n}^\ell \right) z_j^s = 0$$

$$(1 \le j \le \nu) .$$

Rearranging the left side, and on using the convention (2.10), we get

$$\sum_{k=0}^{\nu} \left\{ \sum_{s=0}^{\nu-k} A_{s,n}^\ell z_j^{s+k} + \sum_{s=\nu+1}^{n+\nu} A_{s-k,n}^\ell z_j^s + \alpha^{n+\nu+1} \sum_{s=0}^{\nu-1} A_{s+n+\nu+1-k,n}^\ell z_j^s \right\} \gamma_{kn}^\ell = 0$$

$$j = 1, \ldots, \nu. \tag{2.11}$$

All formulae so far are also true for $1 \le \ell \le \infty$. Now if $1 \le \ell < \infty$, then by (2.4), we have, for $0 \le s \le s + \nu$, on recalling $\gamma := \max(|\alpha|, |\beta|) < \rho$,

$$A_{s,n}^\ell = a_s + \sum_{m=1}^{\infty} a_{m\ell(n+\nu+1)+s} \beta^{m\ell(n+\nu+1)} + \sum_{r=1}^{\ell-1} a_{r(n+\nu+1)+s} \alpha^{r(n+\nu+1)}$$

$$+ \sum_{m=1}^{\infty} \sum_{r=1}^{\ell-1} a_{(m\ell+r)(n+\nu+1)+s} \alpha^{r(n+\nu+1)} \beta^{m\ell(n+\nu+1)} \tag{2.12}$$

$$= a_s + O(1) \left(\left(\frac{|\beta|}{\rho - \varepsilon} \right)^{\ell n} + \left(\frac{|\alpha|}{\rho - \varepsilon} \right)^{n} + \left(\frac{|\alpha| \cdot |\beta|^\ell}{(\rho - \varepsilon)^{\ell+1}} \right)^{n} \right) \frac{1}{(\rho - \varepsilon)^s}$$

$$= a_s + O(1) \left(\frac{\gamma^n}{(\rho - \varepsilon)^{n+s}} \right) = O(1) \left((\rho - \varepsilon)^{-s} \right) , \quad s = 0, 1, \ldots, \quad 1 \le \ell < \infty.$$

Similarly for $\ell = \infty$, we obtain from (2.6)

$$A_{s,n}^\infty = a_s + \sum_{r=1}^{\infty} a_{r(n+\nu+1)+s} \alpha^{r(n+\nu+1)} = a_s + O\left(\frac{\gamma^n}{(\rho - \varepsilon)^{n+s}} \right), \tag{2.13}$$

$$s = 0, 1, \ldots.$$

Substituting these into (2.11), we get by $a_k = O((\rho - \varepsilon)^{-k})$:

$$\sum_{k=0}^{\nu} \left\{ \sum_{s=0}^{\nu-k} a_s z_j^{s+k} + \sum_{s=\nu+1}^{n+\nu} a_{s-k} z_j^s + O(1) \left(\left(\frac{\gamma}{\rho - \varepsilon} \right)^{n} \right) \right\} \gamma_{k,n}^\ell = 0. \tag{2.14}$$

The two inner sums in the above become

$$\sum_{s=0}^{\nu-k} a_s z_j^{s+k} + \sum_{s=\nu+1}^{n+\nu} a_{s-k} z_j^s = z_j^k \sum_{s=0}^{n+\nu-k} a_s z_j^s.$$

Extending the summation over s to ∞, an error of the same order $O\left(\left(\frac{\gamma}{\rho-\varepsilon}\right)^n\right)$ is introduced and from $\sum_{s=0}^{\infty} a_s z_j^s = f(z_j)$, we find from (2.14) that $\gamma_{k,n}^{\ell}$ satisfy the equations

$$\sum_{k=0}^{\nu} \left\{ z_j^k f(z_j) + O\left(\left(\frac{\gamma}{\rho-\varepsilon}\right)^n\right) \right\} \gamma_{k,n}^{\ell} = 0, \quad j = 1, \ldots, \nu, \quad 1 \le \ell \le \infty.$$
(2.15)

Since $f(z_j) \ne 0$ or $j = 1, \ldots, \nu$ by supposition and $\gamma_{\nu,n}^{\ell} = 1$, we have

$$\sum_{k=0}^{\nu} \left\{ z_j^k + O\left(\left(\frac{\gamma}{\rho-\varepsilon}\right)^n\right) \right\} \gamma_{k,n}^{\ell} = -z_j^\nu + O\left(\left(\frac{\gamma}{\rho-\varepsilon}\right)^n\right)$$

$$\text{for} \quad j = 1, \ldots, \nu \text{ and } 1 \le \ell \le \infty. \quad (2.16)$$

In this part of the proof the zeros $(z_j)_{1 \le j \le \nu}$ of $B_{n,\nu}(z)$ are distinct. Therefore for sufficiently large n, this system, in the unknowns $\gamma_{k,n}^{\ell}$, is uniquely solvable, because Det $\{z_j^k\}_{0 \le k \le \nu-1, \, 1 \le j \le \nu} \ne 0$. Now $B_\nu(z_j) = 0$ can be written in the form

$$\sum_{k=0}^{\nu-1} \alpha_k z_j^k = -z_j^\nu, \quad j = 1, \ldots, \nu. \quad (2.17)$$

We can rewrite the difference of (2.16) and (2.17) in the form

$$\sum_{k=0}^{\nu-1} \left\{ \gamma_{k,n}^{\ell} \left(1 + O\left(\frac{\gamma}{\rho-\varepsilon}\right)^n\right) - \alpha_k \right\} z_j^k = O(1) \left(\frac{\gamma}{\rho-\varepsilon}\right)^n, \quad j = 1, \ldots, \nu.$$

Using Cramer's rule, we get

$$\gamma_{k,n}^{\ell} \left(1 + O(1) \left(\frac{\gamma}{\rho-\varepsilon}\right)^n\right) - \alpha_k = \frac{V_k(z_1, \ldots, z_\nu)}{V(z_1, \ldots, z_\nu)} \cdot O(1) \left(\frac{\gamma}{\rho-\varepsilon}\right)^n$$

so that

$$\gamma_{k,n}^{\ell} = \alpha_k + O(1) \left(\frac{\gamma}{\rho-\varepsilon}\right)^n, \quad 0 \le k \le \nu - 1. \quad (2.18)$$

where $V(z_1, \ldots, z_\nu)$ is the Vandermonde determinant and $V_k(z_1, \ldots, z_\nu)$ is obtained from $V(z_1, \ldots, z_\nu)$ on replacing the kth column by $(1, 1 \ldots)^{\text{T}}$. As (2.18) implies that $\lim_{n \to \infty} B_{n,\nu}^{\ell}(z) = B_\nu(z)$, $1 \le \ell \le \infty$, uniformly on every compact set of $\mathbb{C}$. Hence, in particular,

$$\limsup_{n \to \infty} \max_{z \in \mathcal{H}} \left|(B_{n,\nu}^{\ell}(z)\right|^{1/n} = \lim_{n \to \infty} \min_{z \in \mathcal{H}} \left|B_{n,\nu}^{\ell}(z)\right|^{1/n} = 1 \quad (2.19)$$

for every compact set $\mathcal{H}$ provided $z_j \notin \mathcal{H}, j = 1, \ldots, \nu$.

Now (2.12) and (2.13) also imply

$$
A^\ell_{s,n} - A^\infty_{s,n} = \sum_{m=1}^\infty a_{m\ell(n+\nu+1)+s} \beta^{m\ell(n+\nu+1)} - \sum_{r=\ell}^\infty a_{r(n+\nu+1)+s} \alpha^{r(n+\nu+1)}
$$

$$
+ \sum_{m=1}^\infty \sum_{r=1}^{\ell-1} a_{(m\ell+r)(n+\nu+1)+s} \alpha^{r(n+\nu+1)} \beta^{m\ell(n+\nu+1)}
$$

$$
\leq \frac{1}{(\rho-\varepsilon)^s} O(1) \left[\left(\frac{|\beta|}{\rho-\varepsilon}\right)^{\ell n} + \left(\frac{|\alpha|}{\rho-\varepsilon}\right)^{\ell n} + \left(\frac{|\alpha|,|\beta|^\ell}{(\rho-\varepsilon)^{\ell+1}}\right)^n \right]
$$

$$
= O(1) \frac{\gamma^{\ell n}}{(\rho-\varepsilon)^{\ell n+s}}, \quad s = 0, 1, \ldots . \tag{2.20}
$$

Thus (2.11) can be written in the following form (using again also (2.13)):

$$
\sum_{k=0}^\nu \left\{ \sum_{s=0}^{\nu-k} A^\infty_{s,n} z_j^{s+k} + \sum_{s=\nu+1}^{n+\nu} A^\infty_{s-k,n} z_j^s + \alpha^{n+\nu+1} \sum_{s=0}^{\nu-1} A^\infty_{s+n+\nu+1-k,n} z_j^s \right\} \times
$$

$$
\times (\gamma^\ell_{k,n} - \gamma^\infty_{k,n}) + O(1) \left(\frac{\gamma}{\rho-\varepsilon}\right)^{\ell n}
$$

$$
= \sum_{k=0}^\nu \left\{ \sum_{s=0}^{\nu-k} a_s z_j^{s+k} + \sum_{s=\nu+1}^{n+\nu} a_{s-k} z_j^s + O(1) \left(\frac{\gamma}{\rho-\varepsilon}\right)^n \right\} (\gamma^\ell_{k,n} - \gamma^\infty_{k,n})
$$

$$
+ O(1) \left(\frac{\gamma}{\rho-\varepsilon}\right)^{\ell n}
$$

$$
= \sum_{k=0}^\nu \left\{ z_j^k f(z_j) + O(1) \left(\frac{\gamma}{\rho-\varepsilon}\right)^n \right\} (\gamma^\ell_{k,n} - \gamma^\infty_{k,n}) + O(1) \left(\frac{\gamma}{\rho-\varepsilon}\right)^{\ell n} = 0
$$

for $j = 1, \ldots, \nu$. Hence

$$
\sum_{k=0}^\nu \left\{ z_j^k + O(1) \left(\frac{\gamma}{\rho-\varepsilon}\right)^n \right\} (\gamma^\ell_{k,n} - \gamma^\infty_{k,n}) = O(1) \left(\frac{\gamma}{\rho-\varepsilon}\right)^{\ell n} .
$$

This obviously implies that

$$
\gamma^\ell_{k,n} - \gamma^\infty_{k,n} = O(1) \left(\frac{\gamma}{\rho-\varepsilon}\right)^{\ell n}, \quad k = 0, 1, \ldots, \nu \tag{2.21}
$$

whence

$$
\limsup_{n\to\infty} \max_{z\in\mathcal{H}} |B^\ell_{\nu,n}(z) - B^\infty_{\nu,n}(z)|^{1/n} \leq \left(\frac{\gamma}{\rho}\right)^\ell \tag{2.22}
$$

for any compact subset $\mathcal{H} \subset \mathbb{C}$. So far we have dealt only with the solution of (2.9) with respect to the unknowns $\gamma^\ell_{k,n}$. This system (as also in the case $\ell = \infty$) consists of $n + \nu + 1$ equations with the unknowns $\gamma^\ell_{k,n}$ ($k = 0, 1, \ldots, \nu - 1$) and

$p_{j,n}^\ell$ $(j = 0, 1, \ldots, n)$. So the number of unknowns and the number of equations is the same, i.e., $n + \nu + 1$. Hence the rank of the full matrix M is $\leq n + \nu + 1$ where M is the system (2.9).

If we multiply the equations (2.9) by z_j^s and sum on s for $0 \leq s \leq n+\nu$, we get the matrix M_1 (a Vandermonde matrix) to determine $\gamma_{k,n}$ $(k = 0, 1, \ldots, \nu - 1)$, since the unknowns $p_{s,n}^\ell$ drop out as a result of the summation process using the fact that $B_\nu(z_j) = 0$. Since the rank of a matrix does not change by row operations, we may consider the matrix of the system of equations for $\nu \leq s \leq n + \nu$, adjoined to the matrix M_1. A close examination of the sum on the left in (2.9) shows that for $\nu \leq s \leq n + \nu$, the matrix for $p_{s,n}^\ell$ is of the following form:

$$M_2 = \begin{bmatrix} \alpha_\nu & \alpha_{\nu-1} & \alpha_{\nu-2} & \cdots & \alpha_0 & 0 & & \\ & \alpha_\nu & \alpha_{\nu-1} & \cdots & \alpha_1 & \alpha_0 & \cdots & \\ & \cdots & \cdots & \cdots & \cdots & & & \\ & & & \alpha_\nu & \alpha_{\nu-1} & \cdots & & \alpha_0 \\ & 0 & & & \alpha_\nu & \cdots & & \alpha_1 \\ & & & & & \alpha_\nu & \cdots & \alpha_2 \\ & & & & & & \cdots & \\ & & & & & & & \alpha_\nu \end{bmatrix}.$$

M_2 is a triangular matrix of order n with $\alpha_\nu = 1$ in the principal diagonal and so is non-singular. Thus the matrix M for the system of equations (2.9) is row equivalent to $\begin{pmatrix} M_1 & 0 \\ A & M_2 \end{pmatrix}$ where M_2 is the algebraic complement of M_1, and A is some matrix. Clearly the determinant of the system $\begin{pmatrix} M_1 & 0 \\ A & M_2 \end{pmatrix}$ is non-zero. Therefore $\gamma_{k,n}^\ell$ and $p_{s,n}^\ell$ can be uniquely determined for n sufficiently large.

After settling the problem of asymptotic behaviour of the denominators (see (2.19) and (2.22)), we now turn our attention to the numerators. In this respect we prove the following

LEMMA. *We have*

$$\limsup_{n \to \infty} \max_{z \in D_\tau} |U_n^\infty(z) - U_n^\ell(z)|^{1/n} \leq \begin{cases} \left(\frac{\tau}{\rho}\right)^\ell & \text{if } \tau < \rho \\ \left(\frac{\tau}{\rho}\right)^\ell \frac{\tau}{\rho} & \text{if } \tau \geq \rho \end{cases} \tag{2.23}$$

and

$$\limsup_{n \to \infty} \max_{z \in D_\tau} |U_n^\infty(z)|^{1/n} = \begin{cases} 1 & \text{if } \tau < \rho \\ \frac{\tau}{\rho} & \text{if } \tau \geq \rho. \end{cases} \tag{2.24}$$

PROOF. Set $\delta_k := p_{k,n}^\infty - p_{k,n}^\ell, 0 \leq \ell < \infty$ with $p_{k,n}^\ell = p_{k,n}^\infty = 0$ for $k > n$ and $p_{k,n}^0 := 0$. The unknowns $p_{k,n}^\ell$ are given by the equations (2.9). Let $\nu \leq m \leq n$.

We multiply the equations (2.9) by z_j^{s-m} and sum over s from m to $n+\nu$. Then

$$\sum_{s=m}^{n+\nu}\sum_{k=0}^{\nu}\alpha_k p_{s-k,n}^{\ell}z_j^{s-m} = \sum_{k=0}^{\nu}\gamma_{k,n}^{\ell}\sum_{s=m}^{n+\nu}A_{s-k,n}^{\ell}z_j^{s-m}.$$

On putting $s = k + m - r$, the left side can, as $p_{m-r,n}^{\ell} = 0$ for $m - r > n$, be written, since $\sum_{k=0}^{\nu}\alpha_k z_j^k = B_{\nu}(z_j) = 0$, as

$$\sum_{k=0}^{\nu}\sum_{r=m-n+k-\nu}^{k}\alpha_k p_{m-r,n}^{\ell}z_j^{k-r} = \sum_{k=0}^{\nu}\sum_{r=m-n}^{k}\alpha_k p_{m-r,n}^{\ell}z_j^{k-r}$$

$$= \sum_{k=0}^{\nu}\left(\sum_{r=m-n}^{\nu} - \sum_{r=k+1}^{\nu}\alpha_k p_{m-r,n}^{\ell}z_j^{k-r}\right)$$

$$= -\sum_{k=0}^{\nu}\sum_{r=k+1}^{\nu}\alpha_k p_{m-r,n}^{\ell}z_j^{k-r}$$

$$= -\sum_{r=1}^{\nu}\sum_{k=0}^{r-1}\alpha_k p_{m-r,n}^{\ell}z_j^{k-r}$$

$$= \sum_{r=1}^{\nu}\left(\sum_{k=r}^{\nu}\alpha_k z_j^{k-r}\right)p_{m-r,n}^{\ell},$$

So the equations to determine $p_{s,n}^{\ell}$ are:

$$\sum_{r=1}^{\nu}\left(\sum_{k=r}^{\nu}\alpha_k z_j^{k-r}\right)p_{m-r,n}^{\ell} = \sum_{k=0}^{\nu}\left(\sum_{s=m}^{n+\nu}A_{s-k,n}^{\ell}z_j^{s-m}\right)\gamma_{k,n}^{\ell},$$

with a similar equation for $p_{m-r,n}^{\infty}$. Then subtracting we have (on putting $\delta_{m-r} := p_{m-r,n}^{\infty} - p_{m-r,n}^{\ell}$),

$$\sum_{r=1}^{\nu}\left\{\sum_{k=r}^{\nu}\alpha_k z_j^{k-r}\right\}\delta_{m-r} = \sum_{k=0}^{\nu}\gamma_{k,n}^{\infty}\sum_{s=m}^{n+\nu}A_{s-k,n}^{\infty}z_j^{s-m}$$

$$-\sum_{k=0}^{\nu}\gamma_{k,n}^{\ell}\sum_{s=m}^{n+\nu}A_{s-k}^{\ell}z_j^{s-m}, \quad 1 \le j \le \nu.$$

The coefficient matrix of the $\{\delta_{m-r}\}_{r=1}^{\nu}$ has a typical row consisting of the ν elements

$$\left(\sum_{k=1}^{\nu}\alpha_k z_j^{k-1}, \sum_{k=2}^{\nu}\alpha_k z_j^{k-2}, \ldots, \sum_{k=\nu-1}^{\nu}\alpha_k z_j^{k-\nu+1}, \alpha_{\nu}\right).$$

This shows that the coefficient matrix of $\{\delta_{m-r}\}_{r=1}^{\nu}$ is the product of two matrices:

$$V(z_1,\ldots,z_\nu) \quad \text{and} \quad \begin{pmatrix} \alpha_1 & \alpha_2 & \cdots & \alpha_{\nu-1} & \alpha_\nu \\ \alpha_2 & \alpha_3 & \cdots & \alpha_\nu & 0 \\ \cdots & \cdots & \cdots & \cdots & \cdots \\ \alpha_{\nu-1} & \alpha_\nu & \cdots & 0 & \cdots & 0 \\ \alpha_\nu & 0 & \cdots & & 0 & 0 \end{pmatrix}.$$

Thus the determinant of the system of equations is non-zero and using Cramer's rule, we have

$$\delta_{m-r} = \sum_{j=1}^{\nu} d_{r,j} \left\{ \sum_{k=0}^{n+\nu} \gamma_{k,n}^{\infty} \sum_{s=m}^{n+\nu} A_{s-k,n}^{\infty} z_j^{s-m} - \sum_{k=0}^{\nu} \gamma_{k,n}^{\ell} \sum_{s=m}^{n+\nu} A_{s-k,n}^{\ell} z_j^{s-m} \right\}$$

$$\text{for} \quad 1 \le r \le \nu. \qquad (2.25)$$

Here the $d_{r,j}$'s are the minors of the determinants whose elements depend only on the poles of $F(z)$, the α_k's being coefficients of $\prod_{j=1}^{\nu}(z-z_j)$ and so are uniformly bounded, i.e. $|d_{r,j}| < e_5$.

We now estimate the difference in the braces on the right in (2.25). Set

$$\Phi := \sum_{k=0}^{\nu}(\gamma_{k,n}^{\infty} - \gamma_{k,n}^{\ell}) \sum_{s=m}^{n+\nu} A_{s-k,n}^{\infty} z_j^{s-m} + \sum_{k=0}^{\nu} \gamma_{k,n}^{\ell} \sum_{s=m}^{n+\nu}(A_{s-k,n}^{\infty} - A_{s-k,n}^{\ell}) z_j^{s-m}.$$

Write $R_\varepsilon = \max\left(\frac{|\alpha|}{\rho-\frac{\varepsilon}{2}}, \frac{|\beta|}{\rho-\frac{\varepsilon}{2}}\right)$. We have

(i) $|\gamma_{k,n}^{\infty} - \gamma_{k,n}^{\ell}| \le CR_\varepsilon^{\ell n} \le CR_\varepsilon^{\ell(n+\nu+1)}$ \qquad from (2.21),

(ii) $|\gamma_{k,n}^{\ell}| \le C$ \qquad (since $\lim_{n\to\infty}\sum_{k=0}^{\nu}\gamma_{k,n}^{\ell} z^k = B_\nu(z)$),

(iii) $|A_{s-k,n}^{\infty} - A_{s-k,n}^{\ell}| \le \dfrac{1}{(\rho-\frac{\varepsilon}{2})^{s-k}} \left[C_1\left(\dfrac{|\alpha|}{\rho-\frac{\varepsilon}{2}}\right)^{\ell(n+\nu+1)} + \right.$

$$\left. +C_2\left(\dfrac{|\beta|}{\rho-\frac{\varepsilon}{2}}\right)^{\ell(n+\nu+1)} \right]$$

$$\le C_3 \dfrac{(R_\varepsilon)^{\ell(n+\nu+1)}}{(\rho-\frac{\varepsilon}{2})^{s-k}} \qquad \text{for} \quad s \ge k$$

$$|A_{s-k,n}^{\infty} - A_{s-k,n}^{\ell}| \le C_4 \left(\dfrac{|\alpha|}{\rho-\frac{\varepsilon}{2}}\right)^{n+\nu+1} \dfrac{1}{(\rho-\frac{\varepsilon}{2})^{s-k}} \times$$

$$\times \left[\max\left(\dfrac{|\alpha|}{\rho-\frac{\varepsilon}{2}}, \dfrac{|\beta|}{\rho-\frac{\varepsilon}{2}}\right)^{\ell(n+\nu+1)}\right]$$

$$\le C_4\left(\dfrac{|\alpha|}{\rho-\frac{\varepsilon}{2}}\right)^{n+\nu+1} \dfrac{R_\varepsilon^{\ell(n+\nu+1)}}{(\rho-\frac{\varepsilon}{2})^{s-k}}, \qquad \text{for} \quad s < k,$$

(iv) $|A_{s-k,n}^{\infty}| = \begin{cases} \left|\sum_{r=0}^{\infty} \alpha^{r(n+\nu+1)} a_{r(n+\nu+1)+s-k}\right| \le C_2 \dfrac{1}{(\rho-\varepsilon)^{s-k}} \\ \qquad\qquad\qquad\qquad\qquad \text{when} \quad s \ge k, \\[2em] \left|\sum_{r=1}^{\infty} \alpha^{r(n+\nu+1)} a_{r(n+\nu+1)+s-k}\right| < C_1\left(\dfrac{|\alpha|}{\rho-\frac{\varepsilon}{2}}\right)^{n+\nu+1} \times \\[1em] \qquad \times \dfrac{1}{(\rho-\frac{\varepsilon}{2})^{s-k}} \qquad \text{when} \quad s < k \quad \text{and} \quad 0 \le k \le \nu \end{cases}$

by (2.6). Using these estimates, we have

$$|\Phi| \leq C \frac{R_\varepsilon^{\ell(n+\nu+1)}}{(\rho - \frac{\varepsilon}{2})^{s-k}} \left(\frac{|\alpha|}{\rho - \frac{\varepsilon}{2}} \right)^{n+\nu+1}$$

$$\leq C \frac{R_\varepsilon^{\ell(n+\nu+1)}}{(\rho - \frac{\varepsilon}{2})^{m-r}} \qquad \text{for } 0 \leq r \leq \nu \quad \text{and} \quad m \leq s \leq n + \nu.$$

Therefore from (2.23), we get $\delta_k = C \frac{R_\varepsilon^{\ell(n+\nu+1)}}{(\rho - \varepsilon)^k}$, $\quad 0 \leq k \leq n$ where C does not depend on n, k but only on the z_j's. Now with all these estimates, we come to the proof of (2.21). From (2.19)

$$\max_{x \in \mathcal{H}} |U_n^\infty(z) - U_n^\ell(z)| \leq \max_{z \in \mathcal{H}} \left| \sum_{k=0}^n \delta_k z^k \right| \leq \sum_{k=0}^n C \frac{R_\varepsilon^{\ell(n+\nu+1)}}{(\rho - \varepsilon)^k} \tau^k$$

$$\leq C R_\varepsilon^{\ell(n+\nu+1)} \sum_{k=0}^n \left(\frac{\tau}{(\rho - \varepsilon)} \right)^k$$

$$\leq \begin{cases} M R_\varepsilon^{\ell(n+\nu+1)} & \text{if } \tau < \rho - \varepsilon \\ M R_\varepsilon^{\ell(n+\nu+1)} \left(\frac{\tau}{\rho - e} \right)^n & \text{if } \tau \geq \rho. \end{cases}$$

Since $\varepsilon > 0$ may be taken arbitrarily small, this gives

$$\limsup_{n \to \infty} \left\{ \max_{z \in \mathcal{H}} |U_n^\infty(z) - U_n^\ell(z)|^{1/n} \right\} \leq \begin{cases} R^\ell & \text{if } \tau < \rho \\ R^\ell \frac{\tau}{\rho} & \text{if } \tau \geq \rho \end{cases} \qquad (2.26)$$

which is (2.23). Since $U_n^0(z) = 0$, it follows that

$$\limsup_{n \to \infty} \left\{ \max_{z \in \mathcal{H}} |U_n^\infty(z)| \right\}^{1/n} \leq \begin{cases} 1 & \text{if } \tau < \rho \\ < \frac{\tau}{\rho} & \text{if } \tau \geq \rho. \end{cases} \qquad (2.27)$$

$\square$

PROOF OF THEOREM 2. If $\mathcal{H}$ is a compact subset of $|z| < \tau$, $\tau > 0$ and if $\mathcal{H} \subset D_\rho \backslash \overset{\nu}{\underset{1}{\cup}} \{z_i\}$, then recalling that $B_{n,\nu}^\ell(z)$ and $B_{n,\nu}^\infty(z) \to B_\nu(z)$ as $n \to \infty$, we have

$$\limsup_{n \to \infty} \left\{ \max_{z \in \mathcal{H}} |P_{n+\nu+1}^\infty(z) - P_{n+\nu+1}^\ell(z)| \right\}^{1/n}$$

$$= \overline{\lim_{n \to \infty}} \left\{ \max_{z \in \mathcal{H}} |U_n^\infty(z) B_{n,\nu}^\ell(z) - U_n^\ell(z) B_{n,\nu}^\infty(z)| \right\}^{1/n}$$

$$\leq \overline{\lim_{n \to \infty}} \left\{ \max_{z \in \mathcal{H}} |U_n^\infty(z)|^{1/n} |B_{n,\nu}^\ell(z) - B_{n,\nu}^\infty(z)|^{1/n} \right.$$

$$\left. + \max_{z \in \mathcal{H}} |B_{n,\nu}^\infty(z)| \cdot |U_n^\infty(z) - U_n^\ell(z)| \right\}^{1/n}$$

$$\leq \overline{\lim_{n \to \infty}} \left\{ \max_{z \in \mathcal{H}} |U_n^\infty(z)|^{1/n} |B_{n,\nu}^\ell(z) - B_{n,\nu}^\infty(z)|^{1/n} \right\}$$

$$+ \varlimsup_{n\to\infty} \left\{ \max_{z\in\mathcal{H}} |B^\infty_{n,\nu}(z)|^{1/n} |U^\infty_n(z) - U^\ell_n(z)|^{1/n} \right\}$$

$$\leq \left[R^\ell \max\left(1, \frac{\tau}{\rho}\right) + \max\left(R^\ell, \tau\frac{R^\ell}{\rho}\right) \right]$$

$$= \max\left(\frac{\tau R^\ell}{\rho}, R^\ell\right) = \frac{\tau R^\ell}{\rho}, \quad \text{if} \quad \tau > \rho.$$

on using (2.26) and (2.27). $\qquad\qquad\qquad\qquad\qquad\qquad\qquad\qquad\square$

6.3. Historical Remarks

(a) The first generalization of the Walsh equiconvergence theorem for mero-morphic and rational functions (Theorem 1 and Example) is due to Saff, Sharma and Varga [92]. They consider the difference between the (n, ν) rational inter-polating function at the roots of unity to a meromorphic $f(z)$ and the Padé approximation by (n, ν) rational functions to $f(z)$.

(b) Theorem 1 is due to Saff et al [92]. The sharpness result (1.5) gives for $\nu = 0$, the corresponding result of Cavaretta et al [30]. Concerning the monic polynomials in $R_{n,\nu}(F; z)$ and $S_{n,\nu}(F; z)$ they observe that for every compact set $\mathcal{H} \subset \mathbf{C}$,

$$\limsup_{n\to\infty} \max_{z\in\mathcal{H}} |V_{n,\nu}(z) - Q_{n,\nu}(z)|^{1/n} \leq \frac{1}{\rho}$$

follows from Theorem B of Saff [90].

Let E be a closed bounded point set in the z-plane whose complement K (with respect to the extended plane) is connected and regular in the sense that K possesses a Green function $G(z)$ with pole at infinity [114, p. 65]. Let Γ_σ, $\sigma > 1$ denote generically the locus

$$\Gamma_\sigma := \{z \in \mathbf{C} : G(z) = \log\sigma\},$$

and denote by E_σ the interior of Γ_σ. Before giving the proof of Theorem 1, Saff et al [92] consider a more general situation: the rational interpolation in the triangular schemes

$$\beta_1^{(0)}$$
$$\beta_1^{(1)}, \beta_2^{(1)}$$
$$\beta_1^{(2)}, \beta_2^{(2)}, \beta_3^{(2)} \qquad\qquad\qquad (3.1)$$
$$\cdots$$
$$\beta_1^{(n)}, \beta_2^{(n)}, \ldots, \beta_{n+1}^{(n)}$$
$$\cdots$$

and

$$
\begin{array}{c}
\beta_1^{(0)} \\
\beta_1^{(1)}, \beta_2^{(1)} \\
\beta_1^{(2)}, \beta_2^{(2)}, \beta_3^{(2)} \\
\cdots \\
\beta_1^{(\nu)}, \beta_2^{(\nu)}, \ldots, \beta_{n+1}^{(\nu)} \\
\cdots
\end{array}
\tag{3.2}
$$

We assume that no limit points of the tableaux (3.1) or (3.2) lie exterior to E. Let $r_{n,\nu}(F; z)$ be the rational function of the form

$$
r_{n,\nu}(F; z) = r_{n,\nu}(z) = \frac{p_{n,\nu}(z)}{q_{n,\nu}(z)},
$$

$p_{n,\nu} \in \Pi_n$, $q_{n,\nu} \in \Pi_\nu$, $q_{n,\nu}$ monic, which interpolates $F(z)$ in the $n + \nu + 1$ points $\{\beta_j^{(n+\nu)}\}_{j=1}^{n+\nu+1}$, i.e.,

$$
r_{n,\nu}(\beta_j^{(n+\nu)}) = F(\beta_j^{(n+\nu)}), \qquad j = 1, \ldots, n + \nu + 1. \tag{3.3}
$$

Similarly $\tilde{r}(F; z)$ interpolates $F(z)$ in the points $\{\beta_j^{(n+\nu)}\}_{j=1}^{n+\nu+1}$, i.e.,

$$
\tilde{r}_{n,\nu}(\beta_j^{(n+\nu)}) = F(\beta_j^{(n+\nu)}), \qquad j = 1, \ldots, n + \nu + 1. \tag{3.4}
$$

In order to assume that the tableaux (3.1) and (3.2) are, in some sense, close to each other, we set

$$
w_n(z) = \prod_{j=1}^{n+1}(z - \beta_j^{(n)}), \quad \tilde{w}_n(z) = \prod_{j=1}^{n+1}(z - \tilde{\beta}_j^{(n+1)}), \quad \tilde{w}_{-1}(z) = w_{-1}(z) = 1.
\tag{3.5}
$$

Concerning the triangular schemes (3.1) and (3.2), we suppose

$$
\lim_{n \to \infty} |w_n(z)|^{1/n} = \Delta \exp G(z) \tag{3.6}
$$

uniformly in z in each closed bounded subset of K, where Δ is the transfinite diameter (or capacity) [9,Sec. 4.4] of E. The existence of some triangular scheme $\{\beta_j^{(n)}\}$ for E for which (3.6) holds is well known: for example, on defining the tableaux $\{\beta_j^{(n)}\}$ to consist of the Fekete points for E, then (3.6) holds [9, p. 172]. Next, since each $w_j(z)$ and $\tilde{w}_j(z)$ in (3.5) is monic of precise degree $j + 1$, there are unique constants $\gamma_j(n)$, $0 \leq j \leq n$ such that

$$
\tilde{w}_n(z) = w_n(z) + \sum_{j=0}^{n} \gamma_j(n) w_{j-1}(z), \qquad \text{for all } n \geq 1 \tag{3.7}
$$

For fixed ρ we assume (as in Cavaretta et al [30, Sec. 10]) that there exists a constant λ with $-\infty \leq \lambda < 1$, such that

$$\limsup_{n\to\infty} \left\{ \sum_{j=0}^{n} |\gamma_j(n)(\Delta j)^{\rho}| \right\}^{1/n} \leq \Delta\rho^{\lambda} < \Delta\rho. \tag{3.8}$$

With these assumptions it can be shown that for each n sufficiently large, the rational interpolants $r_{n,\nu}(F;z)$ and $\tilde{r}_{n,\nu}(F;z)$ of $F(z)$ in (3.3) and (3.4) do indeed exist and are unique. Then the following theorem holds.

Theorem 3. *Let ρ be fixed with $1 < \rho < \infty$, and suppose that the tableaux (3.1) and (3.2) have no limit points exterior to E and satisfy the conditions (3.3) and (3.4). If $F \in M(E_{\rho};\nu)$, $\nu \geq 0$, and if $\{\alpha_j\}_{j=1}^{\nu}$ are the ν poles of F in $E_{\rho} \setminus E$ (listed according to multiplicities), then the rational interpolants $r_{n,\nu}(F;z)$ of (3.3) and $\tilde{r}_{n,\nu}(F;z)$ of (3.5) satisfy*

$$\lim_{n\to\infty} [\tilde{r}_{n,\nu}(F;z) - r_{n,\nu}(F;z)] = 0, \qquad \text{for all} \quad z \in E_{\rho^{2-\lambda}} \setminus \cup_{j=1}^{\nu}\{\alpha_j\}, \tag{3.9}$$

the convergence being uniform and geometric on any closed subset of

$$E_{\rho^{2-\lambda}} \setminus \cup_{j=1}^{\nu}\{\alpha_j\}.$$

More precisely, on any closed subset $\mathcal{H}$ of any $\overline{E}_{\tau} \setminus \{\alpha_j\}$ with $\rho \leq \tau < \infty$, there holds

$$\limsup_{n\to\infty}\{\max_{z\in\mathcal{H}} |\tilde{r}_{n,\nu}(F;z) - r_{n,\nu}(F;z)|\}^{1/n} \leq \frac{\tau}{\rho^{2-\lambda}}. \tag{3.10}$$

The proof of Theorem 3 depends on the following extension, due to Saff [90], of the de Montessus de Ballore Theorem [39]:

Theorem A. *Suppose that $F \in M(E_{\rho};\nu)$ for some $1 < \rho < \infty$, $\nu \geq 0$, and let $\{\alpha_j\}_{j=1}^{\nu}$ denote the ν poles of F in $E_{\rho} \setminus E$. Suppose further that the points of the triangular scheme*

$$b_1^{(0)}$$
$$b_1^{(1)}, b_2^{(1)}$$
$$b_1^{(2)}, b_2^{(2)}, b_3^{(2)} \tag{3.11}$$
$$\cdots$$
$$b_1^{(n)}, b_2^{(n)}, \ldots, b_{n+1}^{(n)}$$
$$\cdots$$

(which need not be distinct in any row) have no limit points exterior to E, and that

$$\lim_{n\to\infty} \left| \prod_{j=1}^{n+1} (z - zb_j^{(n)}) \right|^{1/n} = \Delta \exp G(z), \qquad (3.12)$$

uniformly on each closed and bounded subset of K. Then, for all n sufficiently large, there exists a uniques rational function

$$s_{n,\nu}(z) = \frac{g_{n,\nu}(z)}{h_{n,\nu}(z)}, \qquad g_{n,\nu} \in \Pi_n,\ h_{n,\nu} \in \Pi_\nu,\ h_{n,\nu} \quad monic \qquad (3.13)$$

which interpolates $F(z)$ in the points $b_1^{(n+\nu)}, b_2^{(n+\nu)}, \ldots, b_{n+\nu+1}^{(n+\nu)}$. Each $s_{n,\nu}(z)$ has precisely ν finite poles, and as $n \to \infty$, these poles approach, respectively, the ν poles of $F(z)$ in $E_\rho \setminus E$. The sequence $\{s_{n,\nu}(z)\}_{n=n_0}^\infty$ converges to $F(z)$ on $E_\rho \setminus \cup_{j=1}^\nu \{\alpha_j\}$, uniformly and geometrically on any closed subset of $E_\rho \setminus \cup_{j=1}^\nu \{\alpha_j\}$. More precisely, on any closed subset $\mathcal{H}$ of any $\overline{E}_\tau \setminus \cup_{j=1}^\nu \{\alpha_j\}$ with $1 < \tau < \rho$, there holds

$$\limsup_{n\to\infty} \{ \max_{z\in\mathcal{H}} |F(z) - s_{n,\nu}(z)|^{1/n} \le \frac{\tau}{\rho}. \qquad (3.14)$$

(c) Theorem 2 is basically the main result from de Bruin and Sharma [35] (for multiple nodes cf. de Bruin and Sharma [36]). Here the idea of Stojanova [101] is used, but the unexplained definition

$$A_{s;n,\nu}^\ell := \sum_{m=0}^{\ell-1} a_{m(n+\nu+1)+s}, \qquad -\nu \le s \le n+\nu; \quad a_{-1} = \cdots = a_{-\nu} := 0$$

there, is replaced by the more reasonable (2.4).

Additional results, obtained by A. Jakimovski and A. Sharma [58], extend equiconvergence theorems for polynomial interpolants to rational Hermite-Padé interpolants are given now. The complete proof of these theorems is out of the scope of this monograph; at the appropriate places we will refer to the literature where the interested reader can complete the argument.

Suppose $f(z) \in A_\rho$, $\rho > 1$, and $0 < |\alpha| \ne |\beta| < \rho$. We recall that $B_\nu(z)$ is a monic polynomial of degree ν with ν zeros, not necessarily simple, in D_ρ. We shall suppose throughout that $f(z)$ does not vanish at the zeros of $B_\nu(z)$. Let the Hermite-Padé interpolant to $\frac{f(z)}{B_\nu(z)}$ on the zeros of $(z^{n+\nu} - \alpha^{n+\nu})^r$ be denoted by $\mathcal{P}_{n,\nu,r}\left(z, \alpha; \frac{f(z)}{B_\nu(z)}\right)$. Then we have

THEOREM 4. *Suppose $f \in A_\rho$, $\rho > 1$ and $0 < |\alpha| \neq |\beta| < \rho$. If $B_\nu(z)$ is a monic polynomial with no common zeros with $f(z)$, then*

$$\mathcal{P}_{n,\alpha+1,1}\left(z,\alpha; \frac{f(z)}{B_\nu(z)}\right) - \mathcal{P}_{n,\nu+1,1}\left(z,\alpha; \frac{L_{n+\nu}(z,\alpha, H^r_{r(n+\nu+1)-1}(\cdot,\beta,f))}{B_\nu(z)}\right)$$

$$= O(1)|z|^{-\nu}\left(\frac{\max(|\alpha|,|\beta|)}{\rho-\epsilon}\right)^{r(n+\nu+1)}\left(\frac{|z|}{\rho-\epsilon}\right)^n \quad as \quad n \to \infty$$

where $O(1)$ is independent of n and z.

The following interesting connection between the Lagrange and Hermite interpolants was used in the proof of the above theorem.

LEMMA 2. *Given a function $f \in A_\rho$, $\rho > 1$, numbers $0 < |\alpha| \neq |\beta| < \rho$ and a positive integer r, then we have*

$$L_{n+\nu}(z,\alpha, H^r_{r(n+\nu+1)-1}(\cdot,\beta,f)) = L_{n+\nu}(z,\alpha,f) - \sum_{j=0}^{n+\nu} c_{j,n+\nu,\alpha,\beta} z^j \quad (7.0)$$

where

$$c_{j,n+\nu,\alpha,\beta} := (\alpha^{n+\nu+1} - \beta^{n+\nu+1})^r \sum_{m=0}^{\infty}\sum_{s=0}^{\infty}\binom{s+r-1}{s}a_{j+(m+s+r)(n+\nu+1)} \times$$

$$\times \alpha^{m(n+\nu+1)}\beta^{s(n+\nu+1)} \quad for \quad 0 \leq j \leq n+\nu,$$

and

$$|c_{j,n+\nu,\alpha,\beta}| \leq \frac{2^r A}{\left(1-\frac{|\alpha|}{\rho-\epsilon}\right)\left(1-\frac{|\beta|}{\rho-\epsilon}\right)^r}\left(\frac{\max(|\alpha|,|\beta|)}{\rho-\epsilon}\right)^{r(n+\nu+1)}\frac{1}{(\rho-\epsilon)^j} . \quad (7.1)$$

$\square$

We assume in this section that $m = rn + s$, $0 \leq s \leq r - 1$ and $pn + q < m$, $p \geq 1$. Let $P_{pn+q-1,(m)}(z)$ denote the unique polynomial $P(z)$ of degree $\leq pn + q - 1$ for which

$$\min_{P \in \pi_{pn+q-1}} \sum_{k=0}^{m-1} |f(\alpha\omega^k) - P(\alpha\omega^k)|^2, \quad \omega^m = 1, \quad 0 < |\alpha| < \rho, \quad \rho > 1, \quad (3.4)$$

is attained. It is known [88] that the polynomial satisfying (3.4) is obtained by truncating the Lagrange interpolant to $f(z)$ on the zeros of $z^m - \alpha^m$. Since

$$L_{m-1}(z,\alpha,f) = \sum_{j=0}^{m-1} c_j z^j, \quad c_j = \sum_{\sigma=0}^{\infty} a_{j+\sigma m}\alpha^{\sigma m},$$

we have

LEMMA 3. *Assume* $m = rn + s$, $0 \leq s \leq r - 1$ *and* $pn + q < m$. *Then we have*

$$P_{pn+q-1,(m)}(z,\alpha,f) = \sum_{j=0}^{pn+q-1} z^j \sum_{\sigma=0}^{\infty} a_{j+\sigma m}\alpha^{\sigma m} = \sum_{j=0}^{pn+q-1} c_j^{(m)} z^j \qquad (3.5)$$

where $c_j^{(m)} = \sum_{\sigma=0}^{\infty} a_{j+\sigma m}\alpha^{\sigma m}$.

THEOREM 5. *Suppose* $f \in A_\rho$, $\rho > 1$ *and* $0 < |\alpha| \neq |\beta| < \rho$. *If* $B_\nu(z)$ *is a monic polynomial with no common zeros with* $f(z)$, *then*

$$\mathcal{P}_{n,0,1}\left(z,\alpha;\frac{f(z)}{B_\nu(z)}\right) - \mathcal{P}_{n,0,1}\left(z,\alpha;\frac{L_{n-1}(z,\alpha,P_{pn+q-1,(m)}(\cdot,\beta,f))}{B_\nu(z)}\right) =$$

$$= O(1)|z|^{-\nu} \max\left(\left(\frac{|\beta|}{\rho-\epsilon}\right)^{rn+s}, \left(\frac{|\alpha|}{\rho-\epsilon}\right)^{pn}\right)\left(\frac{|z|}{\rho-\epsilon}\right)^{n-1-\nu} \quad as\ n \to \infty,$$

where $m = rn + s$, $0 \leq s \leq r - 1$, $pn + q < m$, $p \geq 1$ *and* $P_{pn+q-1}(z,\beta,f)$ *is defined as in* (3.4).

The region of Walsh equiconvergence given in this case agrees with the one in Theorem 4 in [2] with the poles deleted.

THEOREM 6. *Let* $f \in A_\rho$, $\rho > 1$. *Assume* $0 < |\alpha| \neq |\beta| < \rho$ *and that* $k \geq r$ *are two positive integers. Then we have*

$$\mathcal{P}_{n,0,r}\left(z,\alpha;\frac{H_{kn-1}^k(z,\beta,f)}{B_\nu(z)}\right) - \mathcal{P}_{n,0,r}\left(z,0;\frac{f(z)}{B_\nu(z)}\right)$$

$$= O(1)|z|^{-\nu}\left(\frac{|z|}{\rho-\epsilon}\right)^{rn-\nu}\left(\frac{\max(|\alpha|,|\beta|)}{\rho-\epsilon}\right)^n.$$

THEOREM 7. *Suppose* $f \in A_\rho$, $\rho > 1$. *Suppose* $0 < |\alpha| \neq |\beta| < \rho$, $k > r$ *are two positive integers and* $\nu \geq 1$ *is a given integer. Then we have*

$$\mathcal{P}_{n,0,r}\left(z,\alpha;\frac{f}{B_\nu(z)}\right) - \mathcal{P}_{n,0,r}\left(z,\alpha;\frac{H_{km-1}^k(\cdot,\beta,f)}{B_\nu(z)}\right)$$

$$= O(1)|z|^{-\nu}\left(\frac{|z|}{\rho-\epsilon}\right)^{rn-\nu-1}\left(\frac{\max(|\alpha|,|\beta|)}{\rho-\epsilon}\right)^{(k+1-r)n}.$$

The proof of the last theorem requires iteration of two Hermite interpolants and is based on the following lemma.

LEMMA 6. *Given a function $f \in A_\rho$, $\rho > 1$, numbers $0 < |\alpha| \neq |\beta| < \rho$ and positive integers p, r, $p < r$, then we have*

$$H_{pn-1}^p(z, \alpha, H_{rn-1}^r(\cdot, \beta, f)) = H_{pn-1}^p(z, \alpha, f) - \sum_{j=0}^{pn-1} c_{j,n,\alpha,\beta}^* z^j$$

where

$$c_{j,n,\alpha,\beta}^* = O(1) \left(\frac{\max(|\alpha|, |\beta|)}{\rho - \epsilon} \right)^{n(r-p+1)} \frac{1}{(\rho - \epsilon)^j}, \quad 0 \leq j \leq pn - 1.$$

This agrees with the result in Theorem 1 in [2], where $\nu = 0$.

CHAPTER 7

QUANTITATIVE RESULTS IN THE EQUICONVERGENCE OF APPROXIMATION OF MEROMORPHIC FUNCTIONS

The subject of this chapter is similar to that of Chapter 4: we determine the position and number of distinguished points, but now for meromorphic functions and rational interpolants.

7.1. The Main Theorems

If ℓ is a positive integer or ∞, for a function $F(z) \in M_\varrho(\nu)$ we have defined the rational functions

$$P_{n,\nu}^\ell(z, F) := \frac{U_n^\ell(z)}{B_{n,\nu}^\ell(z)} \tag{1.1}$$

and have proved the existence and uniqueness of these rational functions for n sufficiently large. We have equiconvergence of the difference

$$\Delta_{n,\nu}^\ell(z, F) := P_{n,\nu}^\infty(z, F) - P_{n,\nu}^\ell(z, F) \tag{1.2}$$

and have proved (Ch. 6, Theorem 2) that if $\alpha, \beta \in \mathbf{C}$, $|\alpha| \neq |\beta|$, $\gamma := \max(|\alpha|, |\beta|)$ and

$$S_{\ell,\nu}(z, F) = \varlimsup_{n \to \infty} |\Delta_{n,\nu}^\ell(z, F)|^{\frac{1}{n}}, \quad 1 \leq \ell < \infty,$$

then $S_{\ell,\nu}(z, F) \leq K(z)$, where

$$K(z) = \begin{cases} \left(\frac{|\alpha|}{\rho}\right)^\ell & \text{for} \quad |z| < \rho \\ \left(\frac{|\alpha|}{\rho}\right)^\ell \cdot \frac{|z|}{\rho} & \text{for} \quad |z| \geq \rho. \end{cases} \tag{1.3}$$

A point z where $S_{\ell,\nu}(z, F) < K(z)$ will be called *distinguished* or *exceptional* (cf. Ch.4, Section 4.1).

We now shall prove the following four theorems:

THEOREM 1. *For any $F \subset M_\rho(\nu)$, there are at most $\ell - 1$ distinguished points in $|z| < \rho$ (i.e., $D_\rho \backslash \{z_j\}_{j=1}^\nu$).*

THEOREM 2. *For any set of $\ell - 1$ points w_j in $0 < |z| < \rho$ different from $\{z_j\}_{j=1}^{\nu}$, there is a meromorphic function $F(z) \in M_\rho(\nu)$, such that the points $\{w_j\}_{j=1}^{\ell-1}$ are distinguished for $F(z)$.*

THEOREM 3. *For each $F \in M_\rho(\nu)$, $\nu \geq 0$ there are at most $\nu + \ell$ distinguished points in $|z| > \rho$.*

THEOREM 4. *For any set of $\ell + 1 - \nu$ points w_j in $|z| > \rho$, there exists a function $F \in M_\rho(\nu)$, $\nu \geq 1$ such that these $\ell + 1 - \nu$ points are distinguished for the function F.*

7.2. Some Lemmas

We shall need some lemmas for the proof of these theorems. Set

$$\overline{\Delta}_{n,\nu}^{\ell}(z, F) := B_{n,\nu}^{\ell}(z) B_{n,\nu}^{\infty}(z) \Delta_{n,\nu}^{\ell}(z, F).$$

Then we have

LEMMA 1. *For each z satisfying $|z| \neq \rho$ and $B_\nu(z) \neq 0$ the following statements are equivalent:*

(i) $S_{\ell,\nu}(z, F) < K(z)$,

(ii) $\overline{\Delta}_{n,\nu}^{\ell}(z, F) - \left(\frac{z}{\alpha}\right)^{\ell} \overline{\Delta}_{n+1,\nu}^{\ell}(z, F) = O((qK(z))^n)$ *for some q, $0 < q < 1$.*

PROOF. Since $\varliminf_{n\to\infty} |B_{n,\nu}^{\ell}(z)|^{\frac{1}{n}} = 1$ for $B_\nu(z) \neq 0$ (see (2.21) in Ch. 6), part (ii) of the lemma follows from part (i). We shall now show that (ii) implies (i). To see this observe that

$$\overline{\Delta}_{n,\nu}^{\ell}(z, F) - \left(\frac{z}{\alpha}\right)^{s\ell} \overline{\Delta}_{n+s,\nu}^{\ell}(z, F)$$

$$= \sum_{k=0}^{s-1} \left[\left(\frac{z}{\alpha}\right)^{k\ell} \overline{\Delta}_{n+k,\nu}^{\ell}(z, F) - \left(\frac{z}{\alpha}\right)^{(k+1)\ell} \overline{\Delta}_{n+k+1,\nu}^{\ell}(z, F) \right]$$

$$= O\left(\sum_{k=0}^{s-1} \left|\frac{z}{\alpha}\right|^{k\ell} (qK(z))^{n+k} \right). \tag{2.1}$$

(a) *Case 1: $|z| < \rho$.* Using (2.1) with $s = n$ and $n + 1$, respectively, we have

$$\overline{\Delta}_{n,\nu}^{\ell}(z, F) - \left(\frac{z}{\alpha}\right)^{n\ell} \overline{\Delta}_{2n,\nu}^{\ell}(z, F) = O\left(\sum_{k=0}^{n-1} \left(\frac{z}{\alpha}\right)^{k\ell} (qK(z))^{n+k} \right)$$

$$= O\left(\sum_{k=0}^{n-1} \left(\frac{|z|}{|\alpha|}\right)^{k\ell} \left(q\left(\frac{|\alpha|}{\rho}\right)^{\ell}\right)^{n+k} \right)$$

$$= \left(q(\frac{|\alpha|}{\rho})^\ell \right)^n O\left(\sum_{k=0}^{n-1} (\frac{|z|}{|\alpha|})^{k\ell} (q(\frac{|\alpha|}{\rho})^\ell)^k \right)$$

$$= \left(q(\frac{|\alpha|}{\rho})^\ell \right)^n O\left(\sum_{k=0}^{n-1} q^k \right)$$

and

$$\overline{\Delta}^\ell_{n+1,\nu}(z,F) - (\frac{z}{\alpha})^{(n+1)\ell} \overline{\Delta}^\ell_{2n+1,\nu}(z,F) = O\left((q(\frac{|\alpha|}{\rho})^\ell)^n \right).$$

These two relations can be put together and we get

$$(\frac{z}{\alpha})^{[\frac{n+1}{2}]\ell} \overline{\Delta}^\ell_{n,\nu}(z,F) = \overline{\Delta}^\ell_{[\frac{n}{2}],\nu}(z,F) + O\left((\frac{q|\alpha|}{\rho})^\ell)^{\frac{n}{2}} \right).$$

Let us suppose that $S_{\ell,\nu}(z,F) \geq q(\frac{|\alpha|}{\rho})^\ell$. Then

$$\left| \frac{z}{\alpha} \right|^\ell S^2_{\ell,\nu}(z,F) = \varlimsup_{n\to\infty} \left| \overline{\Delta}^\ell_{[\frac{n}{2}],\nu}(z,F) \right|^{\frac{2}{n}} = S_{\ell,\nu}(z,F).$$

Therefore

$$S_{\ell,\nu}(z,F) = \left(\frac{|\alpha|}{|z|} \right)^\ell > \left(\frac{|\alpha|}{\rho} \right)^\ell = K(z)$$

which contradicts Theorem 2 of Ch. 6. Thus $S_{\ell,\nu}(z,f) \leq qK(z) < K(z)$.

(b) *Case 2*: $|z| > \rho$. Using (2.1) with $s = n$ and $n + 1$ we have

$$\overline{\Delta}^\ell_{n,\nu}(z,F) - (\frac{z}{\alpha})^{n\ell} \overline{\Delta}^\ell_{2n,\nu}(z,F) = O\left(q^n |z|^{\ell n} |\alpha|^{\ell n} \left(\frac{|z|}{\rho^{\ell+1}} \right)^{2n} \right).$$

Similarly, we have

$$\overline{\Delta}^\ell_{n,\nu}(z,F) - (\frac{z}{\alpha})^{(n+1)\ell} \overline{\Delta}^\ell_{2n+1,\nu}(z,F) = O\left(q^n |z|^{\ell n} |\alpha|^{\ell n} \left(\frac{|z|}{\rho^{\ell+1}} \right)^{2n} \right).$$

Putting n for $2n$ (or for $2n + 1$), we have

$$\overline{\Delta}^\ell_{[\frac{n}{2}],\nu}(z,F) = (\frac{z}{\alpha})^{[\frac{n+1}{2}]\ell} \overline{\Delta}^\ell_{n,\nu}(z,F) + O\left(q^{1/2} (\frac{|z|}{|\alpha|})^{\frac{\ell}{2}} (\frac{|\alpha|^\ell |z|}{\rho^{\ell+1}}) \right)^n.$$

Let us suppose that $S_\ell(z,F) = K(z)$. Then

$$(K(z))^{1/2} = \varlimsup_{n\to\infty} \left| \overline{\Delta}^\ell_{[\frac{n}{2}],\nu}(z,F) \right|^{\frac{1}{n}}$$

$$= \varlimsup_{n\to\infty} \left| \left(\left| \frac{z}{\alpha} \right| \right)^{[\frac{n+1}{2}]\ell} \cdot \overline{\Delta}{}^{\ell}_{n,\nu}(z,F) \right|^{\frac{1}{n}}$$

$$= \left| \frac{z}{\alpha} \right|^{\frac{\ell}{2}} K(z).$$

This gives

$$K(z) = \left(\frac{|\alpha|}{|z|} \right)^{\ell} = \left| \frac{\rho}{z} \right|^{\ell+1} \cdot \frac{|z|}{\rho} \cdot \left(\left| \frac{\alpha}{\rho} \right| \right)^{\ell} = \left| \frac{\rho}{z} \right|^{\ell+1} K(z) < K(z).$$

From this contradiction and from Theorem 2 of Ch. 6, we see that (i) is also true for $|z| > \rho$.   □

LEMMA 2. *Let* $B_\nu(z) = \sum\limits_{m=1}^{\nu} \alpha_m z^m$. *Then we have the following representation for* $\overline{\Delta}{}^{\ell}_{n,\nu}(z,F)$:

$$\overline{\Delta}{}^{\ell}_{n,\nu}(z,F) = \sum_{s=0}^{n+2\nu} R_{s,n} z^s + \varepsilon_n \tag{2.2}$$

where

$$|\varepsilon_n| \le c \big(qK(z) \big)^n$$

and

$$\sum_{s=0}^{n+2\nu} R_{s,n} z^s = \alpha^{\ell(n+\nu+1)} B_\nu(z) \sum_{k=0}^{\nu} \sum_{s=k}^{n+\nu} \alpha_k a_{\ell(n+\nu+1)+s-k} z^s$$

$$+ z^n \sum_{\tau=1}^{\nu} \sum_{k=0}^{\nu-\tau} \sum_{m=0}^{\nu} (r_{m,n}\alpha_k - r_{k,n}\alpha_m) a_{m+\tau} z^{k+\tau+m} . \tag{2.3}$$

Recall that $r_{s,n}$ $(0 \le s \le \nu)$ *are given by*

$$r_{s,n} := \gamma^{\ell}_{s,n} - \gamma^{\infty}_{s,n}, \quad 1 \le \ell < \infty, \ 0 \le s \le \nu.$$

Here $\gamma^{\ell}_{s,n}, \gamma^{\infty}_{s,n}$ *depend on the coefficients of* $B_\nu(z)$.

The following estimates assumed repeatedly in the sequel are listed below for ready reference:

(a) $\lim\limits_{n\to\infty} \gamma^{\ell}_{s,n} = \alpha_{s,n}, \quad |\gamma^{\ell}_{s,n}| < c$ for all ℓ, s, n ;

(b) $r_{s,n} := \gamma^{\ell}_{s,n} - \gamma^{\infty}_{s,n}, \quad |r_{s,n}| \le c \left(\frac{|\alpha|}{\rho - \varepsilon} \right)^{\ell(n+\nu+1)}$;

(c) $|a_j| \le \dfrac{c}{(\rho - \varepsilon)^j}, \quad j \ge 0$;

(d) $A_{s,n}^{\infty} - A_{s,n}^{\ell} = a_{\ell(n+\nu+1)+s}\alpha^{\ell(n+\nu+1)} + \varepsilon_{s,n}$

where $\varepsilon_{s,n} = \dfrac{c}{(\rho-\varepsilon)^s} \cdot \left(\dfrac{|\alpha|}{\rho-\varepsilon}\right)^{(n+\nu+1)(\ell+1)}$, $s \geq 0$;

(e) $A_{s,n}^{\infty} - A_{s,n}^{\ell} = a_{\ell(n+\nu+1)+n+\nu+1+s}\alpha^{(\ell+1)(n+\nu+1)} + \varepsilon_{s,n}^{*}$

where $\varepsilon_{s,n}^{*} = \left(\dfrac{|\alpha|}{\rho-\varepsilon}\right)^{n+\nu+1}|\varepsilon_{s+n+\nu+1,n}|$, $s < 0$;

(f) $A_{\ell}^{s} = a_s + \delta_{s,n}^{\ell}$, $|\delta_{s,n}^{\ell}| \leq \dfrac{c}{(\rho-\varepsilon)^s} \cdot \left(\dfrac{|\alpha|}{\rho-\varepsilon}\right)^{n+\nu+1}$, $1 \leq \ell \leq \infty$, $s \geq 0$.

PROOF. With $U_n^{\ell}(z) := \sum\limits_{s=0}^{n} p_{s,n}^{\ell} z^s$, the following relation holds between $p_{s,n}^{\ell}$'s and $\gamma_{k,n}^{\ell}$'s :

$$\sum_{k=0}^{\nu} \alpha_k p_{s-k,n}^{\ell} = \sum_{k=0}^{\nu} \gamma_{k,n}^{\ell} A_{s-k,n}^{\ell} \tag{2.4}$$

where $\alpha_k = 0$ for $k < 0$ or $k > \nu$, and $p_{s,n}^{\ell} = 0$ for $s < 0$ or $s > n$. Then

$$B_\nu(z)U_n^{\ell}(z) = \sum_{k=0}^{\nu} \alpha_k z^k \sum_{s=0}^{n} p_{s,n}^{\ell} z^s$$

$$= \sum_{s=0}^{n+\nu} \left(\sum_{k=0}^{\nu} \alpha_k p_{s-k,n}^{\ell}\right) z^s = \sum_{s=0}^{n+\nu} \left(\sum_{k=0}^{\nu} \gamma_{k,n}^{\ell} A_{s-k,n}^{\ell}\right) z^s. \tag{2.5}$$

We shall need the difference (recall that in $A_{s,n}^{\ell}$ the index s can be negative; see (2.12))

$$B_\nu(z)U^{\ell}(z) - B_{n,\nu}^{\ell}(z) \sum_{s=0}^{n} A_{s,n}^{\ell} z^s$$

$$= \sum_{s=0}^{n+\nu} \left(\sum_{k=0}^{\nu} \gamma_{k,n}^{\ell} A_{s-k,n}^{\ell}\right) z^s - \sum_{s=0}^{n} A_{s,n}^{\ell} z^s \sum_{k=0}^{\nu} \gamma_{k,n}^{\ell} z^k .$$

Here the first double sum can be split into three sums and

$$= \sum_{s=0}^{\nu} \left(\sum_{k=0}^{\nu} \gamma_{k,n}^{\ell} A_{s-k,n}^{\ell}\right) z^s + \sum_{s=\nu}^{n} \left(\sum_{k=0}^{\nu} \gamma_{k,n}^{\ell} A_{s-k,n}^{\ell}\right) z^s$$

$$+ \sum_{s=n+1}^{n+\nu} \left(\sum_{k=0}^{\nu} \gamma_{k,n}^{\ell} A_{s-k,n}^{\ell}\right) z^s,$$

and the second double sum can be handled similarly:

$$= \sum_{s=0}^{1} \left(\sum_{k=0}^{s} \gamma_{k,n}^{\ell} A_{s-k,n}^{\ell}\right) z^s + \sum_{s=0}^{n} \left(\sum_{k=0}^{\nu} \gamma_{k,n}^{\ell} A_{s-k,n}^{\ell}\right) z^s$$

$$+ \sum_{s=n+1}^{n+\nu} \left(\sum_{k=s-n}^{\nu} \gamma_{k,n}^{\ell} A_{s-k,n}^{\ell}\right) z^s.$$

The difference of these two can now be seen to be

$$\sum_{s=0}^{\nu-1}\Big(\sum_{k=s+1}^{\nu}\gamma_{k,n}^{\ell}A_{s-k}^{\ell}\Big)z^s + \sum_{s=n+1}^{n+\nu}\Big(\sum_{k=0}^{s-n-1}\gamma_{k,n}^{\ell}A_{s-k,n}^{\ell}\Big)z^s.$$

Hence interchanging the order of summation in s and k in the above two double sums,

$$\sum_{k=1}^{\nu}\sum_{s=0}^{k-1}\gamma_{k,n}^{\ell}A_{s-k,n}^{\ell}z^s + \sum_{k=0}^{\nu-1}\sum_{s=n+k+1}^{n+\nu}\gamma_{k,n}^{\ell}A_{s-k,n}^{\ell}z^s.$$

Replacing s by $s+k$ in the above, we have

$$\sum_{k=1}^{\nu}\sum_{s=-k}^{-1}\gamma_{k,n}^{\ell}A_{s,n}^{\ell}z^{s+k} + \sum_{k=0}^{\nu-1}\sum_{s=n+1}^{n+\nu-k}\gamma_{k,n}^{\ell}A_{s,n}^{\ell}z^{s+k}.$$

Therefore

$$B_\nu(z)U^\ell(z) = B_{n,\nu}^\ell(z)\sum_{s=0}^{n}A_{s,n}^{\ell}z^s + \sum_{k=0}^{\nu-1}\sum_{s=n+1}^{n+\nu-k}\gamma_{k,n}^{\ell}A_{s,n}^{\ell}z^{s+k}$$

$$+ \sum_{k=1}^{\nu}\sum_{s=-k}^{-1}\gamma_{k,n}^{\ell}A_{s,n}^{\ell}z^{s+k}, \quad 1 \le \ell \le \infty. \tag{2.6}$$

If we multiply (2.6) with $\ell = \infty$ by $B_{n,\nu}^\ell(z)$ and from the product subtract (2.6) multiplied by $B_{n,\nu}^\infty(z)$, we get

$$\overline{\Delta}_{n,\nu}^\ell(z,F) = B_{n,\nu}^\ell(z)B_\nu(z)U_n^\infty(z) - B_{n,\nu}^\infty(z)B_\nu(z)U^\ell(z)$$

$$= S_1 + S_2 + S_3 \tag{2.7}$$

where

$$\begin{cases} S_1 &= B_{n,\nu}^\ell(z)B_{n,\nu}^\infty(z)\sum_{s=0}^{n}(A_{s,n}^\infty - A_{s,n}^\ell)z^s \\ S_2 &= B_{n,\nu}^\ell(z)\sum_{k=0}^{\nu-1}\sum_{s=n+1}^{n+\nu-k}\gamma_{k,n}^\infty A_{s,n}^\infty z^{s+k} - B_{n,\nu}^\infty(z)\sum_{k=0}^{\nu-1}\sum_{s=n+1}^{n+\nu-k}\gamma_{k,n}^\ell A_{s,n}^\ell z^{s+k}, \\ S_3 &= B_{n,\nu}^\ell(z)\sum_{k=1}^{\nu}\sum_{s=-k}^{-1}\gamma_{k,n}^\infty A_{s,n}^\infty z^{s+k} - B_{n,\nu}^\infty(z)\sum_{k=1}^{\nu}\sum_{s=-k}^{-1}\gamma_{k,n}^\ell A_{s,n}^\ell z^{s+k}. \end{cases} \tag{2.8}$$

We shall write $S_2 = S_{2,1} + S_{2,2} + S_{2,3}$ where

$$\begin{cases} S_{2,1} &= \Big(B_{n,\nu}^\ell(z) - B_{n,\nu}^\infty(z)\Big)\sum_{k=0}^{\nu-1}\sum_{s=n+1}^{n+\nu-k}\gamma_{k,n}^\infty A_{s,n}^\infty z^{s+k}, \\ S_{2,2} &= B_{n,\nu}^\infty(z)\sum_{k=0}^{\nu-1}\sum_{s=n+1}^{n+\nu-k}\gamma_{k,n}^\infty(A_{s,n}^\infty - A_{s,n}^\ell)z^{s+k}, \\ S_{2,3} &= B_{n,\nu}^\infty(z)\sum_{k=0}^{\nu-1}\sum_{s=n+1}^{n+\nu-k}(\gamma_{k,n}^\infty - \gamma_{k,n}^\ell)A_{s,n}^\ell z^{s+k}. \end{cases} \tag{2.9}$$

We shall prove that

$$S_1 = \alpha^{\ell(n+\nu+1)}\big(B_\nu(z)\big)^2 \sum_{s=0}^n a_{\ell(n+\nu+1)+s}z^s + O\big[(qK(z))^n\big]. \tag{2.10}$$

To see this we use (d) above to get

$$\left|\sum_{s=0}^n (A_{s,n}^\infty - A_{s,n}^\ell)z^s\right| \le C_1\left(\frac{|\alpha|}{\rho-\varepsilon}\right)^{(\ell+1)(n+\nu+1)} \sum_{s=0}^n \left(\frac{|z|}{\rho-\varepsilon}\right)^s = O\Big((qK(z))^n\Big). \tag{2.11}$$

Also, $B_{n,\nu}^\ell(z) - B_\nu(z) = \sum_{s=0}^\nu (\gamma_{s,n}^\ell - \alpha_s)z^s$, where $|\gamma_{s,n}^\ell - \alpha_s| \le Cq^n$ and the same holds for $B_{n,\nu}^\infty(z) - B_\nu(z)$, so that we may replace $B_{n,\nu}^\ell(z)B_{n,\nu}^\infty(z)$ by $\big(B_\nu(z)+\eta\big)^2$ where

$$|\eta| < Cq^n \sum_{s=0}^\nu |z|^s \le Cq^n.$$

It follows that replacing $B_{n,\nu}^\ell(z)B_{n,\nu}^\infty(z)$ by $B_\nu(z)^2$ brings in an error

$$\eta_2 \le 2|B_\nu(z)|q^nC + (Cq^n)^2 \le Cq^n.$$

This proves (2.10) on using (2.11).

We now estimate $S_{2,1}, S_{2,2}, S_{2,3}$ in (2.9) in the same way to get

$$\begin{cases} S_{2,1} = \displaystyle\sum_{m=0}^\nu r_{m,n}z^m \sum_{k=0}^{\nu-1}\sum_{s=n+1}^{n+\nu-k} \alpha_k a_s z^{s+k} + O\big[(qK(z))^n\big], \\[2mm] S_{2,2} = \alpha^{\ell(n+\nu+1)}B_\nu(z)\displaystyle\sum_{k=0}^{\nu-1}\sum_{s=n+1}^{n+\nu-k} \alpha_k a_{\ell(n+\nu+1)+s}z^{s+k} + O\big[(qK(z))^n\big], \\[2mm] S_{2,3} = -B_\nu(z)\displaystyle\sum_{k=0}^{\nu-1}\sum_{s=n+1}^{n+\nu-k} r_{k,n}a_s z^{s+k} + O\big[(qK(z))^n\big]. \end{cases} \tag{2.12}$$

We now split S_3 of (2.8) in three parts as was done for S_2 :

$$S_{3,1} = \big(B_{n,\nu}^\ell(z) - B_{n,\nu}^\infty(z)\big)\sum_{k=s}^\nu\sum_{s=-k}^{-1} \gamma_{k,n}^\infty A_{s,n}^\infty z^{s+k},$$

$$S_{3,2} = B_{n,\nu}^\infty(z)\sum_{k=1}^\nu\sum_{s=-k}^{-1} \gamma_{k,n}^\infty(A_{s,n}^\infty - A_{s,n}^\ell)z^{s+k},$$

$$S_{3,3} = B_{n,\nu}^\infty(z)\sum_{k=1}^\nu\sum_{s=-k}^{-1} (\gamma_{k,n}^\ell - \gamma_{k,n}^\infty)A_{s,n}^\ell z^{s+k}.$$

Using the estimates (a)-(f), we can see that

$$S_{3,i} = O\big[(qK(z))^n\big], \quad i = 1, 2, 3. \tag{2.13}$$

Using (2.10), (2.11) and (2.13) in the expression of $\overline{\Delta}_{n,\gamma}^{\ell}(z,F)$ i.e., in (2.7), we get

$$
\begin{aligned}
\overline{\Delta}_{n,\nu}^{\ell}(z,F) = {}& \alpha^{\ell(n+\nu+1)}\left(B_\nu(z)\right)^2 \sum_{s=0}^{n} a_{\ell(n+\nu+1)+s} z^s \\
& + \sum_{m=0}^{\nu} r_{m,n} z^m \sum_{k=0}^{\nu-1} \sum_{s=n+1}^{n+\gamma-k} \alpha_k a_s z^{s+k} \\
& + \alpha^{\ell(n+\nu+1)} B_\nu(z) \sum_{k=0}^{\nu-1} \sum_{s=n+1}^{n+\nu-k} \alpha_k a_{\ell(n+\gamma+1)+s} z^{s+k} \\
& - B_\nu(z) \sum_{k=0}^{\nu-1} \sum_{s=n+1}^{n+\nu-k} r_{k,n} a_s z^{s+k} + O\big[(qK(z))^n\big].
\end{aligned}
\tag{2.14}
$$

This leads to (2.2) and (2.3), if we can prove that the sums in (2.14) can be rewritten to give (2.3).

Combining the first and third sums on the right in (2.14), we obtain

$$
\begin{aligned}
& \alpha^{\ell(n+\nu+1)} B_\nu(z) \Bigg\{ B_\nu(z) \sum_{s=0}^{n} a_{\ell(n+\nu+1)+s} \alpha^{\ell(n+\nu+1)} z^s \\
& \qquad + \sum_{k=0}^{\nu-1} \sum_{s=n+1}^{n+\nu-k} \alpha_k a_{\ell(n+\nu+1)+s} z^{s+k} \Bigg\} \\
& = \alpha^{\ell(n+\nu+1)} B_\nu(z) \Bigg\{ \sum_{k=0}^{\nu} \sum_{s=0}^{n} \alpha_k a_{\ell(n+\nu+1)+s} z^{s+k} \\
& \qquad + \sum_{k=0}^{\nu-1} \sum_{s=n+1}^{n+\nu-k} \alpha_k a_{\ell(n+\nu+1)s} z^{s+k} \Bigg\} \\
& = \alpha^{\ell(n+\nu+1)} B_\nu(z) \Bigg\{ \sum_{k=0}^{\nu} \sum_{s=0}^{n} \alpha_k a_{\ell(n+\nu+1)+s} z^{s+k} \\
& \qquad + \sum_{k=0}^{\nu} \sum_{s=n+1}^{n+\nu-k} \alpha_k a_{\ell(n+\nu+1)+s} z^{s+k} \Bigg\}.
\end{aligned}
$$

Observe that in the last double sum we have $0 \le k \le \nu$ while in the previous step in the last double sum we have $0 \le k \le \nu - 1$.

This is justified since $\displaystyle\sum_{s=n+1}^{n+\nu-k} \alpha_k a_{\ell(n+\nu+1)+s} z^{s+k} = 0$ for $k = \nu$. Thus on simplifying from the above we have the first and third sums in (2.14) become on adding

$$\alpha^{\ell(n+\nu+1)} B_\nu(z) \sum_{k=0}^{\nu} \sum_{s=0}^{n+\nu-k} \alpha_k a_{\ell(n+\nu+1)+s} z^{s+k}$$

$$= \alpha^{\ell(n+\nu+1)} B_\nu(z) \sum_{k=0}^{\nu} \sum_{s=k}^{n+\nu} \alpha_k a_{\ell(n+\nu+1)+s-k} z^s,$$

which is the first part of (2.3).

Finally the second and fourth sums on the right in (2.14) yield on using the expression for $B_\nu(z)$:

$$\sum_{m=0}^{\nu} r_{m,n} z^m \sum_{k=0}^{\nu-1} \sum_{s=n+1}^{n+\nu-k} \alpha_k a_s z^{s+k} - \sum_{m=0}^{\nu} \alpha_m z^m \sum_{k=0}^{\nu-1} \sum_{s=n+1}^{n+\nu-k} r_{k,n} a_s z^{s+k} =$$

$$= \sum_{m=0}^{\nu} \sum_{k=0}^{\nu-1} \sum_{s=n+1}^{n+\nu-k} (r_{m,n}\alpha_k - r_{k,n}\alpha_m) a_s z^{s+k-m}.$$

Now interchange the summations over s and k. First we have $0 \le k \le \nu - 1$, $n+1 \le s \le n+\nu-k$. This is equivalent to $n+1 \le s \le n+\nu$ and $0 \le k \le n+\nu-s$. Putting $s = \tau + n$, $1 \le \tau \le \nu$, we have

$$z^n \sum_{m=0}^{\nu} \sum_{\tau=1}^{\nu} \sum_{k=0}^{\nu-\tau} (r_{m,n}\alpha_k - r_{k,n}\alpha_m) a_{\tau+n} z^{\tau+k+m}$$

which is the second sum in (2.3). This completes the proof of Lemma 2.   □

7.3 Distinguished Points for $|z| < \rho$ (proof of Theorems 1 and 2)

PROOF OF THEOREM 1. Assume that there is at least one function $F(z) \in M_\rho(\nu)$ for which there are ℓ distinguished points in $|z| < \rho$. Let us denote these points by $\omega_1, \ldots, \omega_\ell$. Then

$$S_{\ell,\nu}(\omega_j, F) < K(\omega_j), \quad j = 1, \ldots, \ell.$$

By Lemma 1, we have

$$\overline{\Delta}_{n,\nu}^\ell(\omega_j) - \left(\frac{\omega_j}{\alpha}\right)^\ell \overline{\Delta}_{n+1,\nu}^\ell(\omega_j) = O\left[(qK(\omega_j))^n\right], \quad 1 \le j \le \ell. \tag{3.1}$$

We shall use Lemma 2 to replace $\overline{\Delta}_{n,\nu}^\ell$ and $\overline{\Delta}_{n+1,\nu}^\ell$ by the corresponding formula (2.3) where the error made by using it is of the same order $(qK(\omega_j))^n$. But the $r_{m,n}$ in formula (2.3) depend upon n and so to distinguish $r_{m,n}$ in the expression for $\overline{\Delta}_{n,\nu}^{\ell}$ from $r_{m,n}$ in $\overline{\Delta}_{n+1,\nu}^{\ell}$, we denote the $r_{m,n}$ in $\overline{\Delta}_{n+1,\nu}^{\ell}$ by $\widehat{r}_{m,n}$. Then by (2.3),

$$\overline{\Delta}_{n,\nu}^{\ell}(\omega_j, F) - \left(\frac{\omega_j}{\alpha}\right)^{\ell}\overline{\Delta}_{n+1,\nu}^{\ell}(\omega_j, F) = T_1 + T_2 + O\left[(qK(\omega_j))^n\right],$$

where

$$T_1 = \alpha^{\ell(n+\nu+1)}B_\nu(\omega_j)\left\{\sum_{k=0}^{\nu}\sum_{s=k}^{n+\nu}\alpha_k a_{\ell(n+\nu+1)+s-k}\omega_j^s \right.$$

$$\left. - \left(\frac{\omega_j}{\alpha}\right)^{\ell}\alpha^{\ell(n+\nu+1)}\sum_{k=0}^{\nu}\sum_{s=k}^{n+1-\nu}\alpha_k a_{\ell(n+\nu+2)+s-k}\omega_j^s\right\} \tag{3.2}$$

and

$$T_2 = \omega_j^n \sum_{\tau=1}^{\nu}\sum_{k=0}^{\nu-\tau}\sum_{m=0}^{\nu}(r_{m,n}\alpha_k - r_{k,n}\alpha_m)a_{\tau+m}\omega_j^{k+\tau+m}$$

$$- \left(\frac{\omega_j}{\alpha}\right)^{\ell}\omega_j^{n+1}\sum_{\tau=1}^{\nu}\sum_{k=0}^{\nu-\tau}\sum_{m=0}^{\nu}(\widehat{r}_{m,n}\alpha_k - \widehat{r}_{k,n}\alpha_m)a_{\tau+m+1}\omega_j^{k+\tau+m}. \tag{3.3}$$

Using the bounds for $\widehat{r}_{m,n}$, $\widehat{r}_{k,n}$ and $a_{\tau+n+1}$, we can see that T_2 is of the same order as $(qK(\omega_j))^n$. Thus from (3.1), we obtain (using T_1) :

$$\alpha^{\ell(n+\nu+1)}B_\nu(\omega_j)\left\{\sum_{k=0}^{\nu}\sum_{s=k}^{n+\nu}\alpha_k a_{\ell(n+\nu+1)+s-k}\omega_j^s \right.$$

$$\left. - \alpha^{\ell}\left(\frac{\omega_j}{\alpha}\right)^{\ell}\sum_{k=0}^{\nu}\sum_{s=k}^{n+\nu+1}\alpha_k a_{\ell(n+\nu+2)+s-k}\omega_j^s\right\} \tag{3.4}$$

$$= O\left[(qK(\omega_j))^n\right].$$

We shall simplify (3.4) further. Thus we get for the left hand side of (3.4)

$$= \alpha^{\ell(n+\nu+1)}B_\nu(\omega_j)\left\{\sum_{k=0}^{\nu}\sum_{s=k}^{n+\nu}\alpha_k a_{\ell(n+\nu+1)+s-k}\omega_j^s \right.$$

$$\left. - \sum_{k=0}^{\nu}\sum_{s=k}^{n+\nu+1}\alpha_k a_{\ell(n+\nu+1)+\ell+s-k}\omega_j^{s+\ell}\right\}$$

$$= \alpha^{\ell(n+\nu+1)}B_\nu(\omega_j)\left\{\sum_{k=0}^{\nu}\sum_{s=k}^{n+\nu}\alpha_k a_{\ell(n+\nu+1)+s-k}\omega_j^s \right.$$

$$\left. - \sum_{k=0}^{\nu}\sum_{s=k+\ell}^{n+\nu+1+\ell}\alpha_k a_{\ell(n+\nu+1)+s-k}\omega_j^s\right\}$$

where we have changed $s + \ell$ to s in the second sum in the braces. Thus we get

$$\alpha^{\ell(n+\nu+1)} B_\nu(\omega_j) \left\{ \sum_{k=0}^{\nu} \sum_{s=k}^{k+\ell-1} \alpha_k a_{\ell(n+\nu+1)+s-k} \omega_j^s \right.$$
$$\left. - \sum_{k=0}^{\nu} \sum_{s=n+\nu+1}^{n+\nu+1+\ell} \alpha_k a_{\ell(n+\nu+1)+s-k} \omega_j^s \right\} = O\big[(qK(\omega_j))^n\big]. \tag{3.5}$$

Since the $B_\nu(\omega_j)$'s are bounded, $|\omega_j| \le q\rho$, $0 < q < 1$, the second sum in the braces in (3.5) is of the same order as $(qK(\omega_j))^n$. Then (3.5) gives with $\tau = s - k$:

$$B_\nu(\omega_j) \sum_{k=0}^{\nu} \sum_{\tau=0}^{\ell-1} \alpha_k a_{\ell(n+\nu+1)+\tau} \omega_j^{\tau+k} = O\left[\left(\frac{qK(\omega_j)}{|\alpha|^\ell}\right)^n\right]$$

or

$$(B_\nu(\omega_j))^2 \sum_{\tau=0}^{\ell-1} a_{\ell(n+\nu+1)+\tau} \omega_j^\tau = O\left(\left(\frac{q(K\omega_j)}{\alpha^\ell}\right)^n\right).$$

Dividing by $B_\nu(\omega_j)^2$ and recalling that $K(\omega_j) = \left(\frac{|\alpha|}{\rho}\right)^\ell$ when $|\omega_j| < \rho$, we get

$$\sum_{\tau=0}^{\ell-1} a_{\ell(n+\nu+1)+\tau} \omega_j^\tau = O\left(\left(\frac{q}{\rho^\ell}\right)^{n+\nu+1}\right), \quad j = 1, \ldots, \ell. \tag{3.6}$$

The above set of ℓ equations in the unknowns $a_{\ell(n+\nu+1)+\tau}$ for $\ell(n+\nu+1) \le \tau \le \ell(n+\nu+1)+\ell-1$ has a non-vanishing Vandermonde determinant and applying Cramer's rule, we obtain

$$a_{\ell(n+\nu+1)+j} = O\left[\left(\frac{q}{\rho^\ell}\right)^{n+\nu+1}\right], \quad 0 \le j \le \ell - 1.$$

As $\ell(n+\nu+1)+j$, $0 \le j \le \ell-1$ runs through all integers $\ge \ell(\nu+1)$ for $n \in \mathbb{N}$, we have

$$\varlimsup_{k\to\infty} |a_k|^{\frac{1}{k}} = \varlimsup_{n\to\infty} |a_{\ell(n+\nu+1)+j}|^{\frac{1}{\ell(n+\nu+1)+j}} = \frac{q^{\frac{1}{\ell}}}{\rho} < \frac{1}{\rho}.$$

This contradicts the hypothesis that $f(z) \in A_\rho$ and completes the proof. $\quad\square$

PROOF OF THEOREM 2. We shall first construct a function $f(z) \in A_\rho$ which is analytic in each of the points ω_j and which satisfies

$$S_{\ell,\nu}\left(\omega_j; \frac{f(z)}{B_\nu(z)}\right) < K(\omega_j), \quad 1 \le j \le \ell - 1.$$

We define the coefficients $\{a_n\}_{n \ge 0}$ by the conditions

$$\begin{cases} a_{\ell m} = \rho^{-\ell m} a_0, & a_0 \ne 0, \quad m = 1, 2, \ldots \\ \displaystyle\sum_{s=0}^{\ell-1} a_{\ell m + s} \omega_j^s = 0, & 1 \le j \le \ell - 1, \quad m = 0, 1, 2, \ldots. \end{cases} \tag{3.7}$$

For $m \geq 0$, we can write

$$\begin{pmatrix} \omega_1 & \omega_1^2 & \cdots & \omega_1^{\ell-1} \\ \omega_2 & \omega_2^2 & \cdots & \omega_2^{\ell-1} \\ \cdots & \cdots & \cdots & \cdots \\ \omega_{\ell-1} & \omega_{\ell-1}^2 & \cdots & \omega_{\ell-1}^{\ell-1} \end{pmatrix} \begin{pmatrix} a_{\ell m+1} \\ a_{\ell m+2} \\ \vdots \\ a_{\ell m+\ell-1} \end{pmatrix} = \rho^{-\ell m} a_0 \begin{pmatrix} 1 \\ 1 \\ \vdots \\ 1 \end{pmatrix}.$$

The determinant of this system of equations is $\omega_1 \omega_2 \ldots \omega_{\ell-1} V(\omega_1, \omega_2, \ldots, \omega_{\ell-1})$ $\neq 0$. By Cramer's rule

$$a_{\ell m+j} = -\rho^{-\ell m} a_0 \frac{V_j(\omega_1, \ldots, \omega_{\ell-1})}{V(\omega_1, \ldots, \omega_{\ell-1})}, \quad 1 \leq j \leq \ell - 1 \tag{3.8}$$

where $V_j(\omega_1, \ldots, \omega_{\ell-1})$ is obtained from $V(\omega_1, \ldots, \omega_{\ell-1})$ by replacing the column containing ω_λ^j by the column $(1\,1 \ldots 1)^{\mathrm{T}}$. From (3.8) it is clear that

$$a_{n+\ell} = \rho^{-\ell} a_n, \quad n \geq 0.$$

Since for an arbitrary m, we can write $m = k\ell + s$, $k \geq 0$ and $1 \leq s \leq \ell - 1$, we see that

$$\varlimsup_{\substack{m \to \infty \\ m \equiv s (\bmod \ell)}} |a_m|^{\frac{1}{m}} = \varlimsup_{k \to \infty} \rho^{-\frac{k\ell}{k\ell+s}} |a_s|^{\frac{1}{k\ell+s}} \leq \frac{1}{\rho}.$$

However for $m = k\ell$, $k \geq 0$, $\lim_{k \to \infty} |a_{k\ell}|^{\frac{1}{k\ell}} = \lim_{k \to \infty} \rho^{-\ell} |a_0|^{\frac{1}{k\ell}} = \frac{1}{\rho^\ell}$. Therefore

$$\varlimsup_{m \to \infty} |a_m|^{\frac{1}{m}} = \frac{1}{\rho},$$

and the function $f(z) = \sum_{m=0}^{\infty} a_m z^m \in A_\rho$ and $F(z) = f(z)/B_\nu(z) \in M_\rho(\nu)$, if we choose a_0 in such a manner that $f(z_j) \neq 0$, $1 \leq j \leq \nu$.

We see on using the representation formula just before (3.6), that for $1 \leq j \leq \ell - 1$,

$$\overline{\Delta}_{n+\nu}^{\ell}(\omega_j) - \left(\frac{\omega_j}{\alpha}\right)^{\ell} \overline{\Delta}_{n+1,\nu}^{\ell}(\omega_j) = \left(B_\nu(\omega_j)\right)^2 \sum_{s=0}^{\ell-1} a_{\ell(n+\nu+1)+s} \omega_j^s + O\left[\left(qK(\omega_j)\right)^n\right]$$
$$= O\left[\left(qK(\omega_j)\right)^n\right],$$

on using the second relation in (3.7). This shows by Lemma 1, that the ω_j's are distinguished for $F(z)$. $\qquad \square$

7.4. Distinguished Points for $|z| \geq \rho$ (proof of Theorem 3)

We shall need some more lemmas before we come to prove Theorems 3 and 4. We begin with

LEMMA 3. *The coefficients $R_{s,n}$ of z^s in (2.3) satisfy the following estimates:*

$$R_{s,n} = O\left[\left(\frac{|\alpha|}{\rho - \varepsilon}\right)^{\ell(n+\nu+1)} \frac{1}{(\rho - \varepsilon)^s}\right], \quad 0 \le s \le n + 2\nu, \qquad (4.1)$$

and

$$R_{s,n} = \alpha^{\ell(n+\nu+1)} \sum_{m=0}^{\nu} \sum_{k=0}^{\nu} \alpha_m \alpha_k a_{\ell(n+\nu+1)+s-m-k} \qquad (4.2)$$

for $2\nu \le s \le n + \nu$.

PROOF. In order to find the coefficient $R_{s,n}$ of z^s in the expansion of $\overline{\Delta}_{n,\nu}^{\ell}(z, F)$ in (2.2), (2.3), we find the coefficient of z^s in

$$\alpha^{\ell(n+\nu+1)} B_\nu(z) \sum_{k=0}^{\nu} \sum_{s=k}^{n+\nu} \alpha_k a_{\ell(n+\nu+1)+s-k} z^s \qquad (4.3)$$

and also the contribution to the coefficient of z^s in the triple sum in (2.3). Now the product (4.3) by interchange of order of summation

$$= \alpha^{\ell(n+\nu+1)} B_\nu(z) \sum_{\tau=0}^{n+\nu} \sum_{k=0}^{\min(\tau,\nu)} \alpha_k a_{\ell(n+\nu+1)+\tau-k} z^\tau$$

$$= \alpha^{\ell(n+\nu+1)} \sum_{s=0}^{n+2\nu} \left\{ \sum_{\tau=\max(0,s-\nu)}^{s} \alpha_{s-\tau} \sum_{k=0}^{\min(\tau,\nu)} \alpha_k a_{\ell(n+\nu+1)+\tau-k} \right\} z^s.$$

The double sum between the curly brackets contains only a finite number of terms $((\nu + 1)^2$ to be precise) which leads to an upper bound for the coefficient of z^s in the above expression. This bound is

$$\left(\frac{|\alpha|}{\rho - \varepsilon}\right)^{\ell(n+\nu+1)} (\rho - \varepsilon)^{-s}.$$

The triple sum in (2.3) also contributes to the coefficient of z^s :

$$z^n \sum_{\tau=1}^{\nu} \sum_{k=0}^{\nu-\tau} \sum_{m=0}^{\nu} (r_{m,n}\alpha_k - r_{k,n}\alpha_m) a_{n+\tau} z^{k+\tau+m}. \qquad (4.4)$$

Putting $\tau + 1$ for τ in the above we have

$$z^n \sum_{\tau=0}^{\nu-1} \sum_{k=0}^{\nu-\tau-1} \sum_{m=0}^{\nu} (r_{m,n}\alpha_k - r_{k,n}\alpha_m) a_{\tau+n+1} z^{k+\tau+1+m} := \sum_{j=1}^{2\nu} C_{n+j} z^{n+j},$$

where we put $j = k + \tau + m + 1$, so that

$$C_{n+j} = \sum (r_{m,n}\alpha_k - r_{k,n}\alpha_m) a_{n+\tau+1}, \quad 1 \le j \le \nu$$

where $\tau + k + m = j - 1$, $\tau, k, m \geq 0$, $\tau \leq \nu - 1$, $m \leq \nu$, $k \leq \nu - 1 - \tau$. When $1 \leq j \leq \nu$, the above conditions can be replaced by the simpler conditions $\tau \geq 0$, $k \geq 0$, $\tau + k \leq j - 1$ and $m = j - 1 - \tau - k$. Therefore we can write for $1 \leq j \leq \nu$:

$$C_{n+j} = \sum_{\tau=0}^{j-1} \sum_{k=0}^{j-\tau-1} (r_{j-1-\tau-k,n}\alpha_k - r_{k,n}\alpha_{j-1-\tau-k})a_{n+\tau+1}.$$

For $0 \leq \tau \leq j - 1 \leq \nu - 1$, the inner sum in the above becomes zero for each τ. This means that the triple sum in (4.4) contributes to the powers of z^s only for $s \geq n + \nu + 1$. For $n + \nu + 1 \leq s \leq n + 2\nu$, there are only a finite number of terms and their order is also $\left(\frac{|\alpha|}{\rho - \varepsilon}\right)^{n+\nu+1}(\rho - \varepsilon)^{-s}$, showing that (4.1) is true for $n + \nu + 1 \leq s \leq n + 2\nu$.

In order to prove (4.2), we have to consider powers of z^s, $2\nu \leq s \leq n + \nu$; these can arise only as a result of (4.3). There we see that the coefficient of z^s is

$$\alpha^{\ell(n+\nu+1)} \sum_{\tau=\max(0,s-\nu)}^{s} \alpha_{s-\tau} \sum_{k=0}^{\min(\tau,\nu)} \alpha_k a_{\ell(n+\nu+1)+\tau-k}$$

where $m + \tau = s$, $0 \leq m \leq \nu$, $0 \leq \tau \leq n + \nu$. Since $\tau = s - m > 2\nu - m \geq \nu$, the above sum can be replaced by

$$\alpha^{\ell(n+\nu+1)} \sum_{m=0}^{\nu} \sum_{k=0}^{\nu} \alpha_m \alpha_k a_{\ell(n+\nu+1)+s-m-k}\alpha^{\ell(n+\nu+1)},$$

which is the value of $R_{s,n}$ in (4.2) for $2\nu \leq s \leq n + \nu$. This completes the proof.
$\square$

LEMMA 4. *Let all the zeros* $\{z_j\}_{j=1}^{\nu}$ *of the monic polynomial* $B_\nu(z) = \sum_{k=0}^{\nu} \alpha_k z^k$, $\alpha_\nu = 1$, *belong to* $|z| < \rho$. *Furthermore let the quantities* $\{A_s\}_{s=0}^{\infty}$ *satisfy*

$$\sum_{k=0}^{\nu} \alpha_k A_{s-k} = O\left[\left(\frac{q}{\rho}\right)^s\right], \quad as \quad s \to \infty \tag{4.5}$$

for some q, $0 < q < 1$. *Then*

$$\limsup_{s \to \infty} |A_s|^{1/s} < \frac{1}{\rho}.$$

PROOF. Consider the formal power series

$$f(z) = \sum_{s=0}^{\infty} A_s z^s.$$

From the order relation (4.5) in the lemma, we see that the coefficients C_s of the formal power series formed by the Cauchy product $B_\nu(z)f(z)$ satisfy

$$\limsup_{s\to\infty}|C_{s,n}|^{\frac{1}{s}} = \frac{q}{\rho} < \frac{1}{\rho}.$$

This implies that $B_\nu(z)f(z)$ is an analytic function at $z = 0$ with Taylor series having a radius of convergence $R_1 > \rho$. As all zeros of $B_\nu(z)$ are in $|z| < \rho$, the function

$$f(z) = \frac{B_\nu(z)f(z)}{B_\nu(z)}$$

is also analytic on $|z| < R_1$ or equivalently we have $\limsup_{s\to\infty}|A_s|^{\frac{1}{s}} < \frac{1}{\rho}$. $\qquad\square$

PROOF OF THEOREM 3. Assume that there is at least one function $F(z) \in M_\rho(\nu)$ with $\ell+\nu+1$ distinguished points ω_j in $|z| > \rho$. We consider throughout the case of simple poles. In other words, this implies that

$$S_{\ell,\nu}(\omega_j, F) < K(\omega_j), \quad 1 \le j \le \ell+\nu+1. \tag{4.6}$$

According to Lemma 1,

$$\Delta := \overline{\Delta}^\ell_{n,\nu}(\omega_j, F) - \left(\frac{\omega_j}{\alpha}\right)^\ell \overline{\Delta}^\ell_{n+1,\nu}(\omega_j, F) = O\big[(qK(\omega_j))^n\big]$$

for some $0 < q < 1$. We first use Lemma 2 (relation (2.2)) to show that for $1 \le j \le \ell+\nu+1$,

$$\begin{aligned}
\Delta = &\sum_{s=0}^{\ell-1} R_{s,n}\omega_j^s + \sum_{s=\ell}^{n+2\nu}\left(R_{s,n} - \frac{1}{\alpha^\ell}R_{s-\ell,n+1}\right)\omega_j^s \\
&- \sum_{s=n+2\nu+1}^{n+2\nu+\ell+1}\frac{1}{\alpha^\ell}R_{s-\ell,m+1}\omega_j^s + O\big[(qK(\omega_j))^n\big].
\end{aligned} \tag{4.7}$$

By Lemma 2, omitting the error terms for simplicity, we have

$$\overline{\Delta}^\ell_{n,\nu}(\omega_j) = \sum_{s=0}^{n+2\nu} R_{s,n}\omega_j^s$$

$$\omega_j^\ell\,\overline{\Delta}^\ell_{n+1,\nu}(\omega_j) = \sum_{s=0}^{n+2\nu+1} R_{s,n+1}\omega_j^{s+\ell} = \sum_{s=\ell}^{n+2\nu+\ell+1} R_{s-\ell,n+1}\omega_j^s.$$

Then

$$\begin{aligned}
\Delta &= \overline{\Delta}^\ell_{n,\nu}(\omega_j) - \left(\frac{\omega_j}{\alpha}\right)^\ell \overline{\Delta}^\ell_{n+1,\nu}(\omega_j) \\
&= \sum_{s=0}^{n+2\nu} R_{s,n}\omega_j^s - \frac{1}{\alpha^\ell}\sum_{s=\ell}^{n+2\nu+\ell+1} R_{s-\ell,n+1}\omega_j^s \\
&= \sum_{s=0}^{\ell-1} R_{s,n}\omega_j^s + \sum_{s=\ell}^{n+2\nu}\left(R_{s,n} - \frac{1}{\alpha^\ell}R_{s-\ell,n+1}\right)\omega_j^s \\
&\quad - \sum_{s=n+2\nu+1}^{n+2\nu+\ell+1}\frac{1}{\alpha^\ell}R_{s-\ell,n+1}\omega_j^s + O\big[(qK(\omega_j))^n\big].
\end{aligned}$$

This proves (4.7).

We shall show that (4.7) can be rewritten as

$$\Delta = \sum_{s=n+\nu+1}^{n+2\nu} \left(R_{s,n} - \frac{1}{\alpha^\ell} R_{s-\ell,n+1} \right) \omega_j^s + \sum_{s=n+2\nu+1}^{n+2\nu+\ell+1} \left(-\frac{1}{\alpha^\ell} R_{s-\ell,n+1} \right) \omega_j^s$$

$$= O\big[(qK(\omega_j))^n \big]. \tag{4.8}$$

To this end we observe that using (4.1), we have

$$\sum_{s=0}^{\ell-1} R_{s,n} \omega_j^s = O\big[(q(K(\omega_j))^n \big] \quad \text{using} \quad q = \frac{\rho - \varepsilon}{\min |\omega_j|} < 1. \tag{4.9}$$

Now

$$\sum_{s=\ell}^{n+\nu} \left(R_{s,n} - \frac{1}{\alpha^\ell} R_{s-\ell,n+1} \right) \omega_j^s = \sum_{s=\ell}^{2\nu+\ell-1} + \sum_{s=2\nu+\ell}^{n+\nu} \left(R_{s,n} - \frac{1}{\alpha^\ell} R_{s-\ell,n+1} \right) \omega_j^s$$

$$= I_1 + I_2.$$

The first sum I_1 is estimated from above using (4.1) by

$$C\left(\frac{|\alpha|}{\rho - \varepsilon} \right)^{\ell(n+\nu+1)} = O\big((qK(\omega_j))^n \big). \tag{4.10}$$

In the second sum we use Lemma 3 (4.2) which holds for $2\nu \le s \le n+\nu$ and

$$\alpha^\ell R_{s-\ell,n+1} = \frac{\alpha^{\ell(n+\nu+2)}}{\alpha^\ell} \sum_{m=0}^{\nu} \sum_{k=0}^{\nu} \alpha_m \alpha_k a_{\ell(n+\nu+1)+\ell+s-\ell-m-k}$$

$$= \alpha^{\ell(n+\nu+1)} \sum_{m=0}^{\nu} \sum_{k=0}^{\nu} \alpha_m \alpha_k a_{\ell(n+\nu+1)+s-k-m}$$

$$= R_{s,n}$$

so that

$$R_{s,n} - \frac{1}{\alpha^\ell} R_{s-\ell,n+1} = 0 \quad \text{for} \quad 2\nu + \ell \le s \le n+\nu. \tag{4.11}$$

Thus combining (4.9), (4.10) and (4.11), we get (4.8).

Put $\tau = s - n - \nu - 1$ in (4.8). Then for $j = 1, 2, \ldots, \ell + \nu + 1$, we have

$$\sum_{\tau=0}^{\nu-1} \left(R_{\tau+n+\nu+1,n} - \frac{1}{\alpha^\ell} R_{\tau+n+\nu+1-\ell,n+1} \right) \omega_j^{\tau+n+\nu+1}$$

$$+ \sum_{\tau=\nu}^{\nu+\ell} \left(-\frac{1}{\alpha^\ell} R_{\tau+n+\nu+1-\ell,n+1} \right) \omega_j^{\tau+n+\nu+1} = O\big[(qK(\omega_j))^n \big]. \tag{4.12}$$

In (4.12) replace $\big(K(\omega_j)\big)^n$ by $\big(K(\omega_j)\big)^{n+\nu+1}$ and then divide (4.12) by $\omega_j^{n+\nu+1}$. This gives a linear system of $\nu+\ell+1$ equations in $\nu+\ell+1$ variables. Applying Cramer's rule, we get

$$R_{\tau+n+\nu+1,n} - \frac{1}{\alpha^\ell} R_{\tau+n+\nu+1-\ell,n+1} = O\Big(\frac{q|\alpha|^\ell}{(\rho-\varepsilon)^{\ell+1}}\Big)^{n+\nu+1}$$

$$\text{for } 0 \le \tau \le \nu-1 \,,$$

$$-\frac{1}{\alpha^\ell} R_{\tau+n+\nu+1-\ell,n+1} = O\Big(\frac{q|\alpha|^\ell}{(\rho-\varepsilon)^{\ell+1}}\Big)^{n+\nu+1}$$

$$\text{for } \nu \le \tau \le n+2\nu+1+\ell.$$

Replacing $\tau+n+\nu+1$ by m, we have

$$R_{m,n} - \frac{1}{\alpha^\ell} R_{m-\ell,n+1} = O\Big(\frac{q|\alpha|^\ell}{(\rho-\varepsilon)^{\ell+1}}\Big)^{n+\nu+1},$$

$$\text{for } n+\nu+1 \le m \le n+2\nu \qquad (4.13)$$

$$-\frac{1}{\alpha^\ell} R_{m-\ell,n+1} = O\Big(\frac{q|\alpha|^\ell}{(\rho-\varepsilon)^{\ell+1}}\Big)^{n+\nu+1},$$

$$\text{for } n+2\nu+1 \le m \le n+2\nu+\ell+1.$$
$$(4.14)$$

Now replace $n+1$ by n in (4.14), multiply by $-\alpha^\ell$ and replace $m-\ell$ by m. Then we get

$$R_{m,n} = O\Big(\frac{q|\alpha|^\ell}{(\rho-\varepsilon)^{\ell+1}}\Big)^{n+\nu+1}, \quad \max\{n+\nu+1,n+2\nu-\ell\} \le m \le n+2\nu. \ (4.15)$$

We now use (4.15) in (4.13) giving $-\frac{1}{\alpha^\ell} R_{m-\ell,n+1}$. Then multiplying by $-\alpha^\ell$ and replacing $m-\ell$ by m, we have

$$R_{m,n+1} = O\Big[\frac{q|\alpha|^\ell}{(\rho-\varepsilon)^{\ell+1}}\Big)^{n+\nu+1}\Big]$$

$$\text{for} \quad \max(n+\nu-\ell+1,n+2\nu-2\ell) \le m \le n+2\nu-\ell.$$

Again replace $n+1$ by n and replace $n+\nu$ in the order term on the right by $n+\nu+1$ to obtain

$$R_{m,n} = O\Big[\Big(\frac{q|\alpha|^\ell}{(\rho-\varepsilon)^{\ell+1}}\Big)^{n+\nu+1}\Big]$$

for

$$\max(n+2\nu-\ell,m+2\nu-2\ell-1) \le m \le n+2\nu-\ell-1. \qquad (4.16)$$

Thus we have been able to get an estimate for $R_{m,n}$ for values of m in (4.16) which have been shifted by $\ell+1$ from the range for m in (4.15). Repeating the process a number of times, we can get the validity of the estimate for $R_{m,n}$ for $n + \nu - \ell < m \leq n + \nu$.

We use the explicit form for $R_{s,n}$ in Lemma 3 (4.2) and combine that form with (4.16) for $\ell + 1$ consecutive values of s and divide by $\alpha^{\ell(n+\nu+1)}$. Then we have

$$\sum_{m=0}^{\nu} \alpha_m \sum_{k=0}^{\nu} \alpha_k a_{\ell(n+\nu+1)+s-m-k} = O\left[\left(\frac{q}{\rho - \varepsilon}\right)^{n+\nu+1} \frac{1}{(\rho - \varepsilon)^{\ell(n+\nu+1)}}\right]$$

for $n + \nu - \ell \leq s \leq n + \nu$.

As $\ell(n + \nu + 1) + s$ runs from $\ell(n + \nu + 1) + n + \nu - \ell = (\ell + 1)(n + \nu)$ to $\ell(n + \nu + 1) + n + \nu = (\ell + 1)(n + \nu) + \ell$, we can let $\{\ell(n + \nu + 1) + s\}_{s,n}$ run through all integers $\geq N_0$ by letting n run through all integers $\geq N_1$. Therefore

$$\sum_{m=0}^{\nu} \alpha_m \sum_{k=0}^{\nu} \alpha_k a_{n-m-k} = O\left[\frac{q^{n+\nu+1}}{(\rho - \varepsilon)^{(\ell+1)(n+\nu+1)}}\right]$$

$$= O\left[\left(\frac{q^{1/\ell}}{\rho - \varepsilon}\right)^n\right].$$

By Lemma 4, we get

$$\sum_{k=0}^{\nu} \alpha_k a_{n-m-k} = O\left(\left(\frac{q_1}{\rho}\right)^n\right) \quad \text{for some} \quad q_1, \quad 0 < q_1 < 1.$$

Applying Lemma 4 again, we get

$$\varlimsup_{n \to \infty} |a_n|^{\frac{1}{n}} < \frac{1}{\rho},$$

a contradiction to $\varlimsup_{n \to \infty} |a_n|^{\frac{1}{n}} = \frac{1}{\rho}$.   □

7.5. A Lemma and Proof of Theorem 4

For the proof of Theorem 4, we shall need the following technical lemma.

LEMMA 5. *Let* $F(z) \in M_\rho(\nu)$ *have simple poles, and let* $\ell \geq \nu$. *If* $|z| > \rho$, *then we have for* $\ell \geq \nu$ *and* n *large, the following relation:*

$$z^{-n}\left\{\overline{\Delta}_{n-\nu-1,\nu}^{\ell}(z, F) - \left(\frac{z}{\alpha}\right)^{\ell} \overline{\Delta}_{n-\nu,\nu}^{\ell}(z, F)\right\}$$

$$= -B_\nu(z)\alpha^{\ell n} \sum_{s=-\nu}^{-1} \sum_{k=-s}^{\nu} \alpha_k a_{(\ell+1)n+s} z^{s+k} \qquad (5.1)$$

$$- (B_\nu(z))^2 \alpha^{\ell n} \sum_{s=0}^{\ell-\nu} a_{(\ell+1)n+s} z^s$$

$$- B_\nu(z)\alpha^{\ell n} \sum_{s=\ell-\nu+1}^{\ell} \sum_{k=0}^{\ell-s} \alpha_k a_{(\ell+1)n+s} z^{s+k}$$
$$+ C\Big(q\big(\frac{|\alpha|}{\rho-\varepsilon}\big)^\ell \frac{|z|}{\rho-\varepsilon}\Big)^n, \quad 0 < q < 1.$$

PROOF. We begin with the explicit form for $\overline{\Delta}^\ell_{n,\nu}(z,F)$ as given in Lemma 2, (2.2) and (2.3). We shall prove that

$$|r_{k,n}| < C\Big[\big(q\,\frac{|\alpha|}{\rho-\varepsilon}\big)^\ell\Big]^{n+\nu+1} \tag{5.2}$$

and show that (5.2) allows us to prove that the triple sum in (2.3) is of the same order as $(qK(z))^n$.

We recall that the system of equations to find the coefficients $\gamma^\ell_{k,n}$, $p^\ell_{k,n}$ can be reduced to a system of equations to determine $\gamma^\ell_{k,n}$ along with a system of equations to determine $p^\ell_{k,n}$ in terms of $\gamma^\ell_{k,n}$. The equations to determine $\gamma^\ell_{k,n}$ ($\gamma^\ell_{\nu,n} = 1$) were given by (2.13) of Ch. 6:

$$\sum_{k=0}^{\nu} \gamma^\ell_{k,n} \sum_{s=0}^{n+\nu} A^\ell_{s-k,n} z_j^s = 0, \quad 1 \le j \le \nu,\ 1 \le \ell \le \infty.$$

Shifting the summation index we get

$$\sum_{k=0}^{\nu} \gamma^\ell_{k,n} z_j^k \sum_{s=-k}^{n+\nu-k} A^\ell_{s,n} z_j^s = 0, \quad 1 \le j \le \nu,\ 1 \le \ell \le \infty. \tag{5.3}$$

This easily yields a system of equations to determine $r_{k,n} = \gamma^\ell_{k,n} - \gamma^\infty_{k,n}$ where $r_{\nu,n} = 0$:

$$\sum_{k=0}^{\nu-1} r_{k,n} z_j^k \sum_{s=-k}^{n+\nu-k} A^\ell_{s,n} z_j^s + \sum_{k=0}^{\infty} \gamma^\infty_{k,n} z_j^k \sum_{s=-k}^{n+\nu-k} (A^\ell_{s,n} - A^\infty_{s,n}) z_j^s = 0. \tag{5.4}$$

We shall now use the following known estimates:

$$\begin{cases} A^\ell_{s,n} = \alpha^{n+\nu+1} A_{n+\nu+1+s}, \quad s < 0 \quad \text{and} \quad A^\ell_{s,n} = a_s + \delta^\ell_{s,n}, \quad s \ge 0 \\ \qquad \text{where} \\ \delta^\ell_{s,n} = \frac{C}{(\rho-\varepsilon)^s}\Big(\frac{|\alpha|}{\rho-\varepsilon}\Big)^{n+\nu+1}, \quad 1 \le s < \infty, \quad s \ge 0. \end{cases} \tag{5.5}$$

Using (5.5), we can write the first double sum in (5.4) as below:

$$\sum_{k=0}^{\nu-1} r_{k,n} z_j^k \Big(\sum_{s=0}^{n+\nu-k} A^\ell_{s,n} z_j^s\Big) + \sum_{k=0}^{\nu-1} r_{k,n} z_j^k \Big\{ \sum_{s=-k}^{-1} \alpha^{n+\nu+1}(a_{n+\nu+1+s} + \delta^\ell_{n+\nu+1+s,n}) z_j^s$$

$$+ \sum_{s=0}^{n+\nu-k} \delta^\ell_{s,n} z_j^s \Big\} = S_1 + S_2. \tag{5.6}$$

Using the estimates on $a_{n+\nu+1+s}$ and of $\delta^\ell_{n+\nu+1+s}$ in (5.5), on the first sum between curly braces, the second of these three double sums is bounded by:

$$\left| \sum_{s=-k}^{-1} \alpha^{n+\nu+1}(a_{n+\nu+1+s} + \delta^\ell_{n+\nu+1+s,n})z_j^s \right| \le$$

$$\le \sum_{s=-k}^{-1} \frac{C|z_j|^s}{(\rho - \varepsilon)^s} \left(\frac{|\alpha|}{\rho - \varepsilon} \right)^{n+\nu+1} \left(\frac{|\alpha|}{\rho - \varepsilon} \right)^{n+\nu+1}$$

$$\le C_1 q^{n+\nu+1}, \quad 0 < q < 1.$$

Similarly,

$$\left| \sum_{s=0}^{n+\nu-k} \delta^\ell_{s,n} z_j^s \right| \le \sum_{s=0}^{n+\nu-k} C\left(\frac{|z_j|}{\rho - \varepsilon} \right)^s \left(\frac{|\alpha|}{\rho - \varepsilon} \right)^{n+\nu+1} \le C_2 q_1^{n+\nu+1}.$$

Thus combining the above we get an upper bound for the second part of (5.6), i.e.,

$$S_2 \le \sum_{k=0}^{\nu-1} r_{k,n}|z_j|^k (C_1 + C_2)q^{n+\nu}, \quad q = \max(q_1, q_2)$$

$$\le C \sum_{k=0}^{\nu-1} \left(\frac{|\alpha|}{\rho - \varepsilon} \right)^{\ell(n+\nu+1)} \rho^k (C_1 + C_2)q^{n+\nu} \qquad (5.7)$$

$$\le C q^{n+\nu} \left(\frac{|\alpha|}{\rho - \varepsilon} \right)^{(n+\nu+1)\ell}.$$

The first double sum in (5.6) can be written as

$$\sum_{k=0}^{\nu-1} r_{k,n} z_j^k \left[f(z_j) - \sum_{s=n+\nu-k+1}^{\infty} a_s z_j^s \right]$$

$$= \sum_{k=0}^{\nu-1} r_{k,n} z_j^k f(z_j)$$

$$+ O\left(\sum_{k=0}^{\nu-1} \left(\frac{|\alpha|}{\rho - \varepsilon} \right)^{\ell(n+\nu+1)} \sum_{s=n+\nu-k+1}^{\infty} \frac{|z_j|^s}{(\rho - \varepsilon)^s} \right)$$

$$= f(z_j) \sum_{k=0}^{\nu-1} r_{k,n} z_j^k + O\left(\left(\frac{|\alpha|}{\rho - \varepsilon} \right)^{\ell(n+\nu+1)} \sum_{k=0}^{\nu-1} \left(\frac{|z|}{\rho - \varepsilon} \right)^{n+\nu+1-k} \right)$$

$$= f(z_j) \sum_{k=0}^{\nu-1} r_{k,n} z_j^n + O\left(\left(\frac{|\alpha|}{\rho - \varepsilon} \right)^{\ell(n+\nu+1)} q^n \right).$$

The second sum in (5.4) can be estimated directly on using $|\gamma^\infty_{k,n}| < C$;

$$|A^\ell_{s,n} - A^\infty_{s,n}| \begin{cases} \le \left(\frac{|\alpha|}{\rho-\varepsilon} \right)^{\ell(n+\nu+1)} + \frac{C}{(\rho-\varepsilon)^s} \cdot \left(\frac{|\alpha|}{\rho-\varepsilon} \right)^{\ell(n+\nu+1)}, & s > 0 \\[2mm] \le \left(\frac{|\alpha|}{\rho-\varepsilon} \right)^{(\ell+1)(n+\nu+1)} \frac{1}{(\rho-\varepsilon)^s} + \frac{C}{(\rho-\varepsilon)^s} \left(\frac{|\alpha|}{\rho-\varepsilon} \right)^{2(n+\nu+1)}, & s < 0 \end{cases}$$

and it gives the same estimate as in (5.7).

Thus the set of equations to determine $r_{k,n}$ can be written as

$$\sum_{k=0}^{\nu-1} r_{k,n} z_j^k f(z_j) = C\Big(\frac{|\alpha|}{\rho-\varepsilon}\Big)^{\ell(n+\nu+1)} q^{n+\nu}, \quad 1 \le j \le \nu. \tag{5.8}$$

Applying Cramer's rule to the above system of equations, we get

$$r_{k,n} = O\Big[\Big(q\Big(\frac{|\alpha|}{\rho-\varepsilon}\Big)^\ell\Big)^n\Big], \quad 0 \le k \le \nu-1 \tag{5.9}$$

since $f(z_j) \ne 0$, and $V(z_1,\ldots,z_\nu) \ne 0$. This proves (5.2).

We now use (5.2) in the triple sum in Lemma 2 (2.3). Since there is only a finite number of terms in the triple sum in (2.3), it is bounded by

$$Cq^{n+\nu}\Big(\frac{|\alpha|}{\rho-\varepsilon}\Big)^{\ell(n+\nu)}\Big(\frac{|z|}{\rho-\varepsilon}\Big)^{n+\nu}, \quad |z| > \rho.$$

From (2.3), we see that

$$\overline{\Delta}^\ell_{n,\nu}(z,F) = \alpha^{\ell(n+\nu+1)} B_\nu(z) \sum_{k=0}^\nu \sum_{s=k}^{n+\nu} \alpha_n a_{\ell(n+\nu+1)+s} + O\big[(qK(z))^{n+\nu}\big].$$

Then

$$\overline{\Delta}^\ell_{n-\nu-1,\nu}(z,F) - \Big(\frac{z}{\alpha}\Big)^\ell \overline{\Delta}^\ell_{n-\nu,\nu}(z,F) = P_B + E \tag{5.10}$$

where P_B has $B_\nu(z)$ as a factor and the error term E is given by

$$C_1 q^{n-1}\Big[\Big(\frac{|\alpha|}{\rho-\varepsilon}\Big)^\ell \frac{|z|}{\rho-\varepsilon}\Big]^{n-1} + C_2 q^n\Big[\Big(\frac{|\alpha|}{\rho-\varepsilon}\Big)^\ell \frac{|z|}{\rho-\varepsilon}\Big]^n \le$$

$$\le Cq^n\Big[\Big(\frac{|\alpha|}{\rho-\varepsilon}\Big)^\ell \frac{|z|}{\rho-\varepsilon}\Big]^n.$$

Now

$$P_B = B_\nu(z)\Big[\alpha^{\ell n} \sum_{k=0}^\nu \sum_{s=k}^{n-1} \alpha_k a_{\ell n+s-k} z^s - \frac{z^\ell \alpha^{\ell(n+1)}}{\alpha^\ell} \sum_{k=0}^\nu \sum_{s=k}^n \alpha_k a_{\ell(n+1)+s-k} z^s\Big]$$

$$= \alpha^{\ell n} B_\nu(z) \sum_{k=0}^\nu \alpha_k\Big\{\sum_{s=k}^{n-1} a_{\ell n+s-k} z^s - \sum_{s=k+\ell}^{n+\ell} a_{\ell n+s-k} z^s\Big\}$$

$$= \alpha^{\ell n} B_\nu(z) \sum_{k=0}^\nu \alpha_k\Big\{\sum_{s=k}^{k+\ell-1} a_{\ell n+s-k} z^s - \sum_{s=n}^{n+\ell} a_{\ell n+s-k} z^s\Big\}. \tag{5.11}$$

The first sum in (5.10) is easily estimated by

$$|\alpha|^{\ell n} \sum_{k=0}^{\nu} C \sum_{s=k}^{k+\ell-1} \frac{|z|^s}{(\rho-\varepsilon)^{\ell n+s-k}} \leq C \Big(\frac{|\alpha|}{\rho-\varepsilon}\Big)^{\ell n} \Big(\frac{|z|}{\rho-\varepsilon}\Big)^{k+\ell-1}$$

$$\leq C \Big(\Big(\frac{|\alpha|}{\rho-\varepsilon}\Big)^{\ell n} \Big(\frac{|z|}{\rho-\varepsilon}\Big)^n \Big(\frac{|\rho-\varepsilon|}{|z|}\Big)^n \Big)$$

$$\leq C \big(K(z)q \big)^n,$$

since $|z| > \rho$. Thus from (5.10) and (5.11), we see that for $|z| > \rho$,

$$\overline{\Delta}^{\ell}_{n-\nu-1,\nu}(z,F) - \Big(\frac{z}{\alpha}\Big)^{\ell} \overline{\Delta}^{\ell}_{n-\nu,\nu}(z,F)$$

$$= -\alpha^{\ell n} B_\nu(z) \sum_{k=0}^{\nu} \alpha_k \sum_{s=n}^{n+\ell} a_{\ell n+s-k} z^s + C \big(q K(z) \big)^n. \tag{5.12}$$

In order to complete the proof of Lemma 5, we examine the double sum S in the above

$$S = \alpha^{\ell n} \sum_{k=0}^{\nu} \sum_{s=n}^{n+\ell} \alpha_k a_{\ell n+s-k} z^s.$$

Set $t = s - n - k$ in the above. Then

$$S = \alpha^{\ell n} z^n \sum_{k=0}^{\nu} \sum_{t=-k}^{\ell-k} \alpha_k a_{\ell n+n+t} z^{t+k}.$$

Here the region of summation is the parallelogram $OABC$ with vertices $O = (0,0)$, $A = (\ell, 0)$, $B = (\ell - \nu, \nu)$, $C = (-\nu, \nu)$ with t along the horizontal axis and k along the vertical axis. This region is split into a triangle $-\nu \leq t \leq -1$, a rectangle $0 \leq t \leq \ell - \nu$ and another triangle $(\ell - \nu + 1 \leq t \leq \ell)$. Thus:

$$S = \alpha^{\ell n} z^n \sum_{t=-\nu}^{-1} \sum_{k=-t}^{\nu} \alpha_k A_{\ell n+n+t} z^{t+k}$$

$$+ \alpha^{\ell n} z^n \sum_{k=0}^{\nu} \sum_{t=0}^{\ell-\nu} a_{\ell n+n+t} z^{t+k}$$

$$+ \alpha^{\ell n} z^n \sum_{t=\ell-\nu+1}^{\ell} \alpha_k \sum_{k=0}^{\ell-k} a_{(\ell+1)n+t} z^{t+k}.$$

The middle sum is

$$\alpha^{\ell n} z^n \sum_{k=0}^{\nu} \alpha_k z^k \sum_{t=0}^{\ell-\nu} a_{(\ell+1)n+t} z^t = \alpha^{\ell n} z^n B_\nu(z) \sum_{t=0}^{\ell-\nu} a_{(\ell+1)n+t} z^t.$$

Inserting this expression in S, we get from (5.12)

$$\overline{\Delta}^{\ell}_{n-\nu-1,\nu}(z,F) - \left(\frac{z}{\alpha}\right)^{\ell}\overline{\Delta}^{\ell}_{n-\nu,\nu}(z,F) = -\alpha^{\ell n}z^n B_\nu(z)^2 \sum_{t=0}^{\ell-\nu} a_{(\ell+1)n+t}z^t$$

$$-\alpha^{\ell n}z^n B_\nu(z)\sum_{t=-\nu}^{-1} a_{\ell n+n+t}z^t$$

$$-\alpha^{\ell n}z^n B_\nu(z)\sum_{t=\ell-\nu+1}^{\ell} a_{\ell n+n+t}z^t,$$

which is the expression in (5.1) and this completes the proof of Lemma 5.   □

PROOF OF THEOREM 4. Let $\{w_j\}$, $1 \le j \le \ell - \nu + 1$ be a set of distinct points in $|z| > \rho$. We shall then construct a function $f(z) \in A_\rho(\nu)$ so that $F(z) = f(z)/B_\nu(z) \in M_\rho(\nu)$. If $f(z) = \sum_{k=0}^{\infty} a_k z^k$, we define the sequence $\{a_k\}_{k=\ell+1}^{\infty}$ by the following scheme:

$$a_{(\ell+1)r+s} = \rho^{-(\ell+1)r-s}, \quad \ell+1-\nu \le s \le \ell, \quad r \ge 1 \tag{5.13}$$

$$a_{(\ell+1)r+s} = 0, \qquad 0 \le s \le \ell - \nu, \quad 1 \le r \le 2\nu - 1. \tag{5.14}$$

This scheme defines the a_k's for $\ell+1 \le k \le 2\nu(\ell+1)-1$. From this point onwards only (5.13) contributes to the construction of the a_k's where $k = (\ell+1)r + s$ and $r \ge 2\nu$ and $\ell+1-\nu \le s \le \ell$. It still remains to construct two sets of a_k's :
(i) the finite set $\{a_k\}$, $0 \le k \le \ell$,
(ii) the infinite set $\{a_{(\ell+1)r+s}\}$, with $r \ge 2\nu$ and $0 \le s \le \ell - \nu$.

The set in (ii) is defined inductively on using the following:

$$\sum_{k=0}^{\ell-\nu} w_j^k a_{(\ell+1)r+k} = -\frac{1}{(B_\nu(w_j))}\left\{\sum_{s=-\nu}^{-1}\sum_{t=-s}^{\nu}\alpha_t a_{(\ell+1)r+s}w_j^{s+t}+\right.$$

$$\left.+\sum_{s=\ell-\nu+1}^{\ell}\sum_{t=0}^{\ell-1}\alpha_t a_{(\ell+1)r+s}w_j^{s+t}\right\} \tag{5.15}$$

for $0 \le s \le \ell - \nu$, $r \ge 2\nu$, $1 \le j \le \ell - \nu + 1$. The first and third sums on the right side above are motivated by the first and third sums on the right in the expression in Lemma 5. On the left in (5.15), the $a_{(\ell+1)r+k}$ are for $0 \le k \le \ell - \nu$, but on the right side the a's have indices of the form $(\ell+1)r + s$ for $\ell - \nu + 1 \le s \le \ell$ and we can use (5.13) for $r \ge 2\nu$. For $-\nu \le s \le -1$, we can write $(\ell+1)r + s = (\ell+1)(r-1) + \ell + 1 + s$ with $\ell - \nu + 1 \le \ell + 1 + s \le \ell$ for $r = 2\nu - 1$ and this again is known by (5.14).

From (5.13), it is clear that once the a_k's are defined, we have

$$\lim_{k \to \infty} |a_k|^{\frac{1}{k}} = \frac{1}{\rho} \tag{5.16}$$

for the sequences in (5.13), i.e., for subsequences of the form $\{a_{(\ell+1)r+s}\}_{r=0}^{\infty}$, $\ell - \nu + 1 \le s \le \ell$ fixed.

We now apply Cramer's rule to (5.15) to solve for $a_{(\ell+1)r+s}$. Then we get

$$a_{(\ell+1)r+t} = d_t \rho^{-(\ell+1)r-t}, \quad 0 \le t \le \ell - \nu, \quad r \ge 2\nu, \tag{5.17}$$

where d_t is a constant which can be given explicitly by

$$d_t = \sum_{j=-\nu}^{-1} C_{t,j} \rho^{t-j} + \sum_{j=\ell-\nu+1}^{\ell} \overline{C}_{t,j} \rho^{t-j},$$

where $C_{t,j}$ and $\overline{C}_{t,j}$ depend only on w_j's and α_k's.

Since t runs through only a finite number of values of (5.16), we see that

$$\overline{\lim_{r \to \infty}} \, |a_{(\ell+1)r+t}|^{\frac{1}{(\ell+1)r+t}} = \frac{1}{\rho}, \quad 0 \le t \le \ell - \nu.$$

Together with (5.16), the subsequences cover all natural numbers and we find that (5.16) holds for the sequence $\{a_k\}_0^{\ell}$ itself as the finite set $a_0, a_1, \ldots, a_{\ell}$ does not change the $\lim \sup$.

The $\ell + 1$ numbers $a_0, a_1, \ldots, a_{\ell}$ are now chosen in such a way that

$$f^*(z) = \sum_{k=0}^{\infty} a_k z^k, \quad f^*(w_j) = 1, \quad 1 \le j \le \nu,$$

and since $\ell \ge \nu$, there is always a solution. The function $F^*(z) = \frac{f^*(z)}{B_\nu(z)}$ then satisfies the conditions that

$$\lim_{k \to \infty} |a_k|^{\frac{1}{k}} = \frac{1}{\rho}.$$

By Lemma 5, for w_j $(1 \le j \le \ell - \nu + 1)$, we have

$$\left[\overline{\Delta}_{n-\nu-1,\nu}^{\ell}(w_j, F^*) - \left(\frac{w_j}{\alpha}\right)^{\ell} \overline{\Delta}_{n-\nu,\nu}^{\ell}(w_j, F^*) \right]$$

$$= w_j^n \left(B_\nu(w_j) \right)^2 \left\{ - \left(B(w_j) \right)^{-1} \sum_{k=-\nu}^{-1} \sum_{t=-k}^{\nu} \alpha_k a_{(\ell+1)n+k} w_j^{k+t} \right.$$

$$- \sum_{k=0}^{\ell-\nu} a_{(\ell+1)n+k} w_j^k - \left(B(w_j) \right)^{-1} \sum_{k=\ell-\nu}^{\ell} \sum_{t=0}^{\ell-k} \alpha_k a_{(\ell+1)n+k} w_j^{k+t} \right\}$$

$$+ C q^n \left(\frac{|\alpha|}{\rho - \varepsilon} \right)^{\ell n} \left(\frac{|w_j|}{\rho - \varepsilon} \right)^n$$

$$= C [q K(w_j)]^{n+\nu+1}$$

by using (5.15). By Lemma 1, ω_j is distinguished for F^* $(1 \leq j \leq \ell - \nu + 1)$. $\square$

7.6. Simultaneous Hermite-Padé Interpolation

Simultaneous Hermite-Padé interpolation has received some attention lately. This section will discuss the setting of the problem and its solution.

Let $d, \nu_0, \nu_1, \nu_2, \ldots, \nu_d$ be natural numbers and let $F_1, F_2, \ldots, F_d$ be d functions, meromorphic in the disks $D_{\rho_j} = \{z \in \mathbb{C} : |z| < \rho\}$, $\rho > 1$, given by

$$F_i(z) := \frac{f_i(z)}{B_i(z)}, \quad f_i(z) := \sum_{k=0}^{\infty} a_{i,k} z^k, \quad \limsup_{k \to \infty} |a_{i,k}| = \frac{1}{\rho} \qquad (6.1)$$

where

$$B_i(z) := \prod_{j=1}^{\mu_i} (z - z_{i,j})^{\lambda_{i,j}} = \sum_{k=0}^{\nu_i} \alpha_{i,k} z^k, \quad \alpha_{i,\nu_i} = 1, \quad \sum_{j=1}^{\mu_i} \lambda_{i,j} = \nu_i. \qquad (6.2)$$

Here it is assumed that the poles given all lie in D_ρ and the poles of F_i are disjoint from those of F_k, $k \neq i$. Let ℓ be an integer and put $n = \sigma + 1$, where $\sigma = \nu_0 + \nu_1 + \nu_2 + \ldots \nu_d$. It is important for the sequel to remember that the ν_i with $i = 1, \ldots, d$ are fixed (thus also $\sigma - \nu_0 = n - \nu_0 - 1$ is fixed) and that ν_0 will go to infinity (or, equivalently, n or σ). Let now $\alpha = \mathbb{C} \backslash \{0\}$ satisfy $|\alpha| < \rho$, such that the zeros of $z^n - \alpha^n$ are different from those of the $B_i(z)$ for all i. Then the Lagrange interpolant to the Taylor sections $\sum_{k=0}^{n\ell-1} a_{i,j} z^k$ on the zeros of $z^n - \alpha^n$ will be denoted by

$$\tilde{f}_{i,\ell}(z) = \sum_{j=0}^{n-1} A_{i,j}^{n-1} A_{i,\ell}^{\ell} z^j.$$

This is the first stage of the problem; explicit formulae for the $A_{i,j}^{\ell}$ are easily derived (cf. [35]). When $\ell = \infty$, $\tilde{f}_{i,0}(z)$ is the full Taylor series of $\tilde{f}_i(z)$. For the sake of completeness the explicit forms for $1 \leq \ell \leq \infty$ are given below:

$$A_{i,s}^{\ell} = \sum_{r=0}^{\ell-1} a_{i,r(\sigma+1)+s} \alpha^{r(\sigma+1)} \quad (s \geq 0), \quad A_{i,s}^{\ell} = \alpha^{\sigma+1} A_{i,\sigma+1+s} \quad (s < 0),$$

$$(6.3)$$

The value of ℓ governs how many packets of n successive coefficients from the Taylor series are used in the Lagrange interpolation; for $\ell = \infty$ the full Taylor series is used. The $A_{i,s}^{\ell}$ with $s < 0$ are needed in the formulae for the second, *simultaneous* stage of the problem which can be stated as:

Find d rational functions $U_i^{\ell}(z)/B^{\ell}(z)$ with the common denominator and
1. *with $U_i^{\ell} = \sum_{s=0}^{\sigma-\nu_i} p_{i,s}^{\ell} z^s$ $(1 \leq i \leq d)$, $B^{\ell}(z) = \sum_{k=0}^{n-\nu_0-1} \gamma_k^{\ell} z^k$,*

2. that interpolate the d rationals $\tilde{f}_{i,\ell}/B_i(z)$ on the zeros of $z^n - \alpha^n$,
3. and B^ℓ is monic: $\gamma^\ell_{n-\nu_0-1} = 1$.

For details the reader is referred to [37]. Define $K(z)$ by

$$K(z) := \left(\frac{|\alpha|}{\rho}\right)^\ell \frac{|z|}{\rho}, \quad |z| \geq \rho; \quad K(z) := \left(\frac{|\alpha|}{\rho}\right)^\ell, \quad |z| < \rho. \qquad (6.4)$$

The two-stage problem has a unique solution for n sufficiently large, a solution that moreover satisfies

$$\lim_{n\to\infty} \gamma^\ell_k = \tilde{\alpha}_k \quad \text{with} \quad \sum_{k=0}^{n-\nu_0-1} \tilde{\alpha}_k z^k = \prod_{i=1}^{d} B_i(z), \quad 1 \leq \ell \leq \infty, \qquad (6.5)$$

$$\limsup_{n\to\infty} \left(\max_{z\in\mathcal{H}} \left|\frac{U^\infty_i(z)}{B^\infty(z)} - \frac{U^\ell_i(z)}{B^\infty(z)}\right|\right)^{1/n} \leq K(\tau), \qquad (6.6)$$

for each compact subset $\mathcal{H}$ of $|z| < \tau$ $(\tau > 0)$, that omits the singularities of the functions $F_i(z)$, $1 \leq i \leq d$. For details the reader is referred to [35].

One of the main results is

THEOREM 5. Let $r, d, \nu_i, f_i, \rho_i, B_i, z_{i,j}$ and ℓ be given as above.
(i)   For n sufficiently large, the interpolating problem stated above has a unique solution that satisfies

$$\lim_{n\to\infty} \gamma^\ell_k = \zeta_k, \quad \text{with} \quad \sum_{k=0}^{n-\nu_0-1} \zeta_k z^k = \prod_{i=1}^{d} B_i(z); \ 1 \leq \ell \leq \infty. \qquad (6.7)$$

(ii)   Let $\mathcal{H}$ be a compact subset of $|z| < \tau$, $\tau > 0$, that omit the singularities of the functions F_i $(1 \leq i \leq d)$. Then

$$\limsup_{n\to\infty} \left(\max_{z\in\mathcal{H}} \left|\frac{U^\infty_i(z)}{B^\infty(z)} - \frac{U^\ell_i(z)}{B^\ell(z)}\right|\right)^{1/n} \leq \begin{cases} R^{(\ell-1)r+1}(\tau/\rho_i)^r & (\tau \geq \rho_i) \\ R^{(\ell-1)r+1} & (\tau < \rho_i), \end{cases} \qquad (6.8)$$

with $R = \max_{1\leq i\leq d} \frac{1}{\rho_i}$.
(iii)   specifically, we have for $|z| < \rho_i R^{-(\ell-1+1/r)}$,

$$\lim_{n\to\infty} \left(\frac{U^\infty_i(z)}{B^\infty(z)} - \frac{U^\ell_i(z)}{B^\ell(z)}\right) = 0 \qquad (6.9)$$

uniformly and geometrically in compact subsets omitting the singularities.

As it is more convenient to study the difference of polynomials than that of rational functions, we can multiply by the denominators - not changing the upper bound because of (6.5).

Define

$$\Delta^{\ell}_{i,\nu_0}(z) := B_i(z) B^{\infty}(z) B^{\ell}(z) \left(\frac{U_i^{\infty}(z)}{B^{\infty}(z)} - \frac{U_i^{\ell}(z)}{B^{\ell}(z)} \right) \qquad (6.10)$$

for z not one of the singularities of the function F_i and the let

$$S^{\ell}(z, F_i) := \limsup_{n \to \infty} |\Delta^{\ell}_{i,\nu_0}(z)|^{1/n}, \qquad (6.11)$$

Then (6.5) takes the form

$$S^{\ell}(z, F_i) \le K(z) \qquad \text{(for } z \text{ with } B_i(z) \neq 0). \qquad (6.12)$$

The question now arises, whether the upper bound $K(z)$ is attained for any z. The following results from [37].

THEOREM 6. *For each $i \in \{1, \ldots, d\}$ there are at most $\ell - 1$ distinguished points for F_i in $|z| < \rho$.*

THEOREM 7. *For any set of $d(\ell - 1)$ points ω_j in $0 < |z| < \rho$, and any subdivision into d sets of $\ell - 1$ points-say $\omega_{i,j}$ ($1 \le j \le \ell - 1, 1 \le i \le d$)- there exists a d-tuple of meromorphic function satisfying (6.1) and (6.2) and such that for each $i \in \{1, \ldots, d\}$ the points $\omega_{i,j}$, ($1 \le j \le \ell - 1$) are distinguished for F_i.*

THEOREM 8. *For each $i \in \{1, \ldots, d\}$ there are at most $\sigma - \nu_0 + \ell$ distinguished points for F_i in $|z| > \rho$.*

THEOREM 9. *For any set of $d(\ell + 1 - (\sigma - \nu_0))$ points ω_j in $|z| > \rho$, and any subdivision into d sets of $\ell + 1 - (\sigma - \nu_0)$ points $-$ say $\omega_{i,j}$ ($1 \le j \le \ell + 1 - (\sigma - \nu_0)$, $1 \le i \le d$) $-$ there exists d-tuple of meromorphic functions satisfying (6.1) and (6.2) and such that for each $i \in \{1, \ldots, d\}$ the points $\omega_{i,j}$ ($1 \le j \le \ell + 1 - (\sigma - \nu_0)$) are distinguished for F_i.*

Remark. The "gap" between Theorems 7 and 8 is obvious; whether this is a matter of the method of proof in [37] or a question of intrinsic behavior of rational interpolation, is a matter for further research.

7.7. Historical Remarks

The method of proof of Theorems 1 to 4 is essentially due to Stojanova [102]. Note that there is a gap between Theorems 3 and 4.

Xin Li [116] gave an interesting extension of Walsh's overconvergence theorem for rational interpolation. Assume that the set $\{\alpha_0, \alpha_1, \ldots\}$ has no limit point on $|z| = 1$, its closure does not separate the complex plane, and for the rational function

$$B_n(z) = \prod_{k=0}^{n} \frac{z - \alpha_k}{1 - \overline{\alpha}_k z}$$

we have

$$\psi(z) := \lim_{n \to \infty} |B_n(z)|^{1/n} \neq \text{const.}$$

locally uniformly in $|z| \geq 1$ except at the closure of the sequence $1/\overline{\alpha}_0, 1/\overline{\alpha}_1, \ldots$. Let $z_{k,n}$, $k = 0, 1, \ldots, n$, be the roots of $B_n(z) = 1$ (these are on the unit circle), and denote

$$R_n(f; z) := \sum_{k=0}^{n} \frac{f(z_{k,n})(B_n(z) - 1)}{B_n'(z_{k,n})(z - z_{k,n})}$$

the rational interpolant to the function $f(z)$ analytic in $\psi(z) < \rho$ $(\rho > 1)$. Further let $r_n(f; z)$ be the least square $(n, n)^{\text{th}}$ degree rational approximation to f on the unit circle. Then

$$\lim_{n \to \infty} [R_n(f; z) - r_n(f; z)] = 0 \qquad \text{for} \quad \psi(z) < \rho^2.$$

CHAPTER 8

EQUICONVERGENCE FOR FUNCTIONS
ANALYTIC IN AN ELLIPSE

8.1. Introduction

In the earlier chapters we have been dealing with equiconvergence problems for functions analytic in the interior of some circle about the origin and its expansion in a power series. It seems natural to consider functions which are analytic in other domains. The simplest extension is to consider equiconvergens results for functions analytic in an ellipse with foci ± 1 and their expansion in terms of Chebyshev series. In this case, similarly to the power series case, there is a largest ellipse in which the Chebyshev expansion converges to the function, and the convergence is uniform on each compact set inside this ellipse. Let E_ρ, $\rho \geq 1$, be the ellipse in the complex z plane which is the image of the circle $|w| = \rho$ in the complex w plane by the mapping $z = \frac{1}{2}(w + w^{-1})$. The ellipse E_ρ has foci ± 1 and the half-axes $a = \frac{1}{2}(\rho + \frac{1}{\rho})$, $b = \frac{1}{2}(\rho - \frac{1}{\rho})$. Notice that an ellipse with foci ± 1 and half axes (in the direction of the x, y-axes) whose sum is $\rho > 1$ is necessarily E_ρ. The set of functions analytic in the interior of E_ρ but not in $E_{\rho'}$, $\rho' > \rho$, will be denoted by $A(E_\rho)$. Let $T_k(z)$ denote the Chebyshev polynomial (of the first kind) of degree k. Each $z \in E_\rho$, $\rho > 1$, has a unique representation $z = \frac{1}{2}(w + \frac{1}{w})$ where $|w| = \rho > 1$. On using the representation

$$T_k(z) = \frac{1}{2}\left[(z + \sqrt{z^2 - 1})^k + (z - \sqrt{z^2 - 1})^k\right]$$

we have

$$T_k(z) = \frac{1}{2}\left(w^k + \frac{1}{w^k}\right), \quad z = \frac{1}{2}\left(w + \frac{1}{w}\right). \tag{1.0}$$

A function $f(z) \in A(E_\rho)$ has a unique Chebyshev expansion inside the ellipse E_ρ of the form

$$f(z) = \frac{1}{2}A_0 + \sum_{k=1}^{\infty} A_k T_k(z). \tag{1.1}$$

Set $S_n(f; z) := \frac{1}{2}A_0 + \sum_{k=0}^{n} A_k T_k(z)$. First we need a result on expansion of functions from the class $A(E_\rho)$.

LEMMA 1. *If $f \in A(E_\rho)$ then*

$$f(z) = \frac{1}{2} A_0 + \sum_{k=1}^{\infty} A_k T_k(z) \tag{1.1}$$

where

$$A_k = \frac{2}{\pi} \int_{-1}^{1} \frac{f(x) T_k(x)}{\sqrt{1 - x^2}} \, dx, \quad k = 0, 1, 2, \ldots \tag{1.2}$$

are the coefficients of the Chebyshev expansion. Here the convergence of the series (1.1) *is uniform in every closed subdomain of E_ρ.*

PROOF. With the substitution $x = \cos t$ and $z = e^{it}$ we obtain

$$A_k = \frac{1}{\pi} \int_{-\pi}^{\pi} f(\cos t) \cos kt \, dt \tag{1.3}$$

$$= \frac{1}{\pi i} \int_{|z|=1} f\left(\frac{z + z^{-1}}{2}\right) \frac{z^k + z^{-k}}{2} \frac{dz}{z}, \quad k = 0, 1, \ldots .$$

Taking an R, $1 < R < \rho$, the function $g(z) := f\left(\frac{z+z^{-1}}{2}\right)$ will be analytic in the closed annulus $G = \{z : \frac{1}{R} \le |z| \le R\}$. When z is on the boundary of G then $w = \frac{1}{2}(z + z^{-1})$ is on the boundary of the ellipse E_R. Thus (1.3) can be written as

$$A_k = \frac{1}{2\pi i} \int_{|z|=1/R} g(z) z^{k-1} dz + \frac{1}{2\pi i} \int_{|z|=R} g(z) z^{-k-1} dz, \quad k = 0, 1, \ldots .$$

Let $M = \max_{z \in G} |g(z)|$, then we get

$$|A_k| \le 2M R^{-k}, \quad k = 0, 1, \ldots . \tag{1.4}$$

Now let r be such that $1 < r < R < \rho$, and let $z \in E_r$. Then with $z = \frac{1}{2}(w + w^{-1})$ by (1.0)

$$|T_k(z)| \le \frac{1}{2}(|w|^k + |w|^{-k}) \le \frac{1}{2}(r^k + r^{-k}) < r^k, \quad k = 0, 1, \ldots . \tag{1.5}$$

(1.4) and (1.5) show that the series (1.1) has a convergent numerical majorant $\sum_{k=0}^{\infty} (\frac{r}{R})^k$, i.e. the Chebyshev series converges in E_r. Since r can be arbitrarily close to ρ, this proves the lemma. □

It is also clear from the proof of Lemma 1 that the order of best polynomial approximation of an $f \in A(E_\rho)$ in any closed subdomain of E_ρ is geometric. We now turn our attention to Lagrange interpolation of functions from the class $A(E_\rho)$. The natural choice for the system of nodes is the roots of the Chebyshev polynomial $T_n(x)$, and it will turn out that such an interpolation also has good convergence properties. First we need a special representation of $L_{n-1}(f; z)$, the Lagrange interpolant to $f(z)$ on zeros of $T_n(x)$.

LEMMA 2. *For any $f \in A(E_\rho)$ with (1.1) we have*

$$L_{n-1}(f;z) = \sum_{k=0}^{n-1}{}' \left\{ A_k + \sum_{j=1}^{\infty} (-1)^j (A_{2nj-k} + A_{2nj+k}) \right\} T_k(z) \qquad (1.6)$$

where the dash indicates that the term $k = 0$ should be halved.

PROOF. From (1.1) we get

$$L_{n-1}(f;z) = \sum_{k=0}^{\infty}{}' A_k L_{n-1}(T_k;z) \qquad (1.7)$$

$$= \sum_{j=0}^{\infty}{}' \sum_{k=0}^{2n-1}{}' A_{2nj+k} L_{n-1}(T_{2nj+k};z)$$

where now the dashes indicate that the term corresponding to $j = k = 0$ should be halved. Using the identity

$$T_\lambda(z) = -T_\mu(z) + 2T_{\frac{\lambda+\mu}{2}}(z) T_{\frac{\lambda-\mu}{2}}(z), \qquad \lambda > \mu, \ \lambda \equiv \mu \, (\mathrm{mod} \, 2) \qquad (1.8)$$

successively, we get

$$L_{n-1}(T_{2jn+k};z) = L_{n-1}\big(-T_{2n(j-1)+k} + 2T_{n(2j-1)+k} T_n; z \big) \qquad (1.9)$$

$$= -L_{n-1}\big(T_{2n(j-1)+k}; z \big) = \cdots = (-1)^j L_{n-1}(T_k;z)$$

$$= \begin{cases} (-1)^j T_k(z) & \text{if } 0 \leq k < n \\ 0 & \text{if } k = n \\ (-1)^{j+1} T_{2n-k}(z) & \text{if } n < k < 2n. \end{cases}$$

Thus we obtain from (1.7)

$$L_{n-1}(f;z) = \sum_{j=0}^{\infty}{}' \left(\sum_{k=0}^{n-1}{}' A_{2nj+k}(-1)^j T_k(z) + \sum_{k=n+1}^{2n-1} A_{2nj+k}(-1)^{j+1} T_{2n-k} \right)$$

$$= \frac{1}{2} A_0 + \sum_{j=1}^{\infty} A_{2nj}(-1)^j + \sum_{k=1}^{n-1} \left\{ A_k + \sum_{j=1}^{\infty} (-1)^j (A_{2nj+k} + A_{2nj-k}) \right\} T_k(z)$$

which proves the lemma. $\qquad \Box$

A similar representation for the Hermite interpolation polynomial $h_{2,2n-1}$ of order 2 on the roots of $T_n(x)$ (that is, function values and first derivatives prescribed) can be established:

LEMMA 3. *For any $f \in A(E_\rho)$ with (1.1), we have*

$$h_{2,2n-1}(f;z) = \sum_{j=0}^{\infty}{}'(-1)^j A_{2nj} +$$

$$+ \sum_{k=1}^{2n-1} \Big\{ \sum_{j=0}^{\infty}(-1)^j[(j+1)A_{2nj+k} + jA_{2n(j+1)-k}]\Big\}T_k(z), \tag{1.10}$$

where the dash means halving the term corresponding to $j = 0$.

PROOF. Again from (1.1) we get

$$h_{2,2n-1}(f;z) = \sum_{k=0}^{\infty}{}'A_k h_{2,2n-1}(T_k(z);z) \tag{1.11}$$

$$= \sum_{j=0}^{\infty}{}' \sum_{k=0}^{2n-1}{}'A_{2nj+k} h_{2,2n-1}(T_{2nj+k};z)$$

where the dashes mean that the term with $j = k = 0$ has to be halved. Here

$$h_{2,2n-1}(T_{2nj+k};z) = \begin{cases} (-1)^j & \text{if } k = 0, \\ (-1)^j[(j+1)T_k(z) + jT_{2n-k}(z)] & \text{if } 1 \le k \le 2n-1. \end{cases} \tag{1.12}$$

We prove this relation by induction on j. For $j = 0$,

$$h_{2,2n-1}(T_k;z) = \begin{cases} 1 & \text{if } k = 0, \\ T_k(z) & \text{if } 1 \le k \le 2n-1 \end{cases} \Bigg\} = T_k(z) \tag{1.13}$$

is obviously true. Assume (1.12) is true for j, then from (1.8) and the identity $T_{2n} = 2T_n^2 - 1$ we obtain

$$h_{2,2n-1}\big(T_{2n(j+1)+k};z\big) = h_{2,2n-1}\big(-T_{2n(j-1)+k} + 2T_{2nj+k}T_{2n},z\big)$$

$$= h_{2,2n-1}\big(-T_{2n(j-1)+k} - 2T_{2nj+k};z\big)$$

$$= \begin{cases} -(-1)^{j-1} - 2(-1)^j \\ (-1)^j[jT_k + (j-1)T_{2n-k}] - 2(-1)^j[(j+1)T_k + jT_{2n-k}] \end{cases}$$

$$= \begin{cases} (-1)^{j+1}, & \text{if } k = 0, \\ (-1)^{j+1}[(j+2)T_k + (j+1)T_{2n-k}, & \text{if } 1 \le k \le 2n-1, \end{cases}$$

i.e., (1.12) holds for $j+1$. Applying this to (1.11) we get

$$h_{2,2n-1}(f;z) = \sum_{j=0}^{\infty}{}'(-1)^j A_{2nj} + \sum_{k=1}^{2n-1}\sum_{j=0}^{\infty}(-1)^j[(j+1)T_k(z) + jT_{2n-k}(z)]A_{2nj+k}$$

whence (1.10) follows by changing k to $2n - k$ in the second term.   □

Lemmas 1 and 3 enable us to prove the convergence of Lagrange and Hermite interpolation.

LEMMA 4. *If $f(z) \in A(E_\rho)$ then both $L_{n-1}(f; z)$ and $h_{2,2n-1}(f; z)$ converge to $f(z)$ as $n \to \infty$, uniformly and geometrically in every closed subdomain of E_ρ.*

PROOF. (1.1) and (1.6) imply

$$f(z) - L_{n-1}(f; z) = \sum_{k=n}^{\infty} A_k T_k(z) - \sum_{k=0}^{n-1}{}' \sum_{j=1}^{\infty} (-1)^j (A_{2nj-k} + A_{2nj+k}) T_k(z)$$

and

$$f(z) - h_{2,2n-1}(f; z) = \sum_{k=2n}^{\infty} A_k T_k(z) - \sum_{j=1}^{\infty} (-1)^j A_{2nj}$$

$$- \sum_{k=1}^{2n-1} \sum_{j=1}^{\infty} (-1)^j [(j+1) A_{2nj+k} + j A_{2n(j+1)-k}] T_k(z).$$

Now choose r and R such that $1 < r < R < \rho$, and let $z \in E_r$. Then by (1.4) and (1.5)

$$|f(z) - L_{n-1}(f; z)| \le 2M \sum_{k=n}^{\infty} \left(\frac{r}{R}\right)^k + 4M \sum_{k=0}^{n-1} \sum_{j=1}^{\infty} R^{-2nj}(rR)^k = O\left(\left(\frac{r}{R}\right)^n\right)$$

and similarly,

$$|f(z) - h_{2,2n-1}(f; z)| \le 2M \sum_{k=2n}^{\infty} \left(\frac{r}{R}\right)^k + 4M \sum_{k=1}^{2n-1} \sum_{j=1}^{\infty} \left[(j+1) R^{-2nj} \left(\frac{r}{R}\right)^k + \right.$$

$$\left. + j R^{-2n(j+1)}(rR)^k \right]$$

$$= O\left(\left(\frac{r}{R}\right)^{2n}\right)$$

which shows that the convergence is geometric.   □

8.2. Equiconvergence (Lagrange Interpolation)

We are interested in exploiting the possible equiconvergence phenomena in connection with interpolating polynomials and partial sums of Chebyshev series. Denoting

$$S_n(f; z) = \sum_{k=0}^{n}{}' A_k T_k(z) \tag{2.1}$$

the nth partial sum of the Chebyshev series (1.1), we have by (1.6)

$$L_{n-1}(f;z) - S_{n-1}(f;z) = \sum_{k=0}^{n-1}{}' \sum_{j=1}^{\infty} (-1)^j (A_{2nj-k} + A_{2nj+k}) T_k(z). \qquad (2.2)$$

Unfortunately, there is no equiconvergence in the classical sense. This can be seen from the following simple example. Let $f_0(z) = \sum_{k=0}^{\infty}{}' \rho^{-k} T_k(z) \in A(E_\rho)$. Then for $z_0 = \frac{1}{2}(\rho + \rho^{-1})$ we obtain from (1.0) and (2.2)

$$L_{n-1}(f_0; z_0) - S_{n-1}(f_0; z_0) = \frac{1}{2} \sum_{k=0}^{n-1}{}' \sum_{j=1}^{\infty} (-1)^j (\rho^{k-2nj} + \rho^{-k-2nj})(\rho^k + \rho^{-k})$$

$$= \frac{1}{2} \sum_{k=0}^{n-1}{}' (\rho^k + \rho^{-k})^2 \sum_{j=1}^{\infty} (-1)^j \rho^{-2nj}$$

$$< \frac{1}{2} \rho^{2n-2} (-\rho^{2n} + \rho^{-2n}) = -\frac{1}{2\rho^2} + O(\rho^{-2n})$$

which shows that the left hand side does not go to zero as $n \to \infty$. Similar phenomenon occurs for the Hermite interpolation. In order to establish equiconvergence we have to modify our operator. We introduce two parameters: a fixed integer $\ell \geq 1$, and an increasing sequence of integers m_n depending on n such that

$$\lim_{n \to \infty} \frac{m_n}{n} =: q \geq 1, \quad \text{and} \quad \lambda := \overline{\lim_{n \to \infty}} \{m_{n+1} - m_n\} < \infty. \qquad (2.3)$$

With these parameters, our operator is defined as

$$\Delta_{\ell-1,n,m_n}(f;z) := S_{n-1}\Big(L_{m_n-1}\big(f - S_{2(\ell-1)m_n+n-1}(f)\big); z\Big) \qquad (2.4)$$

for any $f \in A(E_\rho)$. (Here we use the notation (1.6) - (2.1).) Note that in case $\ell = 1$, $m_n \geq n$ this simplifies to

$$\Delta_{0,n,m_n}(f;z) = S_{n-1}\big(L_{m_n-1}(f) - f; z\big).$$

Our first result gives the precise asymptotic value of this operator.

THEOREM 1. *For any $f \in A(E_\rho)$ and $z \in E_r$ $(r > 1)$ we have*

$$\overline{\lim_{n \to \infty}} |\Delta_{\ell-1,n,m_n}(f;z)|^{1/n} = \frac{r}{\rho^{2\ell q-1}}, \qquad (2.5)$$

except possibly for at most $2\ell\lambda - 2$ points $z_1, \ldots, z_{2\ell\lambda-2}$ outside the interval $[-1,1]$, where

$$\overline{\lim_{n \to \infty}} |\Delta_{\ell-1,n,m_n}(f;z_j)|^{1/n} < \frac{|w_j|}{\rho^{2\ell q-1}}, \quad z_j = \frac{1}{2}(w_j + w_j^{-1}), \quad j = 1, \ldots, 2\ell\lambda-2.$$
$$(2.6)$$

In other words, we have equiconvergence in the ellipse $E_{\rho^{2\ell q-1}}$. Note that (2.5) does not distinguish between the cases $r \leq \rho$ or $r > \rho$, in contrast to the case of functions analytic in a circle.

PROOF. First we prove the upper estimate in (2.5). For this purpose we represent the operator (2.4) in the form

$$\Delta_{\ell-1,n,m_n}(f; z) = \sum_{k=0}^{n-1}{}' \left\{ \sum_{j=\ell}^{\infty} (-1)^j (A_{2jm_n+k} + A_{2jm_n-k}) \right\} T_k(z). \qquad (2.7)$$

To see this, first we write

$$f(z) - S_{2(\ell-1)m_n+n-1}(f; z) = \sum_{k=2(\ell-1)m_n+n}^{\infty} A_k T_k(z),$$

and then apply the Lagrange operator L_{m_n-1} :

$$L_{m_n-1}\left(f - S_{2(\ell-1)m_n+n-1}(f); z \right) = \sum_{k=n}^{2m_n-1} A_{2(\ell-1)m_n+k} L_{m_n-1}\left(T_{2(\ell-1)m_n+k}; z \right)$$

$$+ \sum_{j=\ell}^{\infty} \sum_{k=0}^{2m_n-1} A_{2jm_n+k} L_{m_n-1}\left(T_{2jm+k}; z \right).$$

Using the relation (1.9) we obtain

$$L_{m_n-1}\left(f - S_{2(\ell-1)m_n+n-1}(f); z \right) =$$

$$= (-1)^{\ell} \sum_{k=m_n+1}^{2m_n-1} A_{2(\ell-1)m_n+k} T_{2m_n-k}(z) + (-1)^{\ell-1} \sum_{k=n}^{m_n-1} A_{2(\ell-1)m_n+k} T_k(z)$$

$$+ \sum_{j=\ell}^{\infty} (-1)^j \left\{ \sum_{k=0}^{m_n-1} A_{2jm_n+k} T_k(z) - \sum_{k=m_n+1}^{2m_n-1} A_{2jm_n+k} T_{2m_n-k}(z) \right\}$$

$$= (-1)^{\ell} \sum_{k=1}^{m_n-1} A_{2\ell m_n-k} T_k(z) + (-1)^{\ell-1} \sum_{k=n}^{m_n-1} A_{2(\ell-1)m_n+k} T_k(z)$$

$$+ \sum_{j=\ell}^{\infty} (-1)^j \left\{ \sum_{k=0}^{m_n-1} A_{2jm_n+k} T_k(z) - \sum_{k=1}^{m_n-1} A_{2(j+1)m_n-k} T_k(x) \right\},$$

whence (2.7) follows by replacing $j+1$ by j in the second sum and by applying S_{n-1} to both sides. Now the upper estimate in (2.5) follows by using (1.4) and (1.5):

$$|A_{2jm_n+k}| \leq 2MR^{-2jm_n-k}, \quad |A_{2jm_n-k}| \leq 2MR^{-2jm_n+k}, \qquad 1 < R < \rho, \qquad (2.8)$$

$$|T_k(z)| \leq r^k, \qquad z \in E_r, r > 1,$$

whence by (2.7)

$$|\Delta_{\ell-1,n,m_n}(f;z)| \leq 4M \sum_{k=0}^{n-1} \sum_{j=\ell}^{\infty} R^{k-2jm_n} r^k = O(R^{-2\ell m_n + n} r^n).$$

Thus using (2.3) we get

$$\varlimsup_{n\to\infty} \max_{z\in E_r} |\Delta_{\ell-1,n,m_n}(f;z)|^{1/n} \leq \frac{r}{R^{2\ell q-1}}.$$

Since $R < \rho$ was arbitrary, hence the upper estimate of (2.5). In order to prove the second part of the theorem, we use (1.0) to estimate (2.7) by the help of (1.4) and (2.8):

$$\Delta_{\ell-1,n,m_n}(f;z) = \frac{(-1)^{\ell}}{2} \sum_{k=1}^{n-1} A_{2\ell m_n - k} w^k + O\Big(\sum_{k=0}^{n-1} |A_{2\ell m_n - k}| \cdot |w|^{-k} +$$

$$+ \sum_{k=0}^{n-1} \sum_{j=\ell}^{\infty} |A_{2jm_n + k}| \, |w|^k + \sum_{j=\ell+1}^{\infty} \sum_{k=0}^{n-1} |A_{2jm_n - k}| \cdot |w|^k \Big)$$

$$= \frac{(-1)^{\ell}}{2} \sum_{k=1}^{n-1} A_{2\ell m_n - k} w^k + O\left(\frac{1}{R^{2\ell m_n}} \left(\max\left(\frac{|w|}{R}, \frac{R}{|w|} \right) \right)^n \right)$$

$$= \frac{(-1)^{\ell}}{2} \sum_{k=1}^{n-1} A_{2\ell m_n - k} w^k + O\left(\left(\frac{|w|}{\rho^{2\ell q-1}} - \eta \right)^n \right)$$

with some constant $\eta > 0$, provided $R < \rho$ is close enough to ρ. Hence

$$A := 2(-1)^{\ell} [\Delta_{\ell-1,n,m_n}(f;z) - w^{2\ell(m_n - m_{n+1})} \Delta_{\ell-1,n+1,m_{n+1}}(f;z)]$$

$$= \sum_{k=1}^{n-1} A_{2\ell m_n - k} w^k - \sum_{k=1}^{n} A_{2\ell m_{n+1} - k} w^{k+2\ell(m_n - m_{n+1})} + O\left(\left(\frac{|w|}{\rho^{2\ell q-1}} - \eta \right)^n \right).$$

Here we split the two sums into four:

$$A = \left(\sum_{k=1}^{n-2\ell(m_{n+1}-m_n)} + \sum_{k=n-2\ell(m_{n+1}-m_n)+1}^{n-1} \right) A_{2\ell m_n - k} w^k \tag{2.9}$$

$$- \left(\sum_{k=1}^{2\ell(m_{n+1}-m_n)} + \sum_{k=2\ell(m_{n+1}-m_n)+1}^{n} \right) A_{2\ell m_{n+1} - k} w^{k+2\ell(m_n - m_{n+1})}$$

$$+ O\left(\left(\frac{|w|}{\rho^{2\ell q-1}} - \eta \right)^n \right)$$

$$= w^{n-2\ell(m_{n+1}-m_n)+1} \sum_{k=0}^{2\ell(m_{n+1}-m_n)-2} A_{2\ell m_{n+1} - n - k - 1} w^k +$$

$$+ O\left(\left(\frac{|w|}{\rho^{2\ell q-1}} - \eta \right)^n \right).$$

Namely, changing k to $k + 2\ell(m_{n+1} - m_n)$ in the third sum, it is easily seen to cancel with the first sum. Further, the third sum contains only a bounded number of terms, and is of order $O(A_{2\ell m_n})$ which gives a contribution to the error term. Finally, changing k to $k + n - 2\ell(m_{n+1} - m_n) + 1$ in the second sum, we obtain the above estimate. Now assume that, contrary to (2.6), there exist points $z_1, \ldots, z_{2\ell\lambda - 1}$ outside the interval $[-1, 1]$ such that

$$\varlimsup_{n \to \infty} |\Delta_{\ell-1, n, m_n}(f; z_j)|^{1/n} < \frac{|w_j|}{\rho^{2\ell q - 1}}, \quad j = 1, 2, \ldots, 2\ell\lambda - 1$$

where $z_j = \frac{1}{2}(w_j + w_j^{-1})$, $j = 1, \ldots, 2\ell\lambda - 1$. This means that

$$|\Delta_{\ell-1, n, m_n}(f; z_j)| \leq \left(\frac{|w_j|}{\rho^{2\ell q - 1}} - \eta_1\right)^n, \quad j = 1, \ldots 2\ell\lambda - 1 \tag{2.10}$$

with some constant $\eta_1 > 0$. Since by (2.3), $m_{n+1} - m_n \leq \lambda$ for sufficiently large n's, we can apply the first $2\ell(m_{n+1} - m_n) - 1$ relations in (2.10) to (2.9), and obtain

$$w_j^{n - 2\ell(m_{n+1} - m_n) + 1} \sum_{k=0}^{2\ell(m_{n+1} - m_n) - 2} A_{2\ell m_{n+1} - n - k - 1} w_j^k = O\left(\left(\frac{|w_j|}{\rho^{2\ell q - 1}} - \eta_2\right)^n\right),$$

$$j = 1, \ldots, 2\ell(m_{n+1} - m_n) - 1,$$

where $\eta_2 = \min(\eta_1, \eta)$. Hence by the boundedness of $m_{n+1} - m_n$ we get

$$\sum_{k=0}^{2\ell(m_{n+1} - m_n) - 2} A_{2\ell m_{n+1} - n - k - 1} w_j^k = O\left(\left(\frac{1}{\rho^{2\ell q - 1}} - \eta_3\right)^n\right),$$

$$j = 1, \ldots, 2\ell(m_{n+1} - m_n) - 1,$$

where $\eta_3 = \min_j \frac{\eta_2}{|w_j|}$. The determinant of this system of equations is a Vandermonian. Solving these we obtain

$$|A_{2\ell m_{n+1} - n - k - 1}| = O\left(\left(\frac{1}{\rho^{2\ell q - 1}} - \eta_3\right)^n\right), \quad k = 0, \ldots, 2\ell(m_{n+1} - m_n) - 2.$$

We can see from this relations that the subscripts of the A's attain all integers $s \geq 2\ell m(1)$ (in fact, the set of consecutive integers for fixed n's overlap), whence we have by (2.3) again

$$\varlimsup_{s \to \infty} |A_s|^{1/s} \leq \left(\frac{1}{\rho^{2\ell q - 1}} - \eta_3\right)^{\lim_{s \to \infty} \frac{n}{2\ell m_{n+1} - n - k - 1}}$$

$$= \left(\frac{1}{\rho^{2\ell q - 1}} - \eta_3\right)^{\frac{1}{2\ell q - 1}} < \frac{1}{\rho}.$$

This contradicts the fact that $f(z) \in A(E_\rho)$, and Theorem 1 is proved. $\qquad \square$

8.3. Equiconvergence (Hermite Interpolation)

For Hermite interpolation we modify the operator (2.4) in the following way. Let again $m = m_n$ be an increasing sequence of integers such that (2.3) holds, and set

$$\tilde{\Delta}_{\ell-1,n,m_n}(f;z) := S_{n-1}\left(h_{2,2m_n-1}(f - S_{2\ell m_n-1}(f);z)\right) \qquad (3.1)$$

for any $f \in A(E_\rho)$. Note that in case $\ell = 1$ this simplifies to

$$\tilde{\Delta}_{0,n,m_n}(f,z) = S_{n-1}\left(h_{2,2m_n-1}(f) - f;z\right).$$

In contrast to the case of Lagrange interpolation, here the convergence-divergence behavior of the operator is different if $z \in E_\rho$ or $z \notin E_\rho$. First we consider the case when the point is inside the ellipse of analyticity, and assume that $m_n = qn$ (the case $m_n = qn+$const. could also be treated, but it would complicate the notation).

THEOREM 2. *For any $f \in A(E_\rho)$ and $z \in E_r$, $1 < r < \rho$, we have*

$$\overline{\lim_{n\to\infty}} \, \tilde{\Delta}_{\ell-1,n,qn}(f;z)|^{1/n} = \frac{1}{\rho^{2\ell q}} \qquad (3.2)$$

except at most at $2\ell q$ points in E_ρ, where the equality sign should be changed to "$<$".

PROOF. First we establish an asymptotic expression for the operator (3.1) (In this part, we use only the conditions (2.3) on the sequence m_n.) We have

$$f(z) - S_{2\ell m_n-1}(f;z) = \sum_{k=2\ell m_n}^{\infty} A_k T_k(z),$$

whence, using (1.12), we get

$$h_{2,2m_n-1}\left(f - S_{2\ell m_n-1}(f);z\right) = \sum_{k=2\ell m_n}^{\infty} A_k h_{2m_n-1}(T_k,z)$$

$$= \sum_{j=\ell}^{\infty} \sum_{k=0}^{2m_n-1} A_{2jm_n+k} h_{2,2m_n-1}(T_{2jm_n+k};z)$$

$$= \sum_{j=\ell}^{\infty} (-1)^j \left\{ A_{2jm_n} + \sum_{k=1}^{2m_n-1} A_{2jm_n+k}[(j+1)T_k(z) + jT_{2m_n-k}(z)] \right\}$$

$$= \sum_{j=\ell}^{\infty} (-1)^j \left\{ A_{2jm_n} + \sum_{k=1}^{2m_n-1} [(j+1)A_{2jm_n+k} + jA_{2(j+1)m_n-k}]T_k(z) \right\},$$

i.e., by (3.1)

$$\tilde{\Delta}_{\ell-1,n,m_n}(f;z) = \sum_{j=\ell}^{\infty} (-1)^j \Big\{ A_{2jm_n} + \sum_{k=1}^{n-1}[(j+1)A_{2jm_n+k}+$$

$$+ jA_{2(j+1)m_n-k}]T_k(z) \Big\}. \tag{3.3}$$

This yields

$$\tilde{\Delta}_{\ell-1,n,m_n}(f;z) = O\Big(\frac{1}{(\rho-\varepsilon)^{2\ell m_n}} \Big),$$

i.e.

$$\overline{\lim_{n\to\infty}} \max_{z\in E_r} |\tilde{\Delta}_{\ell-1,n,m_n}(f;z)|^{1/n} \le \frac{1}{(\rho-\varepsilon)^{2\ell q}}$$

follows by (2.3). Since $\varepsilon > 0$ was arbitrary, this yields the upper estimate in (3.2). In order to prove the second part of the theorem, from now on, we assume $m_n = qn$. Using (1.4) we obtain

$$\tilde{\Delta}_{\ell-1,n,qn}(f;z) = (-1)^{\ell}\Big\{ A_{2\ell qn} + (\ell+1)\sum_{k=1}^{n-1} A_{2\ell qn+k}T_k(z) \Big\} + R_n$$

$$\text{for } z \in E_r \text{ and } 1 < r < \rho, \tag{3.4}$$

where

$$R_n = O\Big(\frac{r^n}{(\rho-\varepsilon)^{2(\ell+1)qn-n}} \Big) = O\Big(\frac{1}{(\rho+\varepsilon)^{2[(\ell+1)q-1]n}} \Big) = O\Big(\frac{1}{(\rho+\varepsilon)^{2\ell qn}} \Big),$$

provided $\varepsilon > 0$ is small enough. Hence

$$D(z) := \frac{(-1)^{\ell}}{\ell+1}[\tilde{\Delta}_{\ell-1,n-1,q(n-1)}(f;z) + \tilde{\Delta}_{\ell-1,n+1,q(n+1)}(f;z)-$$

$$- 2T_{2\ell q}(z)\tilde{\Delta}_{\ell-1,n,qn}(f;z)] \tag{3.5}$$

$$= \frac{A_{2\ell q(n-1)} + A_{2\ell q(n+1)} - 2A_{2\ell qn}T_{2\ell q}(z)}{\ell+1} + \sum_{k=1}^{n-2} A_{2\ell q(n-1)+k}T_k(z)$$

$$+ \sum_{k=1}^{n} A_{2\ell q(n+1)+k}T_k(z) - \sum_{k=1}^{n-1} A_{2\ell qn+k}[T_{k+2\ell q}(z) + T_{|k-2\ell q|}(z)] + O(R_n).$$

Here the last sum, after changing the running indices k, can be written in the form

$$\sum_{k=2\ell q+1}^{n+2\ell q-1} A_{2\ell q(n-1)+k}T_k(z) + \sum_{k=1}^{n-2\ell q-1} A_{2\ell q(n+1)+k}T_k(z) + \sum_{k=0}^{2\ell q-1} A_{2\ell q(n+1)-k}T_k(z).$$

Thus we obtain from the previous expression, after taking into account the cancellations,

$$D(z) = \frac{A_{2\ell q(n-1)} - \ell A_{2\ell q(n+1)} + (\ell-1)A_{2\ell qn}T_{2\ell q}(z)}{\ell+1}$$

$$+ \sum_{k=1}^{2\ell q-1} [A_{2\ell q(n-1)+k} - A_{2\ell q(n+1)-k}]T_k(z) - \sum_{k=n-1}^{n+2\ell q-1} A_{2\ell q(n-1)+k}T_k(z)$$

$$+ \sum_{k=n-2\ell q}^{n} A_{2\ell q(n+1)+k}T_k(z) + O(R_n).$$

Here, by (2.8), the last two sums are $O((\rho - \varepsilon)^{-(2\ell q+1)n})$, which is smaller than R_n if $\varepsilon > 0$ is small enough. Thus we obtain

$$D(z) = \frac{A_{2\ell q(n-1)} - \ell A_{2\ell q(n+1)} + (\ell - 1)A_{2\ell qn}T_{2\ell q}(z)}{\ell + 1}$$

$$+ \sum_{k=1}^{2\ell q-1} [A_{2\ell q(n-1)+k} - A_{2\ell q(n+1)-k}]T_k(z) + O(R_n).$$

Now assume that, contrary to the statement of the theorem, for some points $z_j \in E_\rho$, $j = 0, 1, \ldots, 2\ell q$ we have

$$|\tilde{\Delta}_{\ell-1,n,qn}(f; z_j)| \le \frac{1}{(\rho + \varepsilon)^{2\ell qn}} = O(R_n), \qquad j = 1, \ldots, 2\ell q + 1$$

with some $\varepsilon > 0$. Then by (3.4), $|D(z_j)| = O(R_n)$, $j = 0, 1, \ldots, 2\ell q$, whence by (3.5)

$$\frac{A_{2\ell q(n-1)} - \ell A_{2\ell q(n+1)+k} + (\ell - 1)A_{2\ell qn}T_{2\ell q}(z_j)}{\ell + 1} +$$

$$+ \sum_{k=1}^{2\ell q-1} [A_{2\ell q(n-1)+k} - A_{2\ell q(n+1)-k}]T_k(z_j) = O(R_n), \quad j = 0, 1, \ldots, 2\ell q.$$

Consider this as a system of $2\ell q + 1$ linear equations for the unknowns $\frac{A_{2\ell q(n-1)} - \ell A_{2\ell q(n+1)}}{\ell+1}$, $\quad A_{2\ell q(n-1)+k} - A_{2\ell q(n+1)-k}$ (for $k = 1, \ldots, 2\ell q - 1$) and $\frac{(\ell-1)A_{2\ell qn}}{\ell+1}$. The determinant of this system of linear equations consists of the elements $T_k(z_j)$, $k, j = 0, \ldots, 2\ell q$, and because of the linear independence of the Chebyshev polynomials, is different from zero. Evidently, the solution satisfies

$$|A_{2\ell q(n-1)+k} - A_{2\ell q(n+1)-k}| = O(R_n), \qquad k = 1, \ldots, 2\ell q - 1 \tag{3.6}$$

and

$$|A_{2\ell qn}| = O(R_n). \tag{3.7}$$

Now we prove that

$$A_{2\ell qn-k} = \sum_{j=0}^{\infty} \left(A_{2\ell q(n+2[\frac{j}{2}])+(-1)^{j+1}k} - A_{2\ell q(n+2[\frac{j+1}{2}])+(-1)^j k} \right)$$

$$\text{for} \quad k = 1, \ldots, 2\ell q - 1,$$

where $[\cdot]$ means integer part. Evidently, this is a telescoping infinite series. In order to show its convergence, we distinguish two cases. When j is even, the absolute value of the term on the right hand side becomes

$$|A_{2\ell q(n+j)-k} - A_{2\ell q(n+j)+k}| = |A_{2\ell q(n+j-1)+(2\ell q-k)} - A_{2\ell q(n+j+1)-(2\ell q-k)}|$$
$$= O(R_{n+j})$$

upon applying (3.6) with $n+j$ and $2\ell q - k$ instead of n and k, respectively. When j is odd we get similarly

$$|A_{2\ell q(n+j-1)+k} - A_{2\ell q(n+j+1)-k}| = O(R_{n+j})$$

upon applying again (3.6) now with $n+j$ instead of n. Thus we obtain

$$|A_{2\ell qn-k}| = O\left(\sum_{j=0}^{\infty} \frac{1}{(\rho+\varepsilon)^{2\ell q(n+j)}}\right) = O(R_n), \qquad k = 1, \ldots, 2\ell q - 1.$$

This coupled with (3.7) implies

$$\limsup_{s\to\infty} A_s^{1/s} \le \frac{1}{\rho+\varepsilon},$$

which contradicts $f \in A(E_\rho)$. $\qquad\qquad\square$

The next theorem is a similar result for points outside the ellipse of analyticity, but the method of proof is different (and simpler).

THEOREM 3. *For any $f \in A(E_\rho)$ and $z \notin E_\rho$ we have*

$$\overline{\lim_{n\to\infty}} |\tilde{\Delta}_{\ell-1,n,m_n}(f;z)|^{1/n} = \frac{|w|}{\rho^{2\ell q+1}}, \quad z = \frac{1}{2}(w+w^{-1}), \qquad (3.8)$$

except possibly for at most $2\ell\lambda$ points $z_1, \ldots, z_{2\ell\lambda} \notin E_\rho$ where

$$\overline{\lim_{n\to\infty}} |\tilde{\Delta}_{\ell-1,n,m_n}(f;z_j)|^{1/n} < \frac{|w_j|}{\rho^{2\ell q+1}}, \quad j = 1, \ldots, 2\ell\lambda. \qquad (3.9)$$

In other words, we have equiconvergence in the ellipse $E_{\rho^{2\ell q}}$.

PROOF. (2.8) and (3.3) imply for $|w| > \rho$

$$\tilde{\Delta}_{\ell-1,n,m_n}(f;z) = \frac{(-1)^\ell(\ell+1)}{2} \sum_{k=1}^{n-1} A_{2\ell m_n+k}w^k +$$
$$O\left(\frac{1}{(\rho-\varepsilon)^{2\ell m_n}} + \frac{|w|^n}{(\rho-\varepsilon)^{2(\ell+1)m_n-n}}\right).$$

As in (2.9), consider

$$B := \frac{2(-1)^\ell}{\ell+1} \{\tilde{\Delta}_{\ell-1,n,m_n}(f;z) - w^{2\ell(m_{n+1}-m_n)} \Delta_{\ell-1,n+1,m_{n+1}}(f;z)\} \tag{3.10}$$

$$= \sum_{k=1}^{n-1} A_{2\ell m_n+k} w^k - \sum_{k=1}^{n} A_{2\ell m_{n+1}+k} w^{k+2\ell(m_{n+1}-m_n)}$$

$$+ O\Big(\frac{1}{(\rho-\varepsilon)^{2\ell m_n}} + \frac{|w|^n}{(\rho-\varepsilon)^{2(\ell+1)m_n-n}}\Big).$$

Here, splitting the sums again, we obtain

$$B := \left(\sum_{k=1}^{2\ell(m_{n+1}-m_n)} + \sum_{k=2\ell(m_{n+1}-m_n)+1}^{n-1}\right) A_{2\ell m_n+k} w^k$$

$$- \left(\sum_{k=1}^{n-2\ell(m_{n+1}-m_n)-1} + \sum_{k=n-2\ell(m_{n+1}-m_n)}^{n}\right) A_{2\ell m_{n+1}+k} w^{k+2\ell(m_{n+1}-m_n)}$$

$$+ O\Big(\frac{1}{(\rho-\varepsilon)^{2\ell m_n}} + \frac{|w|^n}{(\rho-\varepsilon)^{2(\ell+1)m_n-n}}\Big)$$

$$= -w^{n-1} \sum_{k=1}^{2\ell(m_{n+1}-m_n)+1} A_{2\ell m_n+n+k} w^k$$

$$+ O\Big(\frac{1}{(\rho-\varepsilon)^{2\ell m_n}} + \frac{|w|^n}{(\rho-\varepsilon)^{2(\ell+1)m_n-n}}\Big).$$

This is obtained by changing k to $k+2\ell(m_{n+1}-m_n)$ in the second sum which then will cancel with the third sum. The first sum has a bounded number of terms, thus it can be majorized by $O(A_{2\ell m_n})$ which can be merged with the error term. Finally, in the fourth sum k is replaced by $k+n-2\ell(m_{n+1}-m_n)-1$, thus arriving at the fifth sum. Assume now that, contrary to (3.9), there exist points $z_1, \ldots, z_{2\ell\lambda+1} \notin E_\rho$ such that

$$\overline{\lim_{n\to\infty}} |\tilde{\Delta}_{\ell-1,n,m_n}(f;z_j)|^{1/n} < \frac{|w_j|}{\rho^{2\ell q+1}}, \quad j = 1, \ldots, 2\ell\lambda+1,$$

i.e.

$$|\tilde{\Delta}_{\ell-1,n,m_n}|f;z_j)| \le \Big(\frac{|w_j|}{\rho^{2\ell q+1}} - \eta\Big)^n, \quad j = 1, \ldots, 2\ell\lambda+1 \tag{3.11}$$

with some $\eta > 0$. Using the first $2\ell(m_{n+1}-m_n)+1$ of these inequalities, we get from (3.11)

$$\sum_{k=1}^{2\ell(m_{n+1}-m_n)+1} A_{2\ell m_n+n+k} w_j^k = O\Big(\big(\frac{1}{\rho^{2\ell q+1}} - \frac{\eta}{|w_j|}\big)^n \Big)$$

$$\text{for} \quad j = 1, \ldots, 2\ell(m_{n+1} - m_n) + 1,$$

provided $\eta > 0$ is sufficiently small. The determinant of this system of equations for the $A_{2\ell m_n + k}$'s is again a Vandermonian. Hence we obtain

$$A_{2\ell m_n + n + k} = O\left(\left(\frac{1}{\rho^{2\ell q + 1}} - \eta_1\right)^n\right), \quad k = 0, 1, \ldots, 2\ell(m_{n+1} - m_n), \quad (3.12)$$

where $\eta_1 > 0$ is sufficiently small. This implies

$$\limsup_{s \to \infty} A_s \le \left(\frac{1}{\rho^{2\ell q}} - \eta_1\right)^{\limsup_{n \to \infty} \frac{n}{2\ell m_n + n + k_n}} = \left(\frac{1}{\rho^{2\ell q}} - \eta_1\right)^{\frac{1}{2\ell q + 1}} < \frac{1}{\rho},$$

which contradicts (3.7).   □

8.4. Historical Remarks

Theorem 1 in the case $\lambda = $ constant and $\ell = 1$ was proved by Rivlin [88] without the explicit error estimates and without distinguished points. Theorems 2 and 3 are new. Theorems 1, 2 and 3 are special cases of a more general result of Jakimovski and Sharma [56], [57]. We now give the results of [56] and [57] and some notations used therein. For their proof the interested reader is referred to the original papers. The Hermite interpolant to $f(z)$ in the zeros of $(T_n(z))^p$ (equivalently $(\cos n\theta)^p$, $p \ge 1$) is denoted by

$$h_{p,pn-1}(f; z) = \sum_{\sigma=0}^{pn-1} h_\sigma^{(p,n)} T_\sigma(z). \quad (4.1)$$

The the difference $\Delta_{\ell-1,n,m,p}(f; z).$ is given by

$$\Delta_{l-1,n,m,p}(f; z) := S_{n-1}(h_{p,pm-1}(f, \cdot); z) -$$
$$- \begin{cases} S_{n-1}(h_{p,pm-1}(S_{plm-1}(f, \cdot)); z) & \text{if } p \text{ is even} \\ S_{n-1}(h_{p,pm-1}(S_{(p(2l-1)-1)m+n-1}(f, \cdot)); z) & \text{if } p \text{ is odd.} \end{cases} \quad (4.2)$$

When $l = 1$, $p = 1$, the Hermite interpolant becomes Lagrange interpolant and the above difference reduces to the case treated by Rivlin [88]. In order to state our result we need

$$f_{l,q,p}(R) := \begin{cases} \dfrac{R}{\rho^{plq+1}} & \text{if } p \text{ is even and } R > \rho \\[2mm] \dfrac{1}{\rho^{plq}} & \text{if } p \text{ is even and } \rho > R > 1 \\[2mm] \dfrac{R}{\rho^{(p(2l-1)+1)q-1}} & \text{if } p \text{ is odd and } R \ne \rho. \end{cases} \quad (4.3)$$

Write $\mathbf{N} := \{1, 2, 3, \ldots\}$, $\mathbf{N}^+ := \{0, 1, 2, 3, \ldots\}$, and $e_n := 1$ when n is odd and $e_n := 0$ when n is even. The following relations are easily proved and will be used often:

$$
\begin{cases}
e_p + e_{p+1} = 1 \ , \ e_{ps} = e_p e_s \ , \ e_{(e_p)} = e_p \ , \ e_{ps+2u-e_j} = (-1)^j e_p e_s + e_j \ , \\[2em]
e_{e_p e_s - 1} = 1 - e_p e_s = e_{s-1} + e_s e_{p-1} \ , \ (-1)^p = (-1)^{e_p} \ , \ e_{-p} = e_p \ , \\[2em]
e_{ps+j} = (-1)^j e_p e_s + e_j \ , \ (-1)^{ps} e_p e_s + e_p e_s = 0 \ .
\end{cases}
\tag{4.4}
$$

Let $\boldsymbol{\eta}$ denote the sequence $\eta_0 := 1$, $\eta_k := 2$ for $k \in \mathbf{N}^+$.

THEOREM A. *Assume that p is a positive even integer and s, n are positive integers. For a function $f(z) \in A(E_R)$, let*

$$
h_{p,pn-1}(f;z) = \sum_{\sigma=0}^{pn-1} h_\sigma^{(p)} T_\sigma(z)
\tag{4.5}
$$

be the Hermite interpolant of order p to $f(z)$ at the zeros of $(T_n(z))^p$. Then we have:

(i) For each integer σ, $0 < \sigma < pn$, $2\lambda n < \sigma < (2\lambda+2)n$, $0 \le \lambda \le \frac{1}{2}p - 1$,

$$
h_\sigma^{(p)} = A_\sigma - \binom{p-1}{\frac{1}{2}p+\lambda} \sum_{s=1}^{\infty} (-1)^{\frac{1}{2}p(s+1)} \sum_{\tau=0}^{\frac{1}{2}p-1} (-1)^\tau \binom{\frac{1}{2}p(s+1)+\tau}{p} \times
$$

$$
\times \left(\frac{p}{\frac{1}{2}ps + \tau - \lambda} A_{(ps+2\tau-2\lambda)n+\sigma} + \frac{p}{\frac{1}{2}ps + \tau + \lambda + 1} A_{(ps+2\tau+2(\lambda+1))n-\sigma} \right) ;
\tag{4.6}
$$

(ii) For each even integer ρ, $\rho \in \{0, 2, 4, \dots, p-2\}$,

$$
h_{\rho n}^{(p)} = A'_{\rho n} - \frac{1}{2}\eta_\rho \binom{p}{\frac{1}{2}p+\frac{1}{2}\rho} \sum_{s=1}^{\infty} (-1)^{\frac{1}{2}p(s+1)} \sum_{\tau=0}^{\frac{1}{2}p-1} (-1)^\tau \frac{1}{p} \times
$$

$$
\times \left(\frac{\frac{1}{2}p(s+1)+\tau-1}{p-1} \right) \frac{(p^2-\rho^2)(ps+2\tau)}{(ps+2\tau)^2-\rho^2} A_{(ps+2\tau)n}
\tag{4.7}
$$

REMARK. It should be observed that in case (i) of the above theorem when $\sigma = (2\lambda+1)n$, we have

$$
A_{(ps+2\tau-2\lambda)n+\sigma} = A_{(ps+2\tau+1)n} = A_{(ps+2\tau+2(\lambda+1))n-\sigma}
$$

EXAMPLE 1. Applying Theorem A to the case $p = 4$, we obtain the following:

(i) For $\sigma = 0$ we have

$$h_0^{(4)} = \frac{1}{2}A_0 - \sum_{s=1}^{\infty} \left\{(2s+1)(2s-1)A_{4sn} - (2s+2)2sA_{(4s+2)n}\right\},$$

(ii) For $0 < \sigma < 2n$

$$\begin{aligned} h_\sigma^{(4)} =A_\sigma &- \frac{1}{2}\sum_{s=1}^{\infty}\left\{(2s+2)(2s+1)(2s-1)A_{4sn+\sigma}\right. \\ &+ (2s+2)2s(2s-1)A_{(4s+2)n-\sigma} \\ &- (2s+3)(2s+2)2sA_{(4s+2)n+\sigma} \\ &\left.-(2s+3)(2s+1)2sA_{(4s+4)n-\sigma}\right\}, \end{aligned}$$

(iii) For $\sigma = 2n$ we have

$$h_{2n}^{(4)} = A_{2n} - \frac{1}{4}\sum_{s=1}^{\infty}\left(4s^2 A_{4sn} - (2s+1)^2 A_{(4s+2)n}\right),$$

and

(iv) For $2n < \sigma < 4n$ we have

$$\begin{aligned} h_\sigma^{(4)} =A_\sigma &- \frac{1}{6}\sum_{s=1}^{\infty}\left\{(2s+2)(2s+1)2sA_{(4s-2)n+\sigma}\right. \\ &+ (2s+1)2s(2s-1)A_{(4s+4)n-\sigma} \\ &- (2s+3)(2s+2)(2s+1)A_{4sn+\sigma} \\ &\left.-(2s+2)(2s+1)2sA_{(4s+6)n-\sigma}\right\}. \end{aligned}$$

THEOREM B. *Assume that p is a positive odd integer. For a function $f(z) \in A(E_R)$, let (4.5) give the Hermite interpolant of order p to $f(z)$ at the zeros of $(T_n(z))^p$. Then:*

(i) For each σ satisfying $pn < \sigma < (p+1)n$ $(\rho = 0,\ldots,p-1)$ or for $\sigma = \rho n$ where ρ is an even integer, we have

$$\begin{aligned} h_\sigma^{(p)} =A_\sigma' &+ \left(\frac{p-1}{\frac{1}{2}(p-1)-(-1)^\rho\frac{1}{2}(\rho-e_\rho)+e_\rho}\right)\sum_{s=1}^{\infty}(-1)^s\sum_{\tau=0}^{p-1}(-1)^\tau\binom{ps+\tau}{p}\times \\ &\times\left\{\frac{p}{ps+\tau-\frac{1}{2}(p-1)+\frac{1}{2}(\rho+e_\rho)}A_{2psn+\left((\rho+e_\rho)-(p-1)+2\tau\right)n-\sigma}\right. \\ &\left.+\frac{p}{ps+\tau-\frac{1}{2}(p-1)-\frac{1}{2}(\rho+e_\rho)}A_{2psn+\left(-(\rho+e_\rho)-(p-1)+2\tau\right)n+\sigma}\right\}; \quad (4.8) \end{aligned}$$

(ii) For each $\sigma = \rho n$ where ρ is an odd integer $0 \leq \rho \leq p - 1$, we have

$$h_\sigma^{(p)} = h_{\rho n}^{(p)} = A_{\rho n} + \left(\frac{p}{\frac{1}{2}(p - \rho)}\right) \sum_{s=1}^{\infty} (-1)^s \sum_{\tau=0}^{p-1} (-1)^\tau \binom{ps + \tau}{p} \times$$

$$\times \left\{ \frac{\frac{1}{2}(p + \rho)}{ps - \frac{1}{2}(p - \rho) + \tau} + \frac{\frac{1}{2}(p - \rho)}{ps - \frac{1}{2}(p + \rho) + \tau} - \frac{p}{ps + \tau} \right\} A_{(2ps + 2\tau - p)n}.$$

$$(4.9)$$

REMARK. It should be observed that in case (i) of the above theorem when $\sigma = \rho n$ and ρ is even we have

$$A_{2psn + ((\rho + e_\rho) - (p-1) + 2\tau)n - \sigma} = A_{2psn + (-(p-1) + 2\tau)n}$$
$$= A_{2psn + (-(\rho + e_\rho) - (p-1) + 2\tau)n + \sigma},$$

and (4.8) becomes simpler.

EXAMPLE 2. Applying Theorem B to the case $p = 3$ we get the following:
(i) For $\sigma = 0$

$$h_0^{(3)} = \frac{1}{2} A_0 + 2 \sum_{s=1}^{\infty} (-1)^s \left\{ 3s(3s - 2) A_{(6s-2)n} - (3s + 1)(3s - 1) A_{6sn} \right.$$

$$+ (3s + 2)(3s) A_{(6s+2)n} \bigg\},$$

(ii) For $0 < \sigma < n$,

$$h_\sigma^{(3)} = A_\sigma + \sum_{s=1}^{\infty} (-1)^s \left\{ 3s(3s - 2) \left(A_{(6s-2)n-\sigma} + A_{(6s-2)n+\sigma} \right) \right.$$

$$- (3s + 1)(3s - 1) \left(A_{6sn-\sigma} + A_{6sn+\sigma} \right)$$
$$+ (3s + 2)3s \left(A_{(6s+2)n-\sigma} + A_{(6s+2)n+\sigma} \right) \bigg\},$$

(iii) For $\sigma = n$,

$$h_n^{\{3\}} = A_n + \sum_{s=1}^{\infty} (-1)^s \left\{ (6s - 3) A_{(6s-3)n} - (6s - 1) A_{(6s-1)n} \right.$$

$$+ (6s + 1) A_{(6s+1)n} \bigg\}.$$

(iv) For $n < \sigma < 2n$,

$$
\begin{aligned}
h_\sigma^{(3)} = A_\sigma + \sum_{s=1}^{\infty} (-1)^s &\left\{ \binom{3s-1}{2} A_{6sn-\sigma} + \binom{3s}{2} A_{(6s-4)n+\sigma} - \right. \\
&- \binom{3s}{2} A_{(6s+2)n-\sigma} - \binom{3s+1}{2} A_{(6s-2)n+\sigma} \\
&\left. + \binom{3s+1}{2} A_{(6s+4)n-\sigma} + \binom{3s+2}{2} A_{6sn+\sigma} \right\}.
\end{aligned}
$$

(v) For $\sigma = 2n$,

$$
\begin{aligned}
h_{2n}^{\{3\}} = A_{2n} + \sum_{s=1}^{\infty} (-1)^s &\left\{ (3s-1)^2 A_{(6s-2)n} - (3s)^2 A_{6sn} \right. \\
&\left. + (3s+1)^2 A_{(6s+2)n} \right\}.
\end{aligned}
$$

(vi) For $2n < \sigma < 3n$,

$$
\begin{aligned}
h_\sigma^{(3)} = A_\sigma + \sum_{s=1}^{\infty} (-1)^s &\left\{ \binom{3s-1}{2} A_{6sn-\sigma} + \binom{3s}{2} A_{(6s-4)n+\sigma} \right. \\
&- \binom{3s}{2} A_{(6s+2)n-\sigma} \\
&- \binom{3s+1}{2} A_{(6s-2)n+\sigma} + \binom{3s+1}{2} A_{(6s+4)n-\sigma} \\
&\left. + \binom{3s+2}{2} A_{6sn+\sigma} \right\}.
\end{aligned}
$$

THEOREM C. *Let* $m = m_n$ *be a sequence of positive integers such that* $m/n \to q > 1$ *as* $n \to \infty$. *Suppose* $f \in A(E_\rho)$. *If* $\rho > 1$ *and* l, p *are positive integers, then when* p *is odd and* $R > 1$, $R \neq \rho$, *or, when* p *is even and* $R > \rho$, *the following holds:*

$$
\max_{f \in A(E_\rho)} \limsup_{n \to \infty} \max_{z \in E_R} |\Delta_{l-1,n,m,p}(f;z)|^{1/n} = f_{l,q,p}(R). \tag{4.10}
$$

where $f_{\ell,q,p}(R)$ *is given by (4.3). If*

$$
f(z) = f_*(z) := \frac{\rho - z}{1 - 2\rho z + \rho^2} \in A(E_\rho) \tag{4.11}
$$

then when p is odd and $R > 1, R \neq \rho$, or, when p is even and $R > \rho$, we have

$$\lim_{n\to\infty} \min_{z\in E_R} |\Delta_{l-1,n,m,p}(f_*; z)|^{1/n} = \lim_{n\to\infty} \max_{z\in E_R} |\Delta_{l-1,n,m,p}(f_*; z)|^{1/n} = f_{l,q,p}(R).$$

(4.12)

This shows that the bound in (4.10) is sharp and that

$$\Delta_{l-1,n,m,p}(f; z) \to 0 \quad as \quad n \to \infty$$

Eor each z in the interior of the ellipse E_P where

$$P = \begin{cases} \rho^{plq+1} & \text{if } p \text{ is even,} \\ \rho^{(p(2l-1)+1)q-1} & \text{if } p \text{ is odd.} \end{cases}$$

The next theorems show that instead of (4.10) we have for each function $f \in A(E_\rho)$ and all, but for a finite number of exceptional points $z = z_j$,

$$\limsup_{n\to\infty} |\Delta_{l-1,n,m(n),p}(f, z)|^{1/n} = f_{l,q,p}(R) .$$

When p is *an even positive integer* we have the following two results.

THEOREM D. *Assume p is a positive even integer, l is a positive integer and $\rho > 1$. Let $m \equiv m_n := qn + c$ where $q \geq 2$ and c are non-negative integers. Let $f \in A(E_\rho)$,$\rho > 1$. Then for all complex numbers z outside the ellipse E_ρ (i.e., $z = \frac{1}{2}(w + w^{-1}), |w| = R > \rho$), except perhaps for at most pql points, we have*

$$\limsup_{n\to\infty} |\Delta_{\ell-1,n,m(n),p}(f; z)|^{1/n} = \frac{R}{\rho^{p\ell q+1}}$$

(4.13)

and for at most pql distinguished points z_j, $z_j = \frac{1}{2}(w_j + w_j^{-1})$, $|w_j| > \rho$, outside the ellipse E_ρ we have

$$\limsup_{n\to\infty} |\Delta_{\ell-1,n,m(n),p}(f; z_j)|^{1/n} < \frac{|w_j|}{\rho^{pql+1}}.$$

(4.14)

THEOREM E. *Suppose the assumptions of Theorem D are satisfied. Let z_j , $1 \leq j \leq pql$, $z_j = \frac{1}{2}(w_j + w_j^{-1})$,$|w_j| > \rho$, be arbitrary given pairwise different numbers outside the ellipse E_ρ . Then there exists a function $f^{**}(z) \in A(E_\rho)$ such that*

$$\limsup_{n\to\infty} |\Delta_{\ell-1,n,m(n),p}(f^{**}; z_j)|^{1/n} < f_{\ell,q,p}(R_j) \quad for \quad 1 \leq j \leq pql .$$

where $R_j := |w_j|$.

When p is *a positive odd integer* we have the following two results.

THEOREM F. *Assume p is a positive odd integer and $\rho > 1$ is a real number. Write $m_n := qn + c$, $q > 1$, where q and c are integers. Write $\omega := p(2\ell - 1) + 1$, $\omega_q := \omega q$ and $\omega_c := \omega c$. Let $f \in A(E_\rho)$. Then for all complex numbers $z \notin [-1, 1]$ (i.e., $z = \frac{1}{2}(w + w^{-1})$, $|z| =: R > 1$) except, perhaps, for at most $\omega_q - 2$ numbers, we have*

$$\limsup_{n \to \infty} |\Delta_{\ell-1,n,m(n),p}(f; z)|^{1/n} = \frac{|w|}{\rho^{\omega_q - 1}}.$$

At most at the $\omega_q - 2$ exceptional points, say, z_j, $z_j = \frac{1}{2}(w_j + w_j^{-1})$, $|w_j| = R_j > 1$, $1 \leq j \leq \omega_q - 2$ we have

$$\limsup_{n \to \infty} |\Delta_{\ell-1,n,m(n),p}(f; z_j)|^{1/n} < \frac{|w_j|}{\rho^{\omega_q - 1}}, \quad 1 \leq j \leq \omega_q - 2.$$

THEOREM G. *Assume p is a positive odd integer and $\rho > 1$ is a real number. Assume pairwise different complex numbers $z_1, \ldots, z_{\omega_q-2} \notin [-1, 1]$, $z \notin E_\rho$, are given (where $z_j = \frac{1}{2}(w_j + w_j^{-1})$, $|w_j| > 1$), $|w_j| \neq \rho$. Then there exists a function $f^*(z) \in A(E_\rho)$ such that we have*

$$\limsup_{n \to \infty} |\Delta_{\ell,n,m(n),p}(f^*; z_j)|^{1/n} < \frac{|w_j|}{\rho^{\omega_q - 1}} \quad for \quad j = 1, \ldots, \omega_q - 2.$$

CHAPTER 9

WALSH EQUICONVERGENCE
THEOREMS FOR THE FABER SERIES

1. Introducing Faber polynomial and Faber expansions.

In this chapter we obtain equiconvergence theorems related to expansions of analytic functions into series of polynomials, the Faber polynomials. The Taylor series expansion and the Chebyshev expansion of analytic functions are two examples out of the different families of the Faber expansions.

1.1 Faber polynomials and some of their properties.

Let E be a compact (i.e., closed and bounded) subset of the extended complex plane $\mathbb{C}$, which is not a point, with a complement E^c which is simply connected in the extended complex plane.

According to the Riemann mapping theorem, there exists a conformal map ψ of $\{w \in \mathbb{C} : |w| > 1\}$ onto $E^c = \{z : z \notin E, \ z \in \mathbb{C}\}$ normalized at infinity by $\psi(\infty) = \infty$ and $c := \psi'(\infty) > 0$, where c is called the *capacity* or *infinite diameter* of E. We have the Laurent expansion $z = \psi(w) = cw + c_0 + \frac{c_1}{w} + \cdots$ for $|w| > 1$.

If the boundary ∂E of the set E is a simple Jordan curve, then the conformal map $\psi(w)$ is continuable to a one to one continuous mapping of $\{w \in \mathbb{C} : |w| \geq 1\}$ onto $\mathbb{C} \setminus \text{Int } E$.

When the boundary ∂E is a simple Jordan curve we write $w_{m,k} := \exp(2\pi k i / (m+1))$ $(k = 0, 1, \ldots, m)$. The associated points $z_{n,k} := \psi(w_{m,k})$ are called the *Fejĕr nodes* with respect to E.

Following Pommerenke [84] we say that ∂E is an r_0-analytic curve $(0 \leq r_0 < 1)$, if the conformal map $\psi(w)$ admits a univalent continuation to $\{w \in \mathbb{C} : |w| > r_0\}$.

The function $\psi(w)$ from $\{w \in \mathbb{C} : |w| > 1\}$ onto E^c has an inverse $w = \phi(z)$ that for all sufficiently large $|z|$ has the expansion $w = \phi(z) = dz + d_0 + \frac{d_1}{z} + \cdots$ where $cd=1$. Also, for all sufficiently large z and each non-negative integer n we have the convergent Laurent expansion $\{\phi(z)\}^n = d^n z^n + \sum_{k=-\infty}^{n-1} d_{nk} z^k$.

199

The polynomial part of the Laurent expansion of $\phi(z)^n$, given by $F_n(z) :=$ $d^n z^n + \sum_{k=0}^{n-1} d_{nk} z^k$ is of exact degree n and is called the Faber polynomial associated with the set E.

It is well known (cf. Curtiss [34]) that for an arbitrary compact set E (not one point) for which $\mathbb{C} \backslash E$ is simply connected, the associated Faber polynomials satisfy

$$\lim_{n \to \infty} |F_n(\psi(\zeta))|^{1/n} = |\zeta|, \quad |\zeta| > 1, \tag{1.1.0}$$

uniformly on every closed subset of $\zeta \in \mathbb{C} : |\zeta| > 1$. This implies the following result.

THEOREM 1. *To each fixed $r > 1$ there corresponds some positive integer $N \equiv N(r)$ such that for each $n > N(r)$ all the zeros of $F_n(z)$ are in G_r and, therefore, all accumulation points of the zeros of $(F_n(z))_{n \geq 1}$ must lie in E.*

From the above Theorem 1 it follows that the polynomial $L_n^*(f; z)$, denoting the Lagrange interpolants of f at the zeros of the Faber polynomial $F_{n+1}(z)$, are well defined for all large n. If ∂E is an r_0-analytic curve, $0 \leq r_0 < 1$ then for every sufficiently large n the Faber nodes all lie in the interior of E. It is known [63] that if E is convex, but not a line segment, then all Faber nodes lie in the interior of E. In the case when E is the line segment $[-1, 1]$, it is well known that the Faber polynomials for E coincide with the classical Chebyshev polynomials of the first kind.

The $m + 1$ zeros of the $(m+1)^{\text{st}}$ Faber polynomial with respect to E are called the *Faber nodes*.

For $R > 1$ denote by C_R the outer level curve of E given by $C_R := \{\psi(w) : |w| = R\}$. If ∂E is an r_0-analytic curve $(0 \leq r_0 < 1)$ then C_R and G_R are defined also for $R > r_0$. When ∂E is an r_0-analytic curve we will assume in the following that $R > r_0$.

For the following theorem see V.I. Smirnov and N.A. Lebedev [100].

THEOREM 2 (THE FABER EXPANSION THEOREM). (i) *Assume $f \in A_R^*$ where $R > 1$. Then the function f can be expanded into a series of Faber polynomials*

$$f(z) = \sum_{k=0}^{\infty} a_k F_k(z) \tag{1.1.1}$$

where

$$a_k = \frac{1}{2\pi i} \oint_{|\zeta| = r'} f(\psi(\zeta)) \zeta^{-(k+1)} d\zeta, \quad k = 0, 1, \dots, \tag{1.1.2}$$

for each r', $1 < r' < R$. The Faber series (1.1.1) is uniformly convergent on each compact subset of Int E. *In this case we have*

$$\limsup_{k \to \infty} |a_k|^{1/k} = \frac{1}{R}. \tag{1.1.3}$$

(ii) *If a Faber series $\sum_{k=0}^{\infty} a_k F_k(z)$ is convergent in some domain then it converges in a domain G_σ, uniformly convergent in compact subset of G_σ, and it is not convergent for each point outside g_σ. For this σ (1.1.3) holds.*

(iii) *If a sequence $(a_k)_{k \geq 0}$ satisfies (1.1.3) then the Faber series $\sum_{k=0}^{\infty} a_k F_k(z)$ is uniformly convergent in each compact subset of G_σ.*

Given the Faber expansion (1.1.1) of f, the sequence of partial sums is given by

$$S_n(f; z) := \sum_{k=0}^{n} a_k F_k(z). \tag{1.1.4}$$

Assume that for a non-negative integer n a sequence of pairwise different interpolation nodes $z_{k,n} \in G_R$ ($k = 0, \ldots, n$) is given. To the given interpolating nodes we associate the polynomial $w_n(z) \equiv w(z; z_{k,n}) := \prod_{k=0}^{n}(z - z_{k,n})$.

1.2 Examples of Faber polynomials.

EXAMPLE 1. If E is the closed unit disk $\overline{D}_1$ then $z = w$ maps univalently the outside of the circle $|w| = 1$ onto the outside of E. In this case $w = \phi(z) = z$ so that $F_n(z) = z^n$ and $F_n(\phi(w)) = w^n \equiv w^n + \sum_{k=0}^{\infty} \alpha_{nk} w^{-k}$ where $\alpha_{nk} = 0$ for $n, k = 0, 1, \cdots$. In this case we have $r_0 = 0$. The Faber expansion theorem in this case is the Taylor series expansion.

EXAMPLE 2. Let E be the real closed segment $[-1, 1]$. Then the mapping

$$w = \phi(z) = z + \sqrt{z^2 - 1} = z\left(1 + \sqrt{1 - \frac{1}{z^2}}\right)$$

(where the branch chosen for $\sqrt{1 - \frac{1}{z^2}}$ on E^c is the one that takes the value 1 for $z = \infty$) maps univalently the exterior of E onto the domain $|w| > 1$. The closed segment $[-1, 1]$ is mapped continuously onto $|w| = 1$, but not univalently. In this example we have $z = \psi(w) = \frac{1}{2}\left(w + \frac{1}{w}\right)$ for $|w| \geq 1$ and $c = \frac{1}{2}$. Since the Laurent expansion at infinity of the function

$$\frac{1}{\phi(z)} = z - \sqrt{z^2 - 1} = z\left(1 - \sqrt{1 - \frac{1}{z^2}}\right)$$

contains no nonnegative powers of z, the Laurent expansion at infinity of the function

$$K_n(z) := (\phi(z))^n + \frac{1}{(\phi(z))^n} = \left(z + \sqrt{z^2 - 1}\right)^n + \left(z - \sqrt{z^2 - 1}\right)^n \quad (1.2.1)$$

has the same term with non-negative powers of z as the Laurent expansion at infinity of $[\phi(z)]^n$ for $n = 1, 2, \dots$. But $K_n(z)$ is a polynomial of degree n, and hence

$$F_0(z) = 1, \quad F_n(z) = \left(z + \sqrt{z^2 - 1}\right)^n + \left(z - \sqrt{z^2 - 1}\right)^n, \quad n = 0, 1, \dots$$

and

$$F_n(\psi(w)) = w^n + \frac{1}{w^n} \equiv w^n + \sum_{k=0}^{\infty} \alpha_{nk} w^{-k} \quad \text{for } n = 0, 1, 2, \dots$$

where $\alpha_{nk} = \begin{cases} 1 & \text{when } k = n \\ 0 & \text{otherwise.} \end{cases}$

Setting $z = \cos t$, we obtain

$$F_n(\cos t) = [(\cos t + i \sin t)^n + (\cos t - i \sin t)^n] = 2 \cos nt,$$

or equivalently

$$F_0(z) = 1, \quad F_n(z) = 2 \cos (n \arccos z), \quad n = 1, 2, \dots.$$

Hence the Faber nodes are the zeros of $\cos nt$. We see that in this example the Faber polynomials turn out to be normalized Chebyshev polynomials of the first kind.

EXAMPLE 3. Assume $\delta > 1$. Write $a = \frac{1}{2}(\delta + \frac{1}{\delta})$ and $b = \frac{1}{2}(\delta - \frac{1}{\delta})$. Let E_δ be the ellipse $\frac{x^2}{a^2} + \frac{y^2}{b^2} = 1$ and its interior. This ellipse has ± 1 as its foci. The function $z = \psi(w) := \frac{1}{2}(\delta w + \frac{1}{\delta w})$ maps the exterior of E_δ univalently onto the exterior of $\overline{D}_1$. In this example we have $c = \frac{1}{2}\delta$, and $w = \phi(z) = \frac{1}{\delta}(z + \sqrt{z^2 - 1})$. Since the Laurent expansion at infinity of the function

$$\frac{1}{\phi(z)}) = \delta z \left(1 - \sqrt{1 - \frac{1}{z^2}}\right) = \delta \left(z - \sqrt{z^2 - 1}\right)$$

contains no nonnegative powers of z, the Laurent expansion at infinity of the function

$$K_n(z) := (\phi(z))^n + \frac{1}{(\phi(z))^n} = \left(\frac{1}{\delta}\left(z + \sqrt{z^2 - 1}\right)\right)^n + \left(\frac{1}{\delta}\left(z - \sqrt{z^2 - 1}\right)\right)^n$$

$$(1.2.2)$$

has the same term with non-negative powers of z as the Laurent expansion at infinity of $[\phi(z)]^n$ for $n = 1, 2, \ldots$. But $K_n(z)$ is a polynomial of degree n, and hence

$$F_0(z) = 1, \quad F_n(z) = \left(\frac{1}{\delta}\left(z + \sqrt{z^2 - 1}\right)\right)^n + \left(\frac{1}{\delta}\left(z - \sqrt{z^2 - 1}\right)\right)^n, \quad n = 0, 1, \ldots$$

and

$$F_n(\psi(w)) = w^n + \frac{1}{\delta^{2n}w^n} \equiv w^n + \sum_{k=0}^{\infty} \alpha_{nk}w^{-k} \quad \text{for } n = 0, 1, 2, \ldots ,$$

where

$$\alpha_{nk} = \begin{cases} \frac{1}{\delta^{2n}} & \text{when } k = n \\ 0 & \text{otherwise.} \end{cases}$$

In this example ∂E_δ is an r_0-analytic curve with $r_0 = 1/\delta$.

For the Fejér nodes $w_{m+1,k}$, where $w_{m+1,k}^{m+1} = 1$, for $0 \le k \le m$, we have $F_{m+1}(\psi(w_{m+1,k})) = F_{m+1}(\psi(1))$. Hence, since $F_{m+1}(\psi(w)) - F_{m+1}(\psi(1))$ vanishes at the Fejér nodes $z_{m+1}(\psi(1))$,

$$\omega_m(\psi(w)) = \left(\frac{\delta}{2}\right)^{m+1} (F_{m+1}(\psi(w)) - F_{m+1}(\psi(1)))$$

$$= \left(\frac{\delta}{2}\right)^{m+1} (w^{m+1} - 1)\left(1 - \frac{1}{(\delta^2 w)^{m+1}}\right)$$

EXAMPLE 4. Let E be the closed interior of the lemniscate $|z^2 - 1| = 1$. It is easy to see that the mapping

$$w = \phi(z) = z\left(1 - \frac{1}{z^2}\right)^{1/2} = \sqrt{z^2 - 1},$$

(where we choose the branch of $\left(1 - \frac{1}{z^2}\right)^{\frac{1}{2}}$ which equals 1 at ∞) maps the exterior of E, univalently, onto the exterior of $\overline{D}_1$. The inverse function is given by

$$z = \psi(w) = w\sqrt{1 + \frac{1}{w^2}} = \sqrt{w^2 + 1}.$$

We have

$$\phi(z)^n = z^n\left(1 - \frac{1}{z^2}\right)^n = \sum_{k=0}^{\infty}(-1)^k\binom{n/2}{k}z^{n-2k}.$$

Therefore

$$F_n(z) = \sum_{k=0}^{[n/2]}(-1)^k\binom{n/2}{k}z^{n-2k}.$$

In particular

$$F_0(z) = 1, \; F_1(z) = z, \; F_2(z) = z^2 - 1, \; F_3(z) = z^3 - \frac{3}{2}z, \; \dots .$$

We have

$$F_n(\psi(w)) = \sum_{k=0}^{[n/2]} (-1)^k \binom{n/2}{k} w^{n-2k} \left(1 + \frac{1}{w^2}\right)^{\frac{n}{2}-k}$$

$$= w^n \sum_{k=0}^{[n/2]} (-1)^k \binom{n/2}{k} w^{-2k} \sum_{j=0}^{\infty} \binom{\frac{n}{2}-k}{j} \frac{1}{w^{2j}}$$

$$= w^n \sum_{k=0}^{[n/2]} (-1)^k \binom{n/2}{k} \sum_{j=0}^{\infty} \binom{\frac{n}{2}-k}{j} \frac{1}{w^{2(k+j)}} .$$

Replacing in the last sum the summation index j by r and using the relation $k + j = r$ we get

$$F_n(\psi(w)) = w^n + w^n \sum_{r=1}^{[n/2]} \sum_{k=0}^{r} (-1)^k \binom{n/2}{k} \binom{\frac{n}{2}-k}{r-k} \frac{1}{w^{2r}}$$

$$+ \sum_{r>[n/2]} \frac{1}{w^{2r-n}} \sum_{k=0}^{[n/2]} (-1)^k \binom{n/2}{k} \binom{\frac{n}{2}-k}{r-k} .$$

Since

$$\binom{n/2}{k} \binom{\frac{n}{2}-k}{r-k} = \binom{r}{k} \binom{n/2}{r}$$

we get

$$F_n(\psi(w)) = w^n + w^n \sum_{1 \le r \le [n/2]} \frac{1}{w^{2r}} \binom{n/2}{r} \sum_{k=0}^{r} (-1)^k \binom{r}{k}$$

$$+ \sum_{r>[n/2]} \frac{1}{w^{2r-n}} \binom{n/2}{r} \sum_{k=0}^{[n/2]} (-1)^k \binom{r}{k} .$$

From

$$\sum_{k=0}^{r} (-1)^k \binom{r}{k} = (1-1)^r = 0 \quad \text{for} \;\; r > 0$$

and

$$\sum_{k=0}^{[n/2]} (-1)^k \binom{r}{k} = (-1)^{[n/2]} \binom{r-1}{[n/2]}$$

we get

$$F_n(\psi(w)) = w^n + \sum_{r>[n/2]} \frac{1}{w^{2r-n}} \binom{n/2}{r} (-1)^{[n/2]} \binom{r-1}{[n/2]} .$$

Using the substitution $r = [n/2] + s + 1$ we obtain

$$F_n(\psi(w)) = w^n + (-1)^{[n/2]} \frac{1}{w^{2[n/2]-n+2}} \sum_{s=0}^{\infty} \binom{n/2}{[n/2]+s+1} \binom{[n/2]+s}{s} \frac{1}{w^{2s}}$$

$$= \begin{cases} w^n & \text{when } n \text{ is even} \\[2mm] w^n + (-1)^{[n/2]} \sum_{s=0}^{\infty} \binom{n/2}{[n/2]+s+1} \binom{[n/2]+s}{s} \frac{1}{w^{2s+1}} & \\ & \text{when } n \text{ is odd} \end{cases}$$

$$\equiv w^n + (-1)^{[n/2]} \sum_{s=0}^{\infty} \alpha_{n,s} w^{-s}.$$

So when n is odd and $s = 0, 1, 2, \ldots$, we have

$$\alpha_{n,2s+1} := \begin{cases} (-1)^{[n/2]} \binom{n/2}{[n/2]+s+1} \binom{[n/2]+s}{s} & \text{when } n \text{ is odd and } s \geq 0 \\ 0 & \text{otherwise.} \end{cases}$$

2. Extended equiconvergence theorems for Faber expansions

Let $f \in A_R^*$. Assume that for a non-negative integer n a sequence of pairwise different interpolation nodes $z_{k,n} \in G_R$ $(k = 0, \ldots, n)$ is given. $L_n(f; \cdot)$ denotes the Lagrange interpolant to $f \in A_R^*$ in these nodes. Given a positive integer s then the Hermite interpolant $h_{s,s(n+1)-1}(z)$ is the polynomial of lowest degree interpolating the functions $f, f', \ldots, f^{(s-1)}$ at the interpolation points $(z_{k,n})$. To the given interpolating nodes we associate the polynomial $\omega_n(z) \equiv \omega(z; z_{k,n}) := \prod_{k=0}^{n}(z - z_{k,n})$.

In order to obtain an equiconvergence theorems for the Faber expansion of a function the Lagrange or Hermite interpolants are replaced by their sections

$$S_n\left(L_m(f; \cdot); z\right) \quad \text{or} \quad S_n\left(h_{s,s(m+1)-1}(f; \cdot); z\right).$$

We use also the following expressions. Set, for $m \geq n \geq 0$ and $j \geq 0$

$$S_{m,n,j}(f; z) := \sum_{k=0}^{n} a_{k+j(m+1)} F_k(z). \tag{2.0}$$

Clearly $S_{m,n,0}(f; z) = S_n(f; z)$ and $S_n(L_m(f; z); z)$ is the nth Faber section of the expansion of $L_m(f; z)$ in terms of Faber polynomials.

We state now, but prove in Section 6 the following theorems.

THEOREM 3. *Assume ∂E is an r_0-analytic curve for some $r_0 \in [0, 1)$. Suppose $f \in A_R^*$. Let q be a fixed positive integer. Write $m := q(n + 1) - 1$. Let $D_{m,n}(f; z)$ be given by*

$$D_{m,n,\ell}(f; z) := S_n(L_m(f; \cdot); z) - \sum_{j=0}^{\ell-1} S_{m,n,j}(f; z) \tag{2.1}$$

where $L_m(f;z)$ is the Lagrange interpolant at the Fejĕr nodes. Then,

$$\limsup_{n\to\infty} |D_{m,n,\ell}(f;z)|^{1/n} \leq \max\left(r_0, \frac{|w|}{R^{\ell q+1}}, \frac{|w|r_0^q}{R}, \frac{|w|r_0^{q-1}}{R^q}\right) \quad for \ |w| > R,$$

$$(2.2)$$

where $z = \psi(w)$, with $0^k := 0$ for any nonnegative integer k and $1/0 := \infty$.

An immediate consequence of Theorem 3 is the following

COROLLARY 1. *Assume the assumptions of Theorem 3 are satisfied. Then we have the equiconvergence result that*

$$\limsup_{n\to\infty} D_{m,n,\ell}(f;z) = 0 \quad for \ |w| < \lambda := \min\left(R^{\ell q+1}, Rr_0^{-q}, R^q r_0^{-(q-1)}\right).$$

$$(2.3)$$

The convergence to zero is uniform and geometric for $|w| < \delta < \lambda$ for each $1 < \delta < \lambda$.

In the same way it will be possible to obtain eqiconvergence theorems from the subsequent theorems in the section.

The following theorem is an analogue of Theorem 3, where the $m + 1$ Fejér nodes are replaced by the $m + 1$ zeros of the $(m + 1)^{\text{st}}$ Faber polynomial i.e., *Faber nodes.*

THEOREM 4. *Assume $f \in A_R^*$. Let ℓ and q be given positive integers. Write $m := q(n + 1) - 1$. Let $\Delta_{m,n,\ell}^*(f;z)$ be given by*

$$\Delta_{m,n,\ell}^*(f;z) := S_n(L_m^*(f;\cdot);z) - \sum_{j=0}^{\ell-1} S_{m,n,j}(f;z), \qquad (2.4)$$

where $S_{m,n,j}(f;z)$ is given by (2.0) and $L_m^*(f;z)$ is the Lagrange interpolant of f at the Faber nodes. Then we have

$$\limsup_{n\to\infty} |\Delta_{m,n,\ell}^*(f;z)|^{1/n} \leq \frac{|w|}{R^q} \quad for \ |w| > R \quad where \ z = \psi(w).$$

Adding to the assumption that ∂E is an r_0-analytic curve for some $r_0 \in [0,1)$ we get the following stronger result.

THEOREM 5. *Suppose the assumptions of Theorem 4 are satisfied and that ∂E is an r_0-analytic curve for some $r_0 \in [0,1)$. Then we have*

$$\limsup_{n\to\infty} |\Delta_{m,n,\ell}(f;z)|^{1/n} \leq \begin{cases} \max\left(\frac{1}{R^q}, \frac{|w|r_0^{q-1}}{R^q}\right) & when \ \ell = 1 \\ \max\left(\frac{1}{R^q}, \frac{|w|}{R^{q+1}}, \frac{|w|r_0^{q-1}}{R^q}\right) & when \ \ell > 1 \end{cases} \qquad (2.5)$$

THEOREM 6. *Let ∂E be an r_0-analytic curve for some $r_0 \in [0,1)$. Suppose $f \in A_R^*$. Assume q and s, p are given positive integers. Write $m := q(n+1) - 1$. For $\Delta_{m,n}^{p,s}(f; z)$ given by*

$$\Delta_{m,n}^{p,s}(f; z) := S_{p(n+1)-1}(h_{s,s(m+1)-1}(f; \cdot); z) - S_{p(n+1)-1}(f; z),$$

where $h_{s,s(m+1)-1}(f; z)$ denotes the Hermite interpolation polynomial to f, f', $\ldots, f^{s-1}$ in the $m + 1$ Fejér nodes. Then we have

$$\limsup_{n\to\infty} \left|\Delta_{m,n}^{p,s}(f; z)\right|^{1/n} \le \lambda^*, \quad for \; z = z(w) \; and \; |w| > R$$

where

$$\lambda^* = \begin{cases} \max\left(\dfrac{r_0^q|w|^p}{R^p}, \dfrac{|w|^p}{R^{p-qs}}, \dfrac{|w|^{qs}}{R^{qs}}\right) & when \; qs \le p \\[2ex] \max\left(\dfrac{1}{R^{qs}}, \dfrac{|w|^p r_0^q}{R^p}, \dfrac{|w|^p r_0^{q-p}}{R^{qs}}, \dfrac{|w|^p}{R^{qs+\tau}}\right) & when \; qs \ge p \; and \; q \ge p \\[2ex] \max\left(\dfrac{|w|^p r_0^q}{R^p}, \dfrac{|w|^p}{R^{qs}}\right) & when \; qs \ge p \; and \; q > p \end{cases}$$

(2.6)

with $\tau := q$ when $p = qu$ and $\tau := v$ when $p = qu + v$, $0 < v < q$.

THEOREM 7. *Assume $f \in A_R^*$. Let q, s, p be positive integers. Write $m := q(n + 1) - 1$. Let $\Delta_{m,n}^{*p,s}(f; z)$ be given by*

$$\Delta_{m,n}^{*p,s}(f; z) := S_{p(n+1)-1}\left(h_{s,s(m+1)-1}^*(\cdot; f); z\right) - S_{p(n+1)-1}(f; z),$$

where $h_{s,s(m+1)-1}^*(f; z)$ denotes the Hermite interpolant to $f, f', \cdots, f^{(s-1)}$ at the $m + 1$ Faber nodes. Then we have

$$\lim_{n\to\infty} \left|\Delta_{m,n}^{*p,s}(f; z)\right|^{1/n} \le \lambda^* |w|^p \; for \; z = \psi(w), |w| > R,$$

where

$$\lambda^* := \begin{cases} \dfrac{1}{R^p} & when \; p \ge qs \\[2ex] \max\left(\dfrac{1}{R^{p+q}}, \dfrac{1}{R^{qs}}\right) & when \; p < qs. \end{cases}$$

(2.7)

In the following two theorems we consider the difference of sections of Hermite interpolants and Lagrange interpolants.

THEOREM 8. *Let ∂E be an r_0-analytic curve for some $r_0 \in (0,1)$. Assume $f \in A_R^*$ and that s, p are two given positive integer. Let $D_{p,s,n}(f; z)$ given by*

$$D_{p,s,n}(f; z) := S_{p(n+1)-1}\left(\{h_{s,s(n+1)-1}(f; \cdot) - L_{s(n+1)-1}(f; \cdot)\}; z\right)$$

where $h_{s,s(n+1)-1}(f;z)$ *and* $L_{s(n+1)-1}(f;z)$ *denote the Hermite and Lagrange interpolants at Fejĕr nodes. Then we have*

$$\limsup_{n\to\infty} |D_{p,s,n}(f;z)|^{1/n} \leq \lambda |w|^p \qquad for \ |w| > R,$$

where

$$\lambda := \begin{cases} \frac{r_0}{R^p} & when \ p \geq s = 1, \\ \max\left(\frac{1}{R^{p+1}}, \frac{r_0}{R^p}\right) & when \ p \geq s > 1, \\ \frac{1}{R^s} & when \ 1 \leq p < s. \end{cases} \tag{2.8}$$

The following theorem is an analogue of Theorem 8 for Hermite and Lagrange interpolants at Faber nodes.

THEOREM 9. *Assume* $f \in A_R^*$. *Let* s, p *be positive integers. Let* $D_{p,s,n}^*(f;z)$ *given by*

$$D_{p,s,n}^*(f;z) := S_{p(n+1)-1}\left(h_{s,s(n+1)-1}^*(f;\cdot) - L_{s(n+1)-1}^*(f;\cdot); z\right)$$

where $h_{s,s(n+1)-1}^*(f;z)$ *is the the Hermite interpolant to* f *at the zeros of* $F_{n+1}(z)^s$ *and* $L_{s(n+1)-1}(f;z)$ *is the Lagrange interpolant to* f *at the zeros of* $F_{s(n+1)}(z)$. *Then we have*

$$\limsup_{n\to\infty} |D_{p,s,n}^*(f;z)|^{1/n} \leq \begin{cases} \frac{|w|^p}{R^{p+1}} & when \ s \geq p \ and \ |w| > R, \\ \frac{|w|^p}{R^{2p+1-s}} & when \ s < p \ and \ |w| > R. \end{cases} \tag{2.9}$$

3. Additional properties of Faber polynomials Let $A_R^* \equiv A_R^*(E)$ denote the class of functions $f(w)$ holomorphic in G_R but not in any $G_{R'}$, $R' > R$.

Consider, for a given $R > 1$, and $z \in G_R$, the integrals

$$\frac{1}{2\pi i} \oint_{C_R} \frac{\phi(t)^n}{t-z} dt = \frac{1}{2\pi i} \oint_{|s|=R} \frac{s^n \psi'(s) ds}{\psi(s) - z} \tag{3.1}$$

connected by the substitution $t = \psi(s)$. The path of the integral on the left of (3.1) can be replaced by a circle of radius R^* large enough so that $\phi(z)^n$ has a uniformly convergent Laurent series on this circle and the value of the integral will remain the same. The Laurent series can be integrated term by term, and when this is done the integral reproduces the principal part of the series at infinity and all the terms with negative exponents vanish because we have for $|z| < R^*$

$$\frac{1}{2\pi i} \oint_{|t|=R} \frac{t^k}{t-z} dt = \begin{cases} z^k \ for \ k \geq 0 \\ 0 \ for \ k \leq -1. \end{cases} \tag{3.2}$$

Thus, by (3.1)

$$F_n(z) = \frac{1}{2\pi i} \oint_{|s|=R} \frac{s^n \psi'(s) ds}{\psi(s) - z}, \quad n = 1, 2, 3, \ldots, \ z \in G_R, \ R > \gamma_0^* = 1. \quad (3.3)$$

For each fixed $z \in G_R$ the function of s, $s\psi'(s)/(\psi(s) - z)$, is regular for $\infty \geq |s| \geq R$ and has the value 1 at $s = \infty$, so it has a Laurent series

$$\frac{s\psi'(s)}{\psi(s) - z} = 1 + p_1(z)\frac{1}{s} + p_2(z)\frac{1}{s^2} + \ldots, \quad |s| \geq R > \gamma_0* = 1, \ z \in G_R. \quad (3.4)$$

But the Cauchy formulas for the coefficients in (3.4) are precisely the integrals appearing in (3.3), so these coefficients $p_n(z)$ are indeed $P_n(z)$ for each n. We thus have the following generating function for the Faber polynomials

$$\frac{s\psi'(s)}{\psi(s) - z} = \sum_{k=0}^{\infty} F_k(z) s^{-k} \quad \text{for } |s| > |w| > \gamma_0^* = 1 \quad (z = \psi(w)). \quad (3.5)$$

This implies that

$$\frac{\psi'(s)}{\psi(s) - \psi(w)} = \frac{1}{s - w} + \sum_{k=0}^{\infty} \left(F_k(\psi(w)) - w^k \right) \frac{1}{s^{k+1}} \quad \text{for } |s| > |w| > \gamma_0^* = 1. \quad (3.6)$$

REMARK 1.. It is easy to see that when ∂E is a γ_0–analytic curve $(0 \leq \gamma_0 < 1)$ then in (3.3),(3.5) and (3.6) we can replace $\gamma_0^* = 1$ by $\gamma_0^* = \gamma_0$.

We obtain now some additional properties of $F_n(\psi(w))$. We need the following notations.

(i) If ∂E is a γ_0-analytic curve $(0 \leq \gamma_0 < 1)$ then we choose R_1 and R so that $\gamma_0^* = \gamma_0 < R_1 < R$.

(ii) If nothing is assumed about ∂E then R_1 and R are chosen so that $\gamma_0^* = 1 < R_1 < R$.

Both in case (i) and in case (ii), (3.3) holds for these R_1 and R. Let $z = \psi(w)$ lie in the region $H_{R_1,R}$ bounded by C_{R_1} and C_R. (We observe that for a given w, $|w| > R_1$, we can always choose R sufficiently large so that $z = \psi(w)$ lies in H.) The function $s^n \psi'(s)/[\psi(s) - \psi(w)]$ as a function of s is regular in the closed region except for a simple pole at $s = w$. The residue at this the pole is

$$\lim_{s \to w} (s - w) \frac{s^n \psi'(s)}{\psi(s) - \psi(w)} = w^n.$$

Thus by the residue theorem and (3.3) we get for $z = \psi(w)$, $|w| > R_1$,

$$F_n(z) = F_n(\psi(w)) = w^n + \frac{1}{2\pi i} \oint_{|s|=R_1} \frac{s^n \psi'(s) ds}{\psi(s) - \psi(w)}. \quad z = \psi(w), \ |w| > R_1. \quad (3.7)$$

From (3.7) we get immediately

$$F_n(\psi(w)) - w^n = O(1)R_1^n \qquad (3.8)$$

uniformly in $\gamma_0^* < R_1 < \delta \leq |w|$ when $R_1 > 1$, and $r_0 < R_1 < 1$ when $\gamma_0^* = \gamma_0$, $0 \leq \gamma_0 < 1$.

The integrand on the right of (3.7) is an analytic function of w for $|w| > R$ and has the value 0 at $w = \infty$. Thus it has a Laurent series in w convergent at least for $|w| > R$. We write this series in the form

$$F_n(\psi(w)) = w^n + \sum_{k=1}^{\infty} \alpha_{n,k} w^{-k}. \qquad (3.9)$$

The series is convergent for all w, $|w| > R_1$, and uniformly convergent for $|w| \geq R^* > R_1$, where R^* is now arbitrary. By the formula for the coefficients of the Laurent expansion we have

$$\alpha_{n,k} = -\frac{1}{4\pi^2} \oint_{|w|=R_2} \oint_{|s|=R_1} \frac{w^{k-1} s^n \psi'(s) ds dw}{\psi(s) - \psi(w)}, \qquad (3.10)$$

where $R_2 > R_1 > 1 = \gamma_0^*$ when $\gamma_0^* = 1$ and $\gamma_0^* = \gamma_0 < R_1 < R_2 < 1$ when $\gamma_0^* = \gamma_0 < 1$.

An immediate consequence of (3.10) is

$$|\alpha_{n,k}| \leq O(1)R_2^k R_1^{\ n} \quad \text{for} \ \ n, k \geq 1 \ \ \text{where} \ \begin{cases} 1 < R_1 < R_2 & \text{when} \ \gamma_0^* = 1 \\ \gamma_0 < R_1 < R_2 & < 1 \\ & \text{when} \ \gamma_0^* < 1. \end{cases}$$

When $\gamma_0^* < 1$, the numbers R_1 and R_2 that satisfy $\gamma_0 < R_1 < R_2 < 1$ are arbitrary. So we have

$$\alpha_{n,k} = O(1) \times \begin{cases} R_2^k R_1^n & \text{for} \ \ n, k \geq 1 \ \text{for each pair} \ 1 < R_1 < R_2 \\ & \text{when} \ \gamma_0^* = 1 \\ \beta^{k+n} & \text{for} \ \ n, k \geq 1 \ \text{for each} \ \beta \ \text{satisfying} \ \gamma_0^* < \beta < 1 \\ & \text{when} \ \gamma_0^* < 1. \end{cases}$$

$$(3.11)$$

From (3.11) we get for $|w| > R_2$ (where $1 < R_1 < R_2$ when $\gamma_0^* = 1$ and $\gamma_0 < R_1 < R_2 < 1$ when $\gamma_0^* = \gamma_0 < 1$)

$$\sum_{\nu=1}^{\infty} \alpha_{n,\nu} w^{-\nu} = O(1)R_1^n \sum_{\nu=1}^{\infty} \left(\frac{R_2}{|w|}\right)^{\nu} = O(1)R_1^n \quad \text{for} \ |w| > R > R_2. \quad (3.12)$$

The convergence in (3.12) is uniform for $|w| \geq \delta > R_2$.

For the following inequality see [34, eq.(4.9)]:

$$|\alpha_{n,k}| \leq \sqrt{\frac{n}{k}} \quad \text{for} \quad n, k \in \mathbb{N}. \tag{3.13}$$

From (3.13), it readily follows that

$$\left| \sum_{\nu=1}^{\infty} \alpha_{n,\nu} w^{-\nu} \right| \leq \sqrt{n} \sum_{\nu=1}^{\infty} \frac{|w|^{-\nu}}{\sqrt{\nu}} \leq \sqrt{n} \sum_{\nu=1}^{\infty} |w|^{-\nu} \leq \sqrt{n} \frac{1}{|w|} \cdot \frac{1}{1 - \frac{1}{|w|}} = \frac{\sqrt{n}}{|w| - 1} \tag{3.14}$$

for $|w| > 1$ and $n \in \mathbb{N}$. Also

$$\sum_{k=1}^{n} \sum_{\nu=1}^{\infty} \alpha_{n,\nu} w^{-\nu} t^{-(k+1)} = O(1) \sum_{k=1}^{n} \frac{\sqrt{k}}{|w| - 1} \cdot \frac{1}{|t|^{k+1}} = O(1) \tag{3.15}$$

as $n \to \infty$, uniformly for $|w| > \delta > 1$, $|t| \geq \mu$, for each $\delta > 1$ and $\mu > 1$. From (3.9) and (3.6) we see that

$$\frac{\psi'(s)}{\psi(s) - \psi(w)} = \frac{1}{s - w} + \sum_{j=0}^{\infty} \frac{1}{s^{j+1}} \sum_{k=1}^{\infty} \alpha_{j,k} w^{-k}. \tag{3.16}$$

The series converges for all w, $|w| > R_1$, and uniformly for $|w| \geq R^*$, for each $R^* > R_1$. Here $R_1 > 1$ when $\gamma_0^* = 1$ and when ∂E is an r_0-analytic curve then $\gamma_0^* = r_0$, $r_0 < 1$ and $r_0 < R_1 < 1$. Always $R_1 > r_0$.

4. Estimate of the polynomials $\omega_n(z)$ for Fejér and Faber nodes

4.1 Estimates of the polynomials $\omega_n(z)$ for Faber nodes

For the Faber nodes we have $\omega_n(z) = c^n F_{n+1}(z)$. From (3.9) and (3.14), we get for any t and ζ satisfying $1 < |t| = r < |\zeta| = r'$ and any integer $s \geq 1$,

$$\frac{F_{m+1}(\psi(\zeta))^s - F_{m+1}(\psi(t))^s}{F_{m+1}(\psi(\zeta))^s} =$$

$$= \frac{\left(\zeta^{m+1} + O(1)\frac{\sqrt{m+1}}{\sqrt{\zeta}-1}\right)^s - \left(t^{m+1} + O(1)\frac{\sqrt{m+1}}{\sqrt{t}-1}\right)^s}{\left(\zeta^{m+1} + O(1)\frac{\sqrt{m+1}}{\sqrt{\zeta}-1}\right)^s}$$

$$= \frac{\zeta^{s(m+1)}\left(1 + O(1)\frac{\sqrt{m+1}}{\zeta^{m+1}(\sqrt{\zeta}-1)}\right)^s - t^{s(m+1)}\left(1 + O(1)\frac{\sqrt{m+1}}{t^{m+1}(\sqrt{t}-1)}\right)^s}{\zeta^{s(m+1)}\left(1 + O(1)\frac{\sqrt{m+1}}{\zeta^{m+1}(\sqrt{\zeta}-1)}\right)^s}$$

$$= \frac{\zeta^{s(m+1)} - t^{s(m+1)}}{\zeta^{s(m+1)}} \cdot \left(1 + O(1)\frac{\sqrt{m+1}}{\zeta^{m+1}(\sqrt{\zeta}-1)}\right)^{-s} \times$$

$$\times \left\{ 1 + \frac{\left(1 + O(1)\frac{\sqrt{m+1}}{\sqrt{\zeta}-1}\right)^s - 1}{\zeta^{s(m+1)} - t^{s(m+1)}} - \frac{\left(1 + O(1)\frac{\sqrt{m+1}}{\sqrt{t}-1}\right)^s - 1}{\zeta^{s(m+1)} - t^{s(m+1)}} \right\}$$

$$= \frac{\zeta^{s(m+1)} - t^{s(m+1)}}{\zeta^{s(m+1)}} \left(1 + O(1)\frac{\sqrt{m+1}}{\zeta^{m+1}(\sqrt{\zeta}-1)}\right) \times$$

$$\times \left\{ 1 + \frac{O(1)\frac{\sqrt{m+1}}{\sqrt{\zeta}-1}}{\zeta^{s(m+1)} - t^{s(m+1)}} - \frac{O(1)\frac{\sqrt{m+1}}{\sqrt{t}-1}}{\zeta^{s(m+1)} - t^{s(m+1)}} \right\}$$

$$= \frac{\zeta^{s(m+1)} - t^{s(m+1)}}{\zeta^{s(m+1)}} \left(1 + O(1)\frac{\sqrt{m+1}}{|\zeta|^{m+1}}\right). \tag{4.1}$$

Thus we have

$$\frac{\omega_m(\psi(\zeta))^s - \omega_m(\psi(t))^s}{\omega_m(\psi(\zeta))^s} = \frac{\zeta^{s(m+1)} - t^{s(m+1)}}{\zeta^{s(m+1)}} \left(1 + O(1)\frac{\sqrt{m+1}}{|\zeta|^{m+1}}\right) \tag{4.2}$$

as $m \to \infty$, uniformly in t, ζ for $1 < r = |t| < r' = |\zeta| < R$.

4.2 Estimate of the polynomials $\omega_n(z)$ for Fejér nodes

Write $w_{m,k} := \exp(2\pi ki/(m+1))$ $(k = 0, 1, \ldots, m)$. We recall that the points $z_{m,k} := \psi(w_{m,k})$ are called the *Fejér nodes* with respect to E. We write

$$\begin{cases} \omega(z) \equiv \omega_n(z) := \prod_{k=0}^n (z - z_{n,k}) & \text{and for } z = \psi(w) \text{ we have} \\ \omega_n(z) = \prod_{k=0}^n (\psi(w) - \psi(w_{n,k})). \end{cases} \tag{4.3}$$

The function $\pi_n(w)$ is defined by

$$\pi_n(w) := \frac{\omega_n(z)}{(w^{n+1} - 1)d^{n+1}} \quad \text{for } z \neq z_{n,k}, \quad z = \psi(w),$$

and is defined by continuity for $z = z_k$. We have

$$\pi_n(w) = \prod_{k=0}^n \frac{z - z_{n,k}}{w - w_{n,k}} = \prod_{k=0}^n \frac{\psi(w) - \psi(w_{n,k})}{(w - w_{n,k})d} \tag{4.4}$$

when $w \neq w_k$, $|w| \geq r_0$. If $w = w_k$ then the k-th factor is to be replaced by $\phi'(w_k)/d$.

LEMMA 1. *Assume ∂E is an r_0-analytic curve, $r_0 \in [0, 1)$. Then to each β, $1 > r > r_0$ there corresponds a number $M \equiv M_{\beta,c} > 0$, depending only on r and c, such that for all sufficiently large $n \geq N$*

$$\left| \frac{\omega_m(\psi(w))}{(w^{m+1} - 1)d^{m+1}} - 1 \right| = \left| \frac{1}{\pi_n(w)} - 1 \right|$$

$$\leq M\beta^n, \quad |w| \geq r_0, \quad k = 0, 1, \ldots, n; \ n \geq N, \tag{4.5}$$

and

$$\left| \frac{(w^{m+1} - 1)d^{m+1}}{\omega_m(\psi(w))}(w^{m+1} - 1)d^{m+1} - 1 \right| = |\pi_n(w) - 1| \leq M\beta^n$$

for $|w| \geq r_0$, $k = 0, 1, \ldots, n$; $n \geq N$.

PROOF. The function

$$\chi(w, s) = \begin{cases} \dfrac{\psi(w) - \psi(s)}{(w - s)d}, & w \neq s, \\[2ex] \dfrac{\phi'(s)}{d}, & w = s, \end{cases}$$

is clearly an analytic function of s for each fixed w, $|s| \geq r_0$, $|w| \geq r_0$; and $\chi(w, \infty) = \chi(\infty, s) = 1$. Also, $\phi'(\infty)/d = 1$. The univalence of $\phi(w)$ implies that $\phi'(w)$ cannot vanish for $|w| > r_0$. The roles of w and s can be interchanged in these remarks. Therefore $|\chi(w, s)|$ and $X := \{w : +\infty \geq |w| \geq r_0\} \times \{s : +\infty \geq |s| \geq r_0\}$. Hence $|\chi(w, s)|$ has a finite maximum and a non-zero minimum on this set X.

Let Log $\chi(w, s)$ denote the branch of the logarithm for which Log $\chi(w, \infty) = 0$ on X. For w fixed, $|w| \geq r_0$, this is an analytic function of s for $|s| \geq r_0$ with a Laurent series expansion around infinity of the form

$$\text{Log } \chi(w, s) = \frac{c_1(w)}{s} + \frac{c_2(w)}{s^2} + \cdots .$$

The coefficients of this Laurent expansion are given by

$$c_m(w) = \frac{1}{2\pi i} \oint_{|s|=r_0} s^{m-1} \text{Log } \zeta(w, s)\, ds.$$

The boundedness of $|\chi(w, s)|$ implies that Log $\chi(w, s)$ is bounded and continuous on X. If M_χ is the maximum of Log $\chi(w, s)$, then by the Cauchy estimates we have $|c_m(w)| \leq M_\chi r_0^m$. For the roots of unity $(w_{n,k})$ we have

$$\sum_{k=0}^{n} w_{n,k}{}^j = \begin{cases} 0 & \text{if } j \neq 0 \bmod(n+1), \\ n+1 & \text{if } j = 0 \bmod(n+1). \end{cases}$$

By (4.4) we get

$$\log \pi_n(w) = \sum_{k=0}^{n} \text{Log } \chi(w, w_k) = \sum_{j=1}^{\infty} c_j(w) w_{n,k}{}^{-j}$$

$$= (n+1) \sum_{j=1}^{\infty} c_{j(n+1)}(w) = (n+1) M_\chi \sum_{j=1}^{\infty} r_0^{j(n+1)}$$

so

$$|\log \pi_n(w)| = \left| \log \frac{1}{\pi_n(w)} \right| \leq \frac{(n+1) r_0^{n+1} M_\chi}{1 - r_0^{n+1}} \to 0 \text{ as } n \to \infty \text{ for, } |w| \geq r_0.$$

If $|\log z| < \varepsilon$ with $0 \le \varepsilon \le 1$, then

$$|z - 1| = |\exp(\log z) - 1| \le \varepsilon(1 + \varepsilon/2! + \varepsilon^2/3! + \ldots) \le \frac{\varepsilon}{1 - \varepsilon}.$$

Let β be a given number satisfying $r_0 < \beta < 1$, and choose $M_0 = M_0(\beta, \chi)$ so that

$$\frac{(n + 1)r_0^{n+1}M_\chi}{1 - r_0^{n+1}} \le M_0\beta^n, \qquad n \ge N(r_0, N_\chi).$$

Then for $|w| \ge r_0$,

$$|\pi_n(w) - 1| \le M_0\beta^n, \quad n \ge N(r_0, M_\chi),$$

$$\left|\frac{1}{\pi_n(w)} - 1\right| \le M_0\beta^n, \quad n \ge N(r_0, M_\chi),$$

which completes the proof. $\qquad\qquad\qquad\qquad\qquad\qquad\qquad\qquad\square$

Applying (4.5) of Lemma 1 with some β, $r_0 < \beta < 1$, to each of the terms in the following fraction we get

$$\frac{\omega_m(\psi(\zeta))^s - \omega_m(\psi(t))^s}{\omega_m(\psi(\zeta))^s} =$$

$$= \frac{(\zeta^{m+1} - 1)^s(1 + O(1)\beta^m)^s - (t^{m+1} - 1)^s(1 + O(1)\beta^m)^s}{(\zeta^{m+1} - 1)^s(1 + O(1)\beta^m)^s}$$

$$= \frac{(\zeta^{m+1} - 1)^s(1 + O(1)\beta^m) - (t^{m+1} - 1)^s(1 + O(1)\beta^m)}{(\zeta^{m+1} - 1)^s(1 + O(1)\beta^m)}$$

$$= \frac{(\zeta^{m+1} - 1)^s - (t^{m+1} - 1)^s}{(\zeta^{m+1} - 1)^s} \times$$

$$\times \left(1 + \frac{(\zeta^{m+1} - 1)^s O(1)\beta^m - (t^{m+1} - 1)^s O(1)\beta^m}{(\zeta^{m+1} - 1)^s - (t^{m+1} - 1)^s}\right)$$

$$= \frac{(\zeta^{m+1} - 1)^s - (t^{m+1} - 1)^s}{(\zeta^{m+1} - 1)^s}(1 + O(1)\beta^m) \qquad (4.6)$$

5. Integral representations of Lagrange and Hermite interpolants for Faber expansions

We use in the proofs of the results of this chapter certain integral representations of the Lagrange, Hermite and a combination of these interpolants in a form which is relevant to Faber polynomials.

Given a function $f \in A_R^*$, $R > 1$, we shall assume in all the proofs of this chapter that the numbers r, r' satisfy $1 < r < r' < R$ and that the interpolating points $(z_{k,n})$ satisfy $z_{k,n} \in G_r$ for all sufficiently large n. The last assumption is satisfied by the Faber and Fejér nodes.

Applying the well-known integral formula for the Hermite interpolation (cf. Chapter 2, (1.0)) we obtain for each $r' \in (1, R)$

$$L_m(f; z) = \frac{1}{2\pi i} \oint_{|\zeta|=r'} f(\psi(\zeta)) \frac{\psi'(\zeta)}{\psi(\zeta) - z} \cdot \frac{\omega_m(\psi(\zeta)) - \omega_m(z)}{\omega_m(\psi(\zeta))} d\zeta, \quad z \in \mathbb{C} \quad (5.1)$$

and

$$h_{s,s(m+1)-1}(f; z) = \frac{1}{2\pi i} \oint_{|\zeta|=r'} f(\psi(\zeta)) \frac{\psi'(\zeta)}{\psi(\zeta) - z} \cdot \frac{\omega_m(\psi(\zeta))^s - \omega_m(z)^s}{\omega_m(\psi(\zeta))^s} d\zeta, \quad z \in \mathbb{C}.$$
$$(5.2)$$

By (1.1.2) and (1.1.4) the expansion of

$$S_n(L_m(f; \cdot); z) \quad \text{and} \quad S_n(h_{s,s(m+1)-1}(f; \cdot); z)$$

in terms of Faber polynomials are given by

$$S_n(L_m(f; \cdot); z) = \frac{1}{2\pi i} \oint_{|t|=r} L_m(\psi(\cdot); t) \sum_{k=0}^n F_k(z) t^{-(k+1)} dt$$

$$= \frac{1}{2\pi i} \oint_{|\zeta|=r'} f(\psi(\zeta)) \left(\frac{1}{2\pi i} \oint_{|t|=r} \frac{\psi'(\zeta)}{\psi(\zeta) - \psi(t)} \times \right.$$

$$\left. \times \frac{\omega_m(\psi(\zeta)) - \omega_m(\psi(t))}{\omega_m(\psi(\zeta))} \sum_{k=0}^n F_k(z) t^{-(k+1)} dt \right) d\zeta, \quad z \in \mathbb{C}.$$
$$(5.3)$$

and

$$S_n(h_{s,s(m+1)-1}(f; \cdot); z) = \frac{1}{2\pi i} \oint_{|t|=r} h_{s,s(m+1)-1}(f; \psi(t)) \cdot \sum_{k=0}^n F_k(z) t^{-(k+1)} dt$$

$$= \frac{1}{2\pi i} \oint_{|\zeta|=r'} f(\psi(\zeta)) \left(\frac{1}{2\pi i} \oint_{|t|=r} \frac{\psi'(\zeta)}{\psi(\zeta) - \psi(t)} \times \right.$$

$$\left. \times \frac{\omega_m^s(\psi(\zeta)) - \omega_m^s(\psi(t))}{\omega_m(\psi(\zeta))^s} \sum_{k=0}^n F_k(z) t^{-(k+1)} dt \right) d\zeta, \quad z \in \mathbb{C}.$$
$$(5.4)$$

We set, for $m \geq n \geq 0$ and $j \geq 0$, as in (2.0),

$$S_{m,n,j}(f; z) := \sum_{k=0}^n a_{k+j(m+1)} F_k(z). \quad (5.5)$$

Clearly $S_{m,n,0}(f; z) = S_n(f; z)$ and $S_n(L_m(f; z); z)$ is the nth Faber section of the expansion of $L_m(f; z)$ in terms of Faber polynomials.

Now (2.0) and (1.1.2) imply that

$$S_{m,n,j}(f; z) = \frac{1}{2\pi i} \oint_{|\zeta|=r'} f(\psi(\zeta)) \zeta^{-j(m+1)} \sum_{k=0}^n F_k(z) \zeta^{-(k+1)} d\zeta, \quad z \in \mathbb{C}.$$

Thus we have, for a given integer $\ell \geq 1$

$$\sum_{j=0}^{\ell-1} S_{m,n,j}(f;z) = \frac{1}{2\pi i} \oint_{|\zeta|=r'} f(\psi(\zeta)) \frac{\zeta^{\ell(m+1)} - 1}{\zeta^{(\ell-1)(m+1)}(\zeta^{m+1} - 1)} \sum_{k=0}^{n} F_k(z)\zeta^{-(k+1)} d\zeta$$

$$\text{for } z \in \mathbb{C} \qquad (5.6)$$

The function $\psi(w)$ is univalent for $|w| > 1$ and has a simple pole at infinity with residue c at ∞. Therefore the function $\frac{\psi'(\zeta)}{\psi(\zeta)-\psi(t)}$ of t is regular but for one simple pole at $t = \zeta$, for $|t| > 1$, and $\lim_{t\to\infty} \frac{\psi'(\zeta)}{\psi(\zeta)-\psi(t)} = 0$. The residue at $t = \zeta$ of the function $\frac{\psi'(\zeta)}{\psi(\zeta)-\psi(t)} \cdot \frac{1}{t^s}$ of the variable t, for an integer $s \geq 0$, is $\lim_{t\to\zeta} \frac{(\zeta-t)\psi'(\zeta)}{\psi(\zeta)-\psi(t)} \cdot \frac{1}{t^s} = \frac{1}{\zeta^s}$. Using now the residue theorem, we obtain

$$\frac{1}{2\pi i} \oint_{|t|=r} \frac{\psi'(\zeta)}{\psi(\zeta) - \psi(t)} \cdot \frac{dt}{t^s} = \frac{1}{\zeta^s}, \quad |\zeta| > r > 1. \qquad (5.7)$$

Applying this formula with $s = 1$ and (5.6) for $1 < r < r'$, we get the following double integral:

$$\sum_{j=0}^{\ell-1} S_{m,n,j}(f;z) = \frac{1}{2\pi i} \oint_{|\zeta|=r'} f(\psi(\zeta)) \left(\frac{1}{2\pi i} \oint_{|t|=r} \frac{\psi'(\zeta)}{\psi(\zeta) - \psi(t)} \cdot \frac{\zeta}{t} \times \right.$$

$$\left. \times \frac{\zeta^{\ell(m+1)} - 1}{\zeta^{(\ell-1)(m+1)}(\zeta^{m+1} - 1)} \sum_{k=0}^{n} F_k(z)\zeta^{-(k+1)} dt \right) d\zeta, \quad z \in \mathbb{C}.$$
$$\qquad (5.8)$$

In analogy to (5.4) it is easy to see that for $1 < r < r' < R$

$$S_{p(n+1)-1}(h^*_{s(m+1)-1}(f;\cdot);z) = \frac{1}{2\pi i} \oint_{|\zeta|=r} f(\psi(\zeta)) \times$$

$$\times \left(\frac{1}{2\pi i} \oint_{|t|=r} \frac{\psi'(\zeta)}{\psi(\zeta) - \psi(t)} \frac{w_m^s(\psi(\zeta)) - w_m^s(\psi(t))}{w_m^s(\psi(\zeta))} \sum_{k=0}^{p(n+1)-1} \frac{F_k(z)}{t^{k+1}} dt \right) d\zeta.$$
$$\qquad (5.9)$$

From (5.8) we get for $\ell = 1$ and $1 < r < r' < R$

$$S_{p(n+1)-1}(f;z)) =$$

$$= \frac{1}{2\pi i} \oint_{|\zeta|=r'} f(\psi(\zeta)) \left(\frac{1}{2\pi i} \oint_{|t|=1} \frac{\psi'(\zeta)}{\psi(\zeta) - \psi(t)} \frac{\zeta}{t} \sum_{k=0}^{p(n+1)-1} \frac{F_k(z)}{\zeta^{k+1}} dt \right) d\zeta.$$
$$\qquad (5.10)$$

6. Proofs of the theorems stated in Section 2

6.1 Proof of Theorem 3

In the proof we assume that $0 < r_0 < R_1 < 1 < r < r' < R$, that $|t| = r$ and $|\zeta| = r'$. If $f(z) \in A_R^*$ has the Faber expansion $f(z) = \sum_{k=0}^{\infty} a_k F_k(z)$, then by (1.1.2) and (1.1.4) we have for each r', $1 < r' < R$,

$$S_n(f; z) = \frac{1}{2\pi i} \oint_{|\zeta|=r'} f(\psi(\zeta)) \sum_{k=0}^{n} F_k(z)\zeta^{-(k+1)}d\zeta, \quad z \in \mathbb{C}.$$

By (2.1), (5.3) and (5.8) we get

$$D_{m,n,\ell}(f; z) = \frac{1}{2\pi i} \oint_{|\zeta|=r'} f(\psi(\zeta)) \left(\frac{1}{2\pi i} \oint_{|t|=r} \frac{\psi'(\zeta)}{\psi(\zeta) - \psi(t)} K_{m,n,\ell}(w, \zeta, t)dt \right) d\zeta .$$

$$(6.1.1)$$

where

$$K_{m,n,\ell}(w, \zeta, t) =$$

$$= \frac{\omega_m(\psi(\zeta)) - \omega_m(\psi(t))}{\omega_m(\psi(\zeta))} \left(\sum_{k=1}^{n} \left(F_k(\psi(w)) - w^k \right) t^{-(k+1)} + \sum_{k=0}^{n} w^k t^{-(k+1)} \right)$$

$$- \frac{\zeta}{t} \cdot \frac{\zeta^{\ell(m+1)-1}}{\zeta^{(\ell-1)(m+1)}(\zeta^{m+1} - 1)} \left(\sum_{k=1}^{n} \left(F_k(\psi(w)) - w^k \right) t^{-(k+1)} + \sum_{k=0}^{n} w^k t^{-(k+1)} \right).$$

$$(6.1.2)$$

We fix a number $\rho = \beta$, $r_0 < \beta < 1$. By (4.6) with $s = 1$ we can write

$$\frac{\omega_m(\psi(\zeta)) - \omega_m(\psi(t))}{\omega_m(\psi(\zeta))} = \frac{\zeta^{m+1}}{\zeta^{m+1} - 1} - \frac{t^{m+1}}{\zeta^{m+1} - 1} + O(1)\beta^m \frac{\zeta^{m+1} - t^{m+1}}{\zeta^{m+1} - 1} \quad (6.1.3)$$

uniformly as $n \to \infty$ on sets of points $\{(\zeta, t) : |t| = r < r' = \zeta\}$ for each pair $1 < r < r' < R$. By applying now (6.1.3) and (3.9) the kernel $K_{m,n,\ell}(w, \zeta, t)$ can be broken into six parts, $K_{m,n,\ell}(w, \zeta, t) := \sum_{j=1}^{6} K_{m,n,\ell}(w, \zeta, t, j)$, where

$$K_{m,n,\ell,1}(w, \zeta, t) := \frac{\zeta^{m+1}}{\zeta^{m+1} - 1} \sum_{k=1}^{n} \sum_{\nu=1}^{\infty} \alpha_{k,\nu} w^{-\nu} t^{-(k+1)}$$

$$- \frac{\zeta}{t} \frac{\zeta^{\ell(m+1)} - 1}{\zeta^{(\ell-1)(m+1)}(\zeta^{m+1} - 1)} \sum_{k=1}^{n} \sum_{\nu=1}^{\infty} \alpha_{k\nu} w^{-\nu} \zeta^{-(k+1)}$$

$$K_{m,n,\ell,2}(w, \zeta, t) := -\frac{t^{m+1}}{\zeta^{m+1} - 1} \sum_{k=1}^{n} \sum_{\nu=1}^{\infty} \alpha_{k\nu} w^{-\nu} t^{-(k+1)}$$

$$(6.1.4)$$

$$K_{m,n,\ell,3}(w, \zeta, t) = O(1)\beta^m \frac{\zeta^{m+1} - t^{m+1}}{\zeta^{m+1} - 1} \sum_{k=1}^{n} \sum_{\nu=1}^{\infty} \alpha_{k\nu} w^{-\nu} t^{-(k+1)}$$

$$K_{m,n,\ell,4}(w, \zeta, t) := \frac{\zeta^{m+1}(t^{n+1} - w^{n+1})}{(t - w)(\zeta^{m+1} - 1)t^{n+1}}$$

$$- \frac{\zeta}{t} \frac{(\zeta^{n+1} - w^{n+1})}{(\zeta - w)\zeta^{n+1}} \cdot \frac{\zeta^{\ell(m+1)} - 1}{\zeta^{(\ell-1)(m+1)}(\zeta^{m+1} - 1)}$$

and

$$K_{m,n,\ell,5}(w,\zeta,t) = O(1)\beta^m \frac{(\zeta^{m+1}-t^{m+1})(t^{n+1}-w^{n+1})}{(t-w)t^{n+1}(\zeta^{m+1}-1)}$$
$$K_{m,n,\ell,6}(w,\zeta,t) = -\frac{t^{m+1}(t^{n+1}-w^{n+1})}{(t-w)t^{n+1}(\zeta^{m+1}-1)} .$$

(6.1.5)

By (3.9) the two series on the right hand side of definition of $K_{m,n,\ell}(w,\zeta,t)$ are uniformly convergent for $|w| > 1$ and $|t| \geq r > 1$.

We write $D_{m,n,\ell}(f;z) = \sum_{j=1}^{6} D_{m,n,\ell,j}(f;z)$ where

$$D_{m,n,\ell,j} = \frac{1}{2\pi i} \oint_{|\zeta|=r'} f(\psi(\zeta)) \left(\frac{1}{2\pi i} \oint_{|t|=r} \frac{\psi'(t)}{\psi(\zeta)-\psi(t)} K_{m,n,\ell,j}(w,\zeta,t)dt \right) d\zeta$$

The regularity of the integrands defining $D_{m,n,\ell,j}$ allows us to replace, independently, in each of the integrals defining $D_{m,n,\ell,j}$ any given triple R_1, r_0, r' by any other triple R_1, r_0, r' as long as it satisfies $r_0 < R_1 < 1 < r < r' < R$.

We estimate now integrands of $D_{m,n,\ell,j}$ for $1 \leq j \leq 6$. In these estimates the assumptions are that $r_0 < \beta < 1 < r < r' < R$, $|w| > R$, $|t| = r$, $|\zeta| = r'$.

Estimate of $D_{m,n,\ell,1}$ By (5.7) we have

$$\frac{1}{2\pi i} \oint_{|t|=r} \frac{\psi'(\zeta)}{\psi(\zeta)-\psi(t)} K_{m,n,\ell,1}(w,\zeta,t)dt = \frac{1}{(\zeta^{m+1}-1)\zeta^{(\ell-1)(m+1)}} \times$$
$$\times \sum_{k=1}^{n} \sum_{\nu=1}^{\infty} \alpha_{k,\nu} w^{-\nu} \zeta^{-(k+1)}$$

and by (3.11)

$$\frac{1}{2\pi i} \oint_{|t|=r} \frac{\psi'(\zeta)}{\psi(\zeta)-\psi(t)} K_{m,n,\ell,1}(w,\zeta,t)dt = O(1)\frac{1}{(r')^{\ell(m+1)}} \sum_{k=1}^{n} \beta^k (r')^{-(k+1)}.$$

Hence

$$\limsup_{n\to\infty} \left| \frac{1}{2\pi i} \oint_{|t|=r} \frac{\psi'(\zeta)}{\psi(\zeta)-\psi(t)} K_{m,n,\ell,1}(w,\zeta,t)dt \right|^{1/n} \leq \frac{1}{(r')^{\ell q}}.$$

Therefore

$$\limsup_{n\to\infty} |D_{m,n,\ell,1}(f;z)|^{1/n} \leq \frac{1}{(r')^{\ell q}}.$$

Letting $r' \nearrow R$ we get

$$\limsup_{n\to\infty} |D_{m,n,\ell,1}|^{1/n} \leq \frac{1}{R^{\ell q}} < 1 \quad \text{for } |w| > R.$$

Estimate of $D_{m,n,\ell,2}$ By (3.11) we have

$$|K_{m,n,\ell,2}(w,\zeta,t)| = O(1)\frac{r^{m+1}}{(r')^{m+1}}\sum_{k=1}^{n}\beta^{k}r^{-(k+1)}.$$

Therefore

$$\limsup_{n\to\infty}|K_{m,n,\ell,2}(m,n,\ell)|^{1/n} \le \left(\frac{r}{r'}\right)^{q}.$$

Hence

$$\limsup_{n\to\infty}|D_{m,n,\ell,2}(f;z)|^{1/n} \le \left(\frac{r}{r'}\right)^{q}.$$

Letting $r \searrow 1$ and $r' \nearrow R$ we get

$$\limsup_{n\to\infty}|D_{m,n,\ell,2}|^{1/n} \le \frac{1}{R^{q}} < 1 \quad \text{for } |w| < R.$$

Estimate of $D_{m,n,\ell,3}$ By (3.11) we have

$$K_{m,n,\ell,3}(w,\zeta,t) = O(1)\beta^{m}.$$

Hence

$$\limsup_{n\to\infty}|K_{m,n,\ell,3}|^{1/q} \le \beta^{q}.$$

Therefore

$$\limsup_{n\to\infty}|D_{m,n,\ell,3}(f;z)|^{1/n} \le \beta^{q}.$$

Letting $\beta \searrow r_{0}$ we get

$$\limsup_{n\to\infty}|D_{m,n,\ell,3}(w,\zeta,t)|^{1/n} \le r_{0}^{1} < 1 \quad \text{for } |w| > R.$$

Estimate of $D_{m,n,\ell,4}$ We have $\frac{t^{n+1}-w^{n+1}}{t^{n+1}(t-w)} = \sum_{k=1}^{n}w^{n-k}t^{k-(n+1)}$. By applying (5.7) we see that

$$\frac{1}{2\pi i}\oint_{|t|=r}\frac{\psi'(\zeta)}{\psi(\zeta)-\psi(t)}K_{m,n,\ell,4}(w,\zeta,t)dt = K_{m,n,\ell,4}(w,\zeta,\zeta)$$

$$= \frac{\zeta^{n+1}-w^{n+1}}{(\zeta-w)\zeta^{((\ell-1)q+1)(n+1)}(\zeta^{q(n+1)}-1)}$$

$$= O(1)\left(\frac{|w|}{(r')^{1+\ell q}}\right)^{n}, \quad \text{as } n \to \infty$$

Hence

$$\limsup_{n\to\infty}\left|\frac{1}{2\pi i}\oint_{|t|=r}\frac{\psi'(\zeta)}{\psi(\zeta)-\psi(t)}K_{m,n,\ell,4}(w,\zeta,t)dt\right|^{1/n} \le \frac{|w|}{(r')^{\ell q+1}}.$$

Therefore

$$\limsup_{n\to\infty} |D_{m,n,\ell,4}|^{1/n} \le \frac{|w|}{(r')^{\ell q+1}}.$$

Letting $r' \nearrow R$ we see that

$$\limsup_{n\to\infty} |D_{m,n,\ell,4}(f;z)|^{1/n} \le \frac{|w|}{R^{\ell q+1}} \qquad \text{for } |w| > R.$$

Estimate of $D_{m,n,\ell,5}$ From the definition of $K_{m,n,\ell,5}(w,\zeta,t)$ it follows immediately that we have

$$K_{m,n,\ell,5}(w,\zeta,t) = O(1) \left(\frac{|w|\beta^q}{r} \right)^n \qquad \text{as } n \to \infty. \tag{6.1.6}$$

Therefore

$$\limsup_{n\to\infty} |K_{m,n,\ell,5}(w,\zeta,t)|^{1/n} \le \frac{|w|\beta^q}{r}.$$

Hence

$$\limsup_{n\to\infty} |D_{m,n,\ell,5}(f;z)|^{1/n} \le \frac{|w|\beta^q}{r}.$$

Letting $r \nearrow R$ and $\beta \searrow r_0$ we get

$$\limsup_{n\to\infty} |D_{m,n,\ell,5}(f;z)|^{1/n} \le \frac{|w|r_0^q}{R} \qquad \text{for } |w| > R.$$

Estimate of $D_{m,n,\ell,6}$ From the definition of $K_{m,n,\ell,6}$ and (3.16) we get

$$\frac{1}{2\pi i} \oint_{|t|=r} \frac{\psi'(\zeta)}{\psi(\zeta) - \psi(t)} K_{m,n,\ell,6}(w,\zeta,t)dt =$$

$$= -\frac{1}{2\pi i} \oint_{|t|=r} \left(\frac{1}{\zeta - t} + \sum_{\nu=1}^{\infty}\sum_{k=1}^{\infty} \alpha_{\nu,k} t^{-k}\zeta^{-(\nu+1)} \right) \frac{t^{m+1}}{t^{n+1}} \frac{t^{n+1} - w^{n+1}}{t - w} \times$$

$$\times \frac{1}{\zeta^{m+1} - 1} dt$$

$$= -\frac{1}{\zeta^{m+1} - 1} \frac{1}{\zeta} \frac{1}{2\pi i} \oint_{|t|=r} \left(\sum_{k=0}^{\infty}\sum_{j=0}^{n} \zeta^{-k} w^j t^{k+m-j} + \right.$$

$$\left. + \sum_{\nu=1}^{\infty}\sum_{k=1}^{\infty}\sum_{j=0}^{n} \alpha_{\nu,k}\zeta^{-\nu} w^j t^{m-k-j} \right) dt.$$

Since

$$\frac{1}{2\pi i} \oint_{|t|=r} t^p = \begin{cases} 1 & \text{for } p = -1 \\ 0 & \text{for } p \ne -1 \end{cases} \tag{6.1.7}$$

and $k+m-j = k+q(n+1)-1-j \geq 0$ the integral of the double sum vanishes and the integral of the triple sum gives contibution only when $k = m+1-j$. Hence we get

$$\frac{1}{2\pi i} \oint_{|t|=r} \frac{\psi'(\zeta)}{\psi(\zeta) - \psi(t)} K_{m,n,\ell,6}(w,\zeta,t)dt =$$

$$= \frac{1}{\zeta^{m+1} - 1} \frac{1}{\zeta} \sum_{\nu=1}^{\infty} \sum_{j=0}^{n} \alpha_{\nu,m+1-j} w^j \zeta^{-(\nu+1)}.$$

From (3.11) we get

$$\frac{1}{2\pi i} \oint_{|t|=r} \frac{\psi'(\zeta)}{\psi(\zeta) - \psi(t)} K_{m,n,\ell,6}(w,\zeta,t)dt =$$

$$= O(1) \frac{1}{(r')^{m+2}} \sum_{\nu=1}^{\infty} \sum_{j=0}^{n} \beta^{m+1-j} \beta^{\nu} \zeta^{-(\nu+1)} w^j = O(1) \frac{\beta^{m+2-n}|w|^n}{(r')^{m+2}}.$$

When $r_0 = 0$ we can choose β as near to 0 as we wish. So that

$$\frac{1}{2\pi i} \oint_{|t|=r} \frac{\psi'(\zeta)}{\psi(\zeta) - \psi(t)} K_{m,n,\ell,6}(w,\zeta,t)dt = 0.$$

Hence

$$\limsup_{n\to\infty} |K_{m,n,\ell,6}(w,\zeta,t)|^{1/n} \leq \begin{cases} \frac{|w|\beta^{q-1}}{(r')^q} & \text{when } r_0 > 0 \\ 0 & \text{when } r_0 = 0 \end{cases}$$

and

$$\limsup_{n\to\infty} |D_{m,n,\ell,6}(f;z)|^{1/n} \leq \begin{cases} \frac{|w|\beta^{q-1}}{(r')^q} & \text{when } r_0 > 0 \\ 0 & \text{when } r_0 = 0. \end{cases}$$

Letting $\beta \searrow r_0$ and $r' \nearrow R$ we get

$$\limsup_{n\to\infty} |D_{m,n,\ell,6}(f;z)|^{1/n} \leq \frac{|w|r_0^{q-1}}{R^q} \qquad \text{for } |w| > R \text{ where } 0^0 := 0.$$

Combining the above six estimates we obtain the result

$$\limsup_{n\to\infty} |D_{m,n,\ell}(f;z)|^{1/n} \leq \max\left(\frac{1}{R^{\ell q}}, \frac{1}{R^q}, r_0, \frac{|w|}{R^{\ell q+1}}, \frac{|w|r_0^q}{R}, \frac{|w|r_0^{q-1}}{R^q} \right)$$

$$= \max\left(r_0, \frac{|w|}{R^{\ell q+1}}, \frac{|w|r_0^q}{R}, \frac{|w|r_0^{q-1}}{R^q} \right).$$

□

6.2 Proof of Theorem 4

We proceed along the lines of the proof of Theorem 3. We assume in the proof that $1 < r < r' < R$ and $|w| > R$. From (5.3) and (5.8) we obtain an integral representation for $\Delta^*_{m,n,\ell}(f;z)$ of (2.4) as the difference of two double integrals. From (5.3), (5.8), (3.9) and (4.2) we see that for $1 < \rho < r < r' < R$ (with $z = \psi(w)$) we can write

$$\Delta^*_{m,n,\ell}(f;z) = \sum_{j=1}^{6} \Delta^*_{m,n,\ell,j}(f;z)$$

$$\equiv \frac{1}{2\pi i} \oint_{|\zeta|=r'} f(\psi(\zeta)) \left(\frac{1}{2\pi i} \oint_{|t|=r} \frac{\psi'(\zeta)}{\psi(\zeta) - \psi(t)} \sum_{j=1}^{6} K^{(j)}_{m,n,\ell}(w,\zeta,t) dt \right) d\zeta$$

where

$$K_{m,n,\ell,1}(w,\zeta,t) = \sum_{k=1}^{n} \sum_{\nu=1}^{\infty} \alpha_{k,n} w^{-\nu} t^{-k-1}$$

$$- \frac{\zeta}{t} \frac{\zeta^{\ell(m+1)} - 1}{\zeta^{(\ell-1)(m+1)}(\zeta^{m+1} - 1)} \sum_{k=1}^{n} \sum_{\nu=1}^{\infty} \alpha_{k,\nu} w^{-\nu} \zeta^{-k-1}$$

$$K_{m,n,\ell,2}(w,\zeta,t) = -\frac{t^{m+1}}{\zeta^{m+1}} \sum_{\nu=1}^{\infty} \alpha_{k,\nu} w^{-\nu} t^{-k-1}$$

$$K_{m,n,\ell,3}(w,\zeta,t) = O(1) \frac{\sqrt{m+1}}{|\zeta|^{m+1}} \frac{\zeta^{m+1} - t^{m+1}}{\zeta^{m+1}} \sum_{k=1}^{n} \sum_{\nu=1}^{\infty} \alpha_{k,\nu} w^{-\nu} t^{-k-1}$$

$$K_{m,n,\ell,4}(w,\zeta,t) := \frac{t^{n+1} - w^{n+1}}{(t-w)t^{n+1}} - \frac{\zeta}{t} \cdot \frac{\zeta^{n+1} - w^{n+1}}{(t-w)\zeta^{n+1}} \cdot \frac{\zeta^{\ell(m+1)} - 1}{\zeta^{(\ell-1)(m+1)}(\zeta^{m+1} - 1)}$$

$$\tag{6.2.1}$$

$$K_{m,n,\ell,5}(w,\zeta,t) := O(1) \frac{\sqrt{m+1}}{|\zeta|^{m+1}} t^m \cdot \frac{t^{n+1} - w^{n+1}}{(t-w)t^{n+1}} \tag{6.2.2}$$

and

$$K_{m,n,\ell,6}(w,\zeta,t) := -\frac{t^{m+1}}{\zeta^{m+1}} \cdot \frac{t^{n+1} - w^{n+1}}{(t-w)t^{n+1}} \tag{6.2.3}$$

We estimate now $\Delta^*_{m,n,\ell,j}(f;z)$ for $1 \leq j \leq 6$.

Estimate of $\Delta^*_{m,n,\ell,1}(f;z)$. By (5.7) and (3.14) we have

$$\frac{1}{2\pi i} \oint_{|t|=r} \frac{\psi'(\zeta)}{\psi(\zeta) - \psi(t)} K_{m,n,\ell,1}(w,\zeta,t) dt = -\frac{\zeta^{(\ell-1)(m+1)} - 1}{\zeta^{(\ell-1)(m+1)}(\zeta^{m+1} - 1)} \times$$

$$\times \sum_{k=1}^{n}\sum_{\nu=1}^{\infty}\alpha_{k,\nu}w^{-\nu}\zeta^{-(k+1)} = O(1)\frac{1}{(r')^{m+1}}\left(1 - \frac{1}{r'}\right)^{3/2}$$

Therefore

$$\limsup_{n\to\infty}\left|\frac{1}{2\pi i}\oint_{|t|=r}\frac{\psi'(\zeta)}{\psi(\zeta) - \psi(t)}K_{m,n,\ell,1}(w,\zeta,t)dt\right|^{1/n} \le \frac{1}{(r')^{q}}.$$

Hence

$$\limsup_{n\to\infty}\left|\Delta^{*}_{m,n,\ell,1}(f;z)\right|^{1/n} \le \frac{1}{(r')^{q}}.$$

Letting $r' \nearrow R$ we obtain

$$\limsup_{n\to\infty}\left|\Delta^{*}_{m,n,\ell,1}(f;z)\right|^{1/n} \le \frac{1}{R^{q}} < 1.$$

Estimate of $\Delta^{*}_{m,n,\ell,2}(f;z)$. By (3.14) we have

$$K_{m,n,\ell,2}(w,\zeta,t) = O(1)\left(\frac{r}{r'}\right)^{m+1}\left(1 - \frac{1}{r}\right)^{-3/2}.$$

Hence

$$\limsup_{n\to\infty}|K_{m,n,\ell,2}(w,\zeta,t)|^{1/n} \le \left(\frac{r}{r'}\right)^{q}.$$

Therefore

$$\limsup_{n\to\infty}\left|\Delta^{*}_{m,n,\ell,2}(f;z)\right|^{1/n} \le \left(\frac{r}{r'}\right)^{q}.$$

Letting $r \searrow 1$ and $r' \nearrow R$ we get

$$\limsup_{n\to\infty}\left|\Delta^{*}_{m,n,\ell,2}(f;z)\right|^{1/n} \le \frac{1}{R^{q}} < 1.$$

Estimate of $\Delta^{*}_{m,n,\ell,3}(f;z)$. By (3.14) we have

$$K_{m,n,\ell,3}(w,\zeta,t) = O(1)\frac{\sqrt{m+1}}{|\zeta|^{m+1}}.$$

Hence we get

$$\limsup_{n\to\infty}|K_{m,n,\ell,3}(w,\zeta,t)|^{1/n} \le \frac{1}{(r')^{q}}$$

which implies

$$\limsup_{n\to\infty}\left|\Delta^{*}_{m,n,\ell,3}(f;z)\right|^{1/n} \le \frac{1}{(r')^{q}}.$$

Letting $r' \nearrow R$ we get

$$\limsup_{n\to\infty}\left|\Delta^{*}_{m,n,\ell,3}(f;z)\right|^{1/n} \le \frac{1}{R^{q}} < 1 \qquad \text{for } |w| > R.$$

Estimate of $\Delta^*_{m,n,\ell,4}(f;z)$. Again using the residue theorem, we see that for $|w| > R$,

$$\frac{1}{2\pi i} \oint_{|t|=r} \frac{\psi'(\zeta)}{\psi(\zeta) - \psi(t)} K_{m,n,\ell,4}(w,\zeta,t)dt = K_{m,n,\ell,4}(w,\zeta,\zeta)$$

$$= \frac{\zeta^{n+1} - w^{n+1}}{(\zeta - w)\zeta^{n+1}} \cdot \frac{1 - \zeta^{(\ell-1)(m+1)}}{\zeta^{(\ell-1)(m+1)}(\zeta^{m+1} - 1)}$$

$$= \begin{cases} 0, & \ell = 1 \\ O(1) \left(\frac{|w|}{(r')^{1+q}} \right)^n, & \ell \geq 2. \end{cases}$$

Hence we get

$$\limsup_{n \to \infty} |K_{m,n,\ell,4}(w,\zeta,t)|^{1/n} \leq \begin{cases} 0 & \text{when} & \ell = 1 \\ \frac{|w|}{(r')^{q+1}} & \text{when} & \ell \geq 2. \end{cases}$$

Therefore

$$\limsup_{n \to \infty} \left| \Delta^*_{m,n,\ell,4}(f;z) \right|^{1/n} \leq \begin{cases} 0 & \text{when} & \ell = 1 \\ \frac{|w|}{(r')^{q+1}} & \text{when} & \ell \geq 2. \end{cases}$$

Letting $r' \nearrow R$ we get

$$\limsup_{n \to \infty} \left| \Delta^*_{m,n,\ell,4}(f;z) \right|^{1/n} \leq \begin{cases} 0 & \text{when} & \ell = 1 \\ \frac{|w|}{R^{q+1}} & \text{when} & \ell \geq 2. \end{cases}$$

Estimate of $\Delta^*_{m,n,\ell,5}(f;z)$. It is obvious from the expression for $K_{m,n,\ell,5}$ (w,ζ,t) that for $|w| > R > r'$,

$$\limsup_{n \to \infty} |K_{m,n,\ell,5}(w,\zeta,t)|^{1/n} \leq \frac{|w|}{r|\zeta|^q} = \frac{|w|}{r(r')^q}.$$

Hence

$$\limsup_{n \to \infty} \left| \Delta^*_{m,n,\ell,5}(f;z) \right|^{1/n} \leq \frac{|w|}{r((r')^q}.$$

Letting $r' \nearrow R$, $\varepsilon \searrow 0$ and then $r \nearrow R$ we get

$$\limsup_{n \to \infty} \left| \Delta^*_{m,n,\ell,5}(f;z) \right|^{1/n} \leq \frac{|w|}{R^{q+1}} \quad \text{for } |w| > R.$$

Estimate of $\Delta^*_{m,n,\ell,6}(f;z)$. We have

$$|K_{m,n,\ell,6}(w,\zeta,t)| \leq \frac{r^{m+1}(|w|^{n+1} + r^{n+1})}{(r')^{m+1}(|w| - r)r^{n+1}}.$$

Therefore

$$\limsup_{n \to \infty} |K_{m,n,\ell,6}(w,\zeta,t)|^{1/n} \leq \frac{|w|r^{q-1}}{(r')^q}.$$

Hence

$$\limsup_{n\to\infty} \left|\Delta^*_{m,n,\ell,6}(f;z)\right|^{1/n} \le \frac{|w|r^{q-1}}{(r')^q}.$$

Letting $r \searrow 1$ and $r' \nearrow R$ we get

$$\limsup_{n\to\infty} \left|\Delta^*_{m,n,\ell,6}(f;z)\right|^{1/n} \le \frac{|w|}{R^q}.$$

On combining the above estimates we see that

$$\limsup_{n\to\infty} \left|\Delta^*_{m,n,\ell}(f;z)\right|^{1/n} \le \frac{|w|}{R^q}.$$

$\square$

Proof of Theorem 5 Since ∂E is an r_0-analytic curve for some $0 \le r_0 < 1$, (3.12) leads, for $\beta < |t| < |\zeta|$, to

$$\frac{\omega_m(\psi(\zeta)) - \omega_m(\psi(t))}{\omega_m(\psi(\zeta))} = t^m \left(1 + O(1)\left(\frac{\beta}{|\zeta|}\right)^m\right), \quad m \to \infty. \qquad (6.2.4)$$

An examination of the proof of Theorem 4 shows that by applying (6.2.4) in the above proof, the estimates of $\Delta^*_{m,n,\ell,j}(f;z)$, $1 \le j \le 6$ can be improved. Applying (3.11) and (6.2.4) we get

$$\limsup_{n\to\infty} \left|\Delta_{m,n,\ell,1}(f;z)\right|^{1/n} \le \frac{1}{R^q},$$

$$\limsup_{n\to\infty} \left|\Delta_{m,n,\ell,2}(f;z)\right|^{1/n} \le \frac{1}{R^q},$$

$$\limsup_{n\to\infty} \left|\Delta_{m,n,\ell,3}(f;z)\right|^{1/n} \le \frac{r_0^q}{R^q},$$

and

$$\limsup_{n\to\infty} \left|\Delta_{m,n,\ell,4}(f;z)\right|^{1/n} \le \begin{cases} 0 & \text{when } \ell = 1 \\ \frac{|w|}{R^{q+1}} & \text{when } \ell > 1. \end{cases}$$

Using now (3.11) and (6.2.4) we obtain

$$\limsup_{n\to\infty} \left|\Delta_{m,n,\ell,5}(f;z)\right|^{1/n} \le \frac{|w|r_0^q}{R^{q+1}}.$$

Applying (6.2.3) and (3.16) we get

$$F_{m,n}(w,\zeta) = \frac{1}{2\pi i} \oint_{|t=r|} \frac{\psi'(\zeta)}{\psi(\sigma) - \psi(t)} K_{m,n,\ell,6}(w,\zeta,t)dt =$$

$$= -\frac{1}{2\pi i} \oint |t| = r \left(\frac{1}{\zeta - t} + \sum_{\nu=1}^{\infty}\sum_{k=1}^{\infty} \alpha_{\nu,k} t^{-k}\zeta^{-(\nu+1)}\right) \frac{t^{m+1}}{\zeta^{m+1}} \frac{t^{n+1} - w^{n+1}}{(t-w)t^{n+1}}$$

$$= -\frac{1}{2\pi i} \oint_{|t|=r} \frac{1}{\zeta^{m+2}} \left(\sum_{k=0}^{\infty} \sum_{j=0}^{n} \zeta^{-k} w^j t^{k+m-j} + \right.$$

$$\left. + \sum_{\nu=1}^{\infty} \sum_{k=1}^{\infty} \sum_{j=0}^{n} \alpha_{\nu,k} \zeta^{-\nu} w^j t^{m-k-j} \right) dt .$$

Since $k + m - j = k + q(n+1) - 1 - j \geq 0$, the integral of the double sum above vanishes and the triple sum gives a contribution for the integral only when $k = m + 1 - j$, so that

$$F_{m,n}(w,\zeta) = \frac{1}{\zeta^{m+2}} \sum_{\nu=1}^{\infty} \sum_{j=0}^{n} \alpha_{\nu,m+1-j} w^j \zeta^{-\nu}.$$

Applying now (3.11) for $|\zeta| = r'$, we have

$$F_{m,n}(w,\zeta) = O(1) \frac{1}{(r')^m} \sum_{\nu=1}^{\infty} \sum_{j=0}^{n} \beta^{\nu+m-j} |w|^j (r')^{-\nu}$$

$$= O(1) \left(\frac{\beta}{r'} \right)^m \sum_{\nu=1}^{\infty} \left(\frac{\beta}{r'} \right)^\nu \sum_{j=0}^{n} \left(\frac{|w|}{\beta} \right)^j = O(1) \left(\frac{|w|\beta^{q-1}}{(r')^q} \right)^n .$$

Thus,

$$\lim_{n\to\infty} |\Delta_{m,n,\ell,6}(f;z)|^{1/n} \leq \frac{|w| r_0^{q-1}}{R^q} .$$

Combining the six estimates we get when $\ell = 1$

$$\limsup_{n\to\infty} |\Delta_{m,n,\ell}(f;z)|^{1/n} \leq \max \left(\frac{1}{R^q}, \frac{r_0^q}{R^q}, 0, \frac{|w| r_0^q}{R^{q+1}}, \frac{|w| r_0^{q-1}}{R^q} \right)$$

$$= \max \left(\frac{1}{R^q}, \frac{|w| r_0^{q-1}}{R^q} \right)$$

and when $\ell > 1$

$$\limsup_{n\to\infty} |\Delta_{m,n,\ell}(f;z)|^{1/n} \leq \max \left(\frac{1}{R^q}, \frac{r_0^q}{R^q}, \frac{|w|}{R^{q+1}}, \frac{|w| r_0^q}{R^{q+1}}, \frac{|w| r_0^{q-1}}{R^q} \right)$$

$$= \max \left(\frac{1}{R^q}, \frac{|w|}{R^{q+1}}, \frac{|w| r_0^{q-1}}{R^q} \right). \tag{6.2.5}$$

$\square$

6.3 Proof of Theorem 6

Observe that the formula for $\Delta_{m,n}^{p,s}(f;z)$ remains the same as the difference of (5.9) and (5.10). We have to keep in mind that $w_n(\psi(t))$ is based now on Fejér nodes. When ∂E is an r_0-analytic curve, with $r_0 \in [0,1)$ we have by (4.6)

$$\frac{w_m^s(\psi(\zeta)) - w_m^s(\psi(t))}{w_m^s(\psi(\zeta))} = \frac{(\zeta^{m+1} - 1)^s - (t^{m+1} - 1)^s}{(\zeta^{m+1} - 1)^s} (1 + O(1)\beta^m), \quad r_0 < \beta < 1,$$

$$\tag{6.3.1}$$

uniformly on sets $\{(\zeta,t): |t|/|\zeta| \le \alpha < 1\}$, for each $0 < \zeta < 1$. Here again, we can write

$$\Delta_{m,n}^{p,s}(f;z) = \sum_{j=1}^{6} \Delta_{m,n,j}^{p,s}(f;z),$$

where for $j = 1,2,3,4,5,6$ we have

$$\Delta_{m,n,j}^{p,s}(f;z) = \frac{1}{2\pi i} \oint_{|\zeta|=r'} f(\psi(\zeta)) \left(\frac{1}{2\pi i} \oint_{|t|=r} \frac{\psi'(\zeta)}{\psi(\zeta) - \psi(t)} K_{m,n,j}(w,\zeta,t) dt \right) d\zeta.$$

The kernels $K_{m,n,j}(w,\zeta,t)$ are defined as follows:

$$K_{m,n,1}(w,\zeta,t) := \frac{(\zeta^{m+1}-1)^s}{(\zeta^{m+1}-1)^s} \sum_{k=1}^{p(n+1)-1} \sum_{\nu=1}^{\infty} \frac{\alpha_{k,\nu}}{w^\nu t^{k+1}} - \frac{\zeta}{t} \sum_{k=1}^{p(n+1)-1} \sum_{\nu=1}^{\infty} \frac{\alpha_{k,\nu}}{w^\nu \zeta^{k+1}}$$

$$K_{m,n,2}(w,\zeta,t) := -\frac{(t^{m+1}-1)^s}{(\zeta^{m+1}-1)^s} \sum_{k=1}^{p(n+1)-1} \sum_{\nu=1}^{\infty} \frac{\alpha_{k,\nu}}{w^\nu t^{k+1}}$$

$$K_{m,n,3}(w,\zeta,t) := O(1)\beta^m \frac{(\zeta^{m+1}-1)^s - (t^{m+1}-1)^s}{(\zeta^{m+1}-1)^s} \sum_{k=1}^{p(n+1)-1} \sum_{\nu=1}^{\infty} \frac{\alpha_{k,\nu}}{w^\nu t^{k+1}}$$

$$K_{m,n,4}(w,\zeta,t) := \frac{t^{p(n+1)} - w^{p(n+1)}}{(t-w)t^{p(n+1)}} - \frac{\zeta}{t} \cdot \frac{\zeta^{p(n+1)} - w^{p(n+1)}}{(t-w)\zeta^{p(n+1)}}$$

$$K_{m,n,5}(w,\zeta,t) := O(1)\beta^m \frac{(\zeta^{m+1}-1)^s - (t^{m+1}-1)^s}{(\zeta^{m+1}-1)^s} \cdot \frac{t^{p(m+1)} - w^{p(m+1)}}{(t-w)t^{p(n+1)}},$$

and

$$K_{m,n,6}(w,\zeta,t) := -\frac{(t^{m+1}-1)^s}{(\zeta^{m+1}-1)^s} \cdot \frac{t^{p(n+1)} - w^{p(n+1)}}{(t-w)t^{p(n+1)}}.$$

We begin now the estimation of $\Delta_{m,n,j}^{p,s}(f;z)$ for $1 \le j \le 6$. In the proof it is assumed that $0 < r_0 < R_1 < 1 < r < r' < R$, that $z = \psi(w)$ and that $|w| > R$.

Estimate of $\Delta_{m,n,1}^{p,s}(f;z)$. By (1.1.4) and (3.14) we have

$$\frac{1}{2\pi i} \oint_{|t|=r} \frac{\psi'(\zeta)}{\psi(\zeta) - \psi(t)} K_{m,n,1}(w,\zeta,t) = 0.$$

Therefore $\limsup_{n\to\infty} |\Delta_{m,n,1}^{p,s}(f;z)|^{1/n} = 0 < 1$.

Estimate of $\Delta_{m,n,2}^{p,s}(f;z)$. We have by (3.12)

$$K_{m,n,2}(w,\zeta,t) = O(1) \left(\frac{r}{r'}\right)^{s(m+1)} \sum_{k=1}^{p(n+1)-1} \frac{1}{r} \left(\frac{R_1}{r}\right)^k = O(1) \left(\frac{r}{r'}\right)^{s(m+1)}.$$

Hence $\limsup_{n\to\infty} |K_{m,n,2}(w,\zeta,t)|^{1/n} \le \left(\frac{r}{r'}\right)^{qs}$. Therefore

$$\limsup_{n\to\infty} \left|\Delta^{p,s}_{m,n,2}(f;z)\right|^{1/n} \le \left(\frac{r}{r'}\right)^{qs}.$$

Letting $r \searrow 1$ and $r' \nearrow$ we get

$$\limsup_{n\to\infty} \left|\Delta^{p,s}_{m,n,2}(f;z)\right|^{1/n} \le \frac{1}{R^{qs}} < 1.$$

Estimate of $\Delta^{p,s}_{m,n,3}(f;z)$. We have by (3.12)

$$K_{m,n,3}(w,\zeta,t) = O(1)\beta^m \sum_{k=1}^{p(n+1)-1} R_1^k r^{-(k+1)} = O(1)\beta^m.$$

Hence

$$\limsup_{n\to\infty} |K_{m,n,3}(w,\zeta,t)|^{1/n} \le \beta^q.$$

Therefore

$$\limsup_{n\to\infty} \left|\Delta^{p,s}_{m,n,3}(f;z)\right|^{1/n} \le \beta^q.$$

Letting $\beta \searrow r_0$ we get

$$\limsup_{n\to\infty} \left|\Delta^{p,s}_{m,n,3}(f;z)\right|^{1/n} \le r_0^q < 1.$$

Estimate of $\Delta^{p,s}_{m,n,4}(f;z)$. We have by (5.7)

$$\frac{1}{2\pi i} \oint_{|t|=r} K_{m,n,4}(w,\zeta,t)dt = K_{m,n,4}(w,\zeta,t)(w,\zeta,\zeta) = 0.$$

Hence

$$\limsup_{n\to\infty} \left|\Delta^{p,s}_{m,n,4}(f;z)\right|^{1/n} = 0 < 1.$$

Estimate of $\Delta^{p,s}_{m,n,5}(f;z)$. By (3.12) we have

$$K_{m,n,5}(w,\zeta,t) = O(1)\beta^m |w|^{p(n+1)} r^{-(p(n+1))}.$$

Hence

$$\limsup_{n\to\infty} |K_{m,n,5}(w,\zeta,t)|^{1/n} \le \frac{\beta^q}{r^p}|w|^{pq}.$$

Therefore

$$\limsup_{n\to\infty} \left|\Delta^{p,s}_{m,n,5}(f;z)\right|^{1/n} \le \frac{\beta^q}{r^p}|w|^{pq}.$$

Letting $r \nearrow R$ and $\beta \searrow r_0$ we get

$$\limsup_{n \to \infty} \left| \Delta^{p,s}_{m,n,5}(f;z) \right|^{1/n} \leq \frac{r_0^q |w|^p}{R^p}.$$

Estimate of $\Delta^{p,s}_{m,n,6}(f;z)$. The argument used in the proof now is similar to that used in part (i.6) of the proof of Theorem 3. From the definition of $K_{n,m,\ell,6}(w,\zeta,t)$ and (3.16) we get

$$\frac{1}{2\pi i} \oint_{|t|=r} \frac{\psi'(\zeta)}{\psi(\zeta)-\psi(t)} K_{m,n,6}(w,\zeta,t)dt =$$

$$= -\frac{1}{2\pi i} \oint_{|t|=r} \left(\sum_{k=0}^{\infty} t^k \zeta^{-(k+1)} + \sum_{\mu=1}^{\infty}\sum_{k=1}^{\infty} \alpha_{\mu,k} t^{-k}\zeta^{-\mu-1} \right) \times$$

$$\times \frac{(t^{m+1}-1)^s}{(\zeta^{m+1}-1)^s} \frac{t^{p(n+1)} - w^{p(n+1)}}{t-w} \frac{1}{t^{p(n+1)}} dt$$

$$= -\frac{1}{(\zeta^{m+1}-1)^s} \frac{1}{2\pi i} \oint_{|t|=r} \left(\sum_{k=0}^{\infty} t^k \zeta^{-(k+1)} \right) \left(\sum_{j=0}^{s}(-1)^{s-j}\binom{s}{j} t^{j(m+1)} \right) \times$$

$$\times \left(\sum_{\nu=0}^{p(n+1)-1} t^{-(\nu+1)}w^{\nu} \right) dt$$

$$- \frac{1}{(\zeta^{m+1}-1)^s} \frac{1}{2\pi i} \oint_{|t|=r} \left(\sum_{\mu=1}^{\infty}\sum_{k=1}^{\infty} \alpha_{\mu,k} t^{-k}\zeta^{-\mu-1} \right) \times$$

$$\times \left(\sum_{j=0}^{s}(-1)^{s-j}\binom{s}{j} t^{j(m+1)} \right) \times$$

$$\times \left(\sum_{\nu=0}^{p(n+1)-1} t^{-(\nu+1)}w^{\nu} \right) dt$$

$$\equiv I_1 + I_2. \tag{6.3.2}$$

The integrands in the integral with respect to t in I_1 will vanish (by 6.1.7) except when $k = \nu - j(n+1) \geq 0$. Therefore we have

$$I_1 = -\frac{1}{\zeta(\zeta^{m+1}-1)^s} \sum_{j=0}^{\min\left(\left[\frac{p(n+1)-1}{q(n+1)}\right],s\right)} (-1)^{s-j}\binom{s}{j}\zeta^{jq(n+1)} \sum_{\nu \geq jq(n+1)}^{q(n+1)-1} \left(\frac{w}{\zeta}\right)^{\nu}.$$

We consider now the two possibilities: $p \leq qs$ and $p > qs$.

Estimating I_1 *when* $p \leq qs$. We have in this case

$$\min\left(\left[\frac{p(n+1)-1}{q(n+1)}\right],s\right) = \left[\frac{p(n+1)-1}{q(n+1)}\right]$$

Therefore

$$I_1 = O(1) \frac{1}{(r')^{qs(n+1)+1}} \frac{|w|^{p(n+1)-1}(r')^{q(n+1)\left[\frac{p(n+1)-1}{q(n+1)}\right]}}{(r')^{p(n+1)-1}}.$$

Write $p = qu + v$ where $u \in \mathbb{N}^+$ and $0 \le v < q$. Then we have

$$q(n+1)\left[\frac{p(n+1)-1}{q(n+1)}\right] = \begin{cases} q(n+1)(u-1) & \text{when } v = 0 \\ q(n+1)u & \text{when } 0 < v < q. \end{cases}$$

Hence

$$I_1 = O(1)|w|^{p(n+1)-1} \times \begin{cases} (r')^{-q(s+1)(n+1)-2} & \text{when } v = 0 \\ (r')^{-(qs+v)(n+1)-2} & \text{when } 0 \le v < q. \end{cases}$$

Hence

$$\text{When } qs \ge p \qquad \limsup_{n \to \infty} |I_1|^{1/n} = |w|^p \times \begin{cases} (r')^{-q(s+1)} & \text{when } v = 0 \\ (r')^{-qs-v} & \text{when } 0 < v < q. \end{cases} \tag{6.3.3}$$

Estimating I_1 when $p > qs$. We have $\min\left\{\left[\frac{p}{q} - \frac{1}{q(n+1)}\right], s\right\} = s$. Then

$$I_1 = O(1) \frac{|w|^{p(n+1)-1}}{(r')^{(p-qs)(n+1)-1}}.$$

Therefore

$$\limsup_{n \to \infty} |I_1|^{1/n} \le \frac{|w|^p}{(r')^{p-qs}} \qquad \text{when} \quad sq < p. \tag{6.3.4}$$

Estimating I_2. The integrands in the integral with respect to t in I_2 will vanish (by 6.1.7) except when $k = j(m+1) - \nu \ge 1$. Therefore we have, by applying (3.11)

$$I_2 = O(1) \frac{1}{(r')^{sq(n+1)+1}} \sum_{\mu=1}^{\infty} \left(\frac{\beta}{r'}\right)^{\mu} \sum_{\nu=0}^{\min(p(n+1)-1,sq(n+1)-1)} \left(\frac{|w|}{\beta}\right)^{\nu} \sum_{j=1}^{s} \beta^{jq(n+1)}.$$

Write $\nu = v + u(n+1)$ where $0 \le v \le n$ and $0 \le u \le \min(p, sq) - 1$. Then we get

$$I_2 = O(1) \frac{1}{(r')^{sq(n+1)}} \sum_{u=0}^{\min(p,sq)-1} \sum_{v=0}^{n} \left(\frac{|w|}{\beta}\right)^{v+u(n+1)} \sum_{\substack{j=1 \\ jq(n+1) \ge v+u(n+1)+1}}^{s} \beta^{jq(n+1)}$$

$$= O(1) \frac{1}{(r')^{sq(n+1)}} \sum_{u=0}^{\min(p,sq)-1} \left(\frac{|w|}{\beta}\right)^{u(n+1)} \sum_{v=0}^{n} \left(\frac{|w|}{\beta}\right)^{v} \times$$

$$\times \sum_{\substack{j=1 \\ jq(n+1) \ge v+u(n+1)+1}}^{s} \beta^{jq(n+1)}$$

$$= O(1) \frac{1}{(r')^{sq(n+1)}} \sum_{u=0}^{\min (p,sq)-1} |w|^{u(n+1)} \sum_{v=0}^{n} |w|^v \times$$

$$\times \sum_{\substack{j=1 \\ jq(n+1) \geq v+u(n+1)+1}}^{s} \beta^{jq(n+1)-u(n+1)-v}$$

$$= O(1) \frac{1}{(r')^{sq(n+1)}} \sum_{u=0}^{\min (p,sq)-1} |w|^{u(n+1)} \sum_{v=0}^{n} |w|^v \sum_{\substack{j=1 \\ jq(n+1) \geq v+u(n+1)+1}}^{s} \beta ,$$

and since $\nu = v + u(n+1)$

$$= O(1) \frac{1}{(r')^{sq(n+1)}} \sum_{\nu=0}^{(n+1)\cdot(\min (p,sq)) -1} |w|^{\nu} \sum_{v=0}^{n} |w|^v \sum_{\substack{j=1 \\ jq(n+1) \geq v+u(n+1)+1}}^{s} \beta$$

$$= O(1) \frac{|w|^{(n+1)\cdot\min (p.sq)}}{(r')^{sq(n+1)}}.$$

Therefore

$$\limsup_{n\to\infty} |I_2|^{1/n} \leq \frac{|w|^{\min (p,sq)}}{(r')^{qs}} \tag{6.3.5}$$

An improved estimate of I_2 when $q \geq p$. Assume $q \geq p$ (which implies $qs \geq p$). In this case (6.3.5) can be improved. We have under this assumption $jq(n+1) \geq jp(n + 1) \geq v + 1 + u(n + 1)$, when $0 \leq v \leq n$ and $0 \leq u \leq \min (p, sq) - 1$. Hence $jq(n + 1) - u(n + 1) - v \geq jp(n + 1) - u(n + 1) - v \geq 1$. Therefore

$$\sum_{\substack{j=0 \\ jq(n+1) \geq v+u(n+1)+1}}^{s} \beta^{jq(n+1)-u(n+1)-v} =$$

$$= \beta^{(q-p)(n+1)} \sum_{\substack{j=1 \\ jq(n+1) \geq v+u(n+1)+1}}^{s} \beta^{jp(n+1)-u(n+1)-v}$$

$$= O(1) \beta^{(q-p)(n+1)}.$$

Hence

$$I_2 = O(1) \frac{|w|^{p(n+1)} \beta^{j(q-p)(n+1)}}{(r')^{qs(n+1)}} \qquad \text{when} \quad q \geq p.$$

Therefore

$$\limsup_{n\to\infty} |I_2|^{1/n} \leq \frac{|w|^p \beta^{q-p}}{(r')^{qs}} \qquad \text{when} \quad q \geq p. \tag{6.3.6}$$

this improves the estimate of (6.3.5) when $q \geq p$.

Final estimate of $\Delta^{p,s}_{m,n,6}(f;z)$. Combining (6.3.2), (6.3.3), (6.3.4) (6.3.6), and using the notation $\tau = q$ when $p = qu$ and $\tau = v$ when $p = qu + v$, $0 < v < q$, we get

$$\limsup_{n \to \infty} \left| \frac{1}{2\pi i} \oint_{|t|=r} \frac{\psi'(\zeta)}{\psi(\zeta) - \psi(t)} K_{m,n,6}(w,\zeta,t) dt \right|^{1/n} \le$$

$$\le \begin{cases} \max \left(\frac{|w|^p}{(r')^{p-qs}}, \frac{|w|^{qs}}{(r')^{qs}} \right) & \text{when } qs < p \\[2mm] \max \left(\frac{|w|^p \beta^{q-p}}{(r')^{qs}}, \frac{|w|^p}{(r')^{qs+\tau}} \right) & \text{when } qs \ge p \text{ and } q \ge p \\[2mm] \max \left(\frac{|w|^p}{(r')^{qs}}, \frac{|w|^p}{(r')^{qs+\tau}} \right) & \text{when } qs \ge p \text{ and } q < p. \end{cases}$$

Hence

$$\limsup_{n \to \infty} \left| \Delta^{p,s}_{m,n,6}(f;z) \right|^{1/n} \le \begin{cases} \max \left(\frac{|w|^p}{(r')^{p-qs}}, \frac{|w|^{qs}}{(r')^{qs}} \right) & \text{when } qs < p \\[2mm] \max \left(\frac{|w|^p \beta^{q-p}}{(r')^{qs}}, \frac{|w|^p}{(r')^{qs+\tau}} \right) \\[1mm] & \text{when } qs \ge p \text{ and } q \ge p \\[2mm] \frac{|w|^p}{(r')^{qs}} & \text{when } qs \ge p \text{ and } q < p. \end{cases}$$

Letting $r' \nearrow R$ and $\beta \searrow r_0$ we get

$$\limsup_{n \to \infty} \left| \Delta^{p,s}_{m,n,6}(f;z) \right|^{1/n} \le \begin{cases} \max \left(\frac{|w|^p}{R^{p-qs}}, \frac{|w|^{qs}}{R^{qs}} \right) & \text{when } qs < p \\[2mm] \max \left(\frac{|w|^p r_0^{q-p}}{R^{qs}}, \frac{|w|^p}{R^{qs+\tau}} \right) \\[1mm] & \text{when } qs \ge p \text{ and } q \ge p \\[2mm] \frac{|w|^p}{R^{qs}} & \text{when } qs \ge p \text{ and } q < p. \end{cases}$$

Combining the estimates of $\Delta^{p,s}_{m,n,j}(f;z)$ for $1 \le j \le 6$ we obtain (2.6). $\square$

6.4 Proof of Theorem 7

We assume in the proof that $1 < r < r' < R$. For Faber nodes $w_m(z) = c^{m+1} F_{m+1}(z)$. Apply (4.1) for $|t| = r$, $|\zeta| = r'$. Then for each choice $\rho = (1-\varepsilon)r'$, $0 < \varepsilon < 1$, we have

$$\frac{w^s_m(\psi(\zeta)) - w^s_m(\psi(t))}{w^s_m(\psi(\zeta))} = \frac{\zeta^{s(m+1)} - t^{s(m+1)}}{\zeta^{s(m+1)}} \left(1 + O(1) \frac{\sqrt{m}}{|\zeta|^m} \right) \tag{6.4.1}$$

uniformly on $\{(\zeta,t) : |t| = r, |\zeta| = r'\}$. Apply now (5.8) with $\ell = 1$ and n given the value $p(n+1) - 1$, respectively, together with (5.3) with n again replaced by $p(n+1) - 1$. Then we see (since $z = \psi(w)$) that

$$\Delta^{*p,s}_{m,n,j}(f;z) := \frac{1}{2\pi i} \oint_{|\zeta|=r'} f(\psi(\zeta)) \left(\frac{1}{2\pi i} \oint_{|t|=r} \frac{\psi'(\zeta)}{\psi(\zeta) - \psi(t)} K_{m,n}(w,\zeta,t) dt \right) d\zeta,$$

where

$$K_{m,n}(w,\zeta,t) = \frac{\omega_m(\psi(\zeta))^s - \omega_m(\psi(t))^s}{\omega_m(\psi(\zeta))^s} \sum_{k=1}^{n} F_k(\psi(w)) t^{-(k+1)}$$

$$- \frac{\zeta}{t} \cdot \sum_{k=1}^{n} F_k(\psi(w)) \zeta^{-(k+1)}$$

$$= \frac{\omega_m(\psi(\zeta))^s - \omega_m(\psi(t))^s}{\omega_m(\psi(\zeta))^s} \left(\sum_{k=1}^{n} \left(F_k(\psi(w)) - w^k \right) t^{-(k+1)} \right.$$

$$\left. + \sum_{k=0}^{n} w^k t^{-(k+1)} \right)$$

$$- \frac{\zeta}{t} \cdot \left(\sum_{k=1}^{n} \left(F_k(\psi(w)) - w^k \right) \zeta^{-(k+1)} + \sum_{k=0}^{n} w^k \zeta^{-(k+1)} \right)$$

$$= \frac{\omega_m(\psi(\zeta))^s - \omega_m(\psi(t))^s}{\omega_m(\psi(\zeta))^s} \left(\sum_{k=1}^{p(n+1)-1} \sum_{\nu=1}^{\infty} \alpha_{k,\nu} w^{-\nu} t^{-(k+1)} + \right.$$

$$\left. + \frac{w^{p(m+1)} - t^{p(n+1)}}{(w-t) t^{p(n+1)}} \right)$$

$$- \frac{\zeta}{t} \left(\sum_{k=1}^{p(n+1)-1} \sum_{\nu=1}^{\infty} \alpha_{k,\nu} w^{-\nu} \zeta^{-(k+1)} + \frac{w^{p(n+1)} - \zeta^{p(n+1)}}{(w-z) \zeta^{p(n+1)}} \right)$$

$$= \frac{\zeta^{s(m+1)} - t^{s(m+1)}}{\zeta^{s(m+1)}} \left(1 + O(1) \frac{\sqrt{m+1}}{|\zeta|^{m+1}} \right) \times$$

$$\times \left(\sum_{k=1}^{p(n+1)-1} \sum_{\nu=1}^{\infty} \alpha_{k,\nu} w^{-\nu} t^{-(k+1)} + + \frac{w^{p(n+1)} - t^{p(n+1)}}{(w-t) t^{p(n+1)}} \right)$$

$$- \frac{\zeta}{t} \left(\sum_{k=1}^{p(n+1)-1} \sum_{\nu=1}^{\infty} \alpha_{k,\nu} w^{-\nu} \zeta^{-(k+1)} + \frac{w^{p(n+1)} - \zeta^{p(n+1)}}{(w-\zeta) \zeta^{p(n+1)}} \right)$$

$$= \frac{\zeta^{s(m+1)} - t^{s(m+1)}}{\zeta^{s(m+1)}} \cdot 1 \cdot \frac{w^{p(n+1)} - t^{p(n+1)}}{(w-t) t^{p(n+1)}} + \frac{\zeta^{s(m+1)} - t^{s(m+1)}}{\zeta^{s(m+1)}} \times$$

$$\times \sum_{k=1}^{p(n+1)-1} \sum_{\nu=1}^{\infty} \zeta_{k,\nu} w^{-\nu} t^{-(k+1)}$$

$$+ O(1) \frac{\sqrt{m+1}}{|\zeta|^{m+1}} \frac{\zeta^{s(m+1)} - t^{s(m+1)}}{\zeta^{s(m+1)}} \frac{w^{p(n+1)} - \zeta^{p(n+1)}}{(w-z) \zeta^{p(n+1)}}$$

$$+ O(1) \frac{\sqrt{m+1}}{|\zeta|^{m+1}} \cdot \frac{\zeta^{s(m+1)} - t^{s(m+1)}}{\zeta^{s(m+1)}} \sum_{k=1}^{p(n+1)-1} \sum_{\nu=1}^{\infty} \alpha_{k,\nu} w^{-\nu} t^{-()k+1}$$

$$- \frac{\zeta}{t} \sum_{k=1}^{p(n+1)-1} \sum_{\nu=1}^{\infty} \alpha_{k,\nu} w^{-\nu} \zeta^{-(k+1)} - \frac{\zeta}{t} \frac{w^{p(n+1)} - \zeta^{p(n+1)}}{(w-z) \zeta^{p(n+1)}}$$

$$
= \frac{\zeta^{s(m+1)}}{\zeta^{s(m+1)}} \frac{w^{p(n+1)} - t^{p(n+1)}}{(w-t)t^{p(n+1)}} - \frac{t^{s(m+1)}}{\zeta^{s(m+1)}} \frac{w^{p(n+1)} - t^{p(n+1)}}{(w-t)t^{p(n+1)}}
$$

$$
- \frac{t^{s(m+1)}}{\zeta^{s(m+1)}} \sum_{k=1}^{p(n+1)-1} \sum_{\nu=1}^{\infty} \alpha_{k,\nu} w^{-\nu} t^{-(k+1)}
$$

$$
+ \frac{\zeta^{s(m+1)}}{\zeta^{s(m+1)}} \sum_{k=1}^{p(n+1)-1} \sum_{\nu=1}^{\infty} \zeta_{k,\nu} w^{-\nu} t^{-(k+1)}
$$

$$
+ O(1) \frac{\sqrt{m+1}}{|\zeta|^{m+1}} \frac{\zeta^{s(m+1)} - t^{s(m+1)}}{\zeta^{s(m+1)}} \frac{w^{p(n+1)} - t^{p(n+1)}}{(w-z)t^{p(n+1)}}
$$

$$
+ O(1) \frac{\sqrt{m+1}}{|\zeta|^{m+1}} \cdot \frac{\zeta^{s(m+1)} - t^{s(m+1)}}{\zeta^{s(m+1)}} \sum_{k=1}^{p(n+1)-1} \sum_{\nu=1}^{\infty} \alpha_{k,\nu} w^{-\nu} t^{-()k+1}
$$

$$
- \frac{\zeta}{t} \sum_{k=1}^{p(n+1)-1} \sum_{\nu=1}^{\infty} \alpha_{k,\nu} w^{-\nu} \zeta^{-(k+1)} - \frac{\zeta}{t} \frac{w^{p(n+1)} - \zeta^{p(n+1)}}{(w-\zeta)\zeta^{p(n+1)}} \qquad (6.4.2)
$$

Applying (6.4.1) and (3.9) we split $\Delta_{m,n}^{*p,s}(f;z)$ as

$$
\Delta_{m,n,\ell}^{*p,s}(f;z) = \sum_{j=1}^{6} \Delta_{m,n,\ell,j}^{*p,s}(f;z), \qquad (6.4.3)
$$

where

$$
\Delta_{m,n,j}^{*p,s}(f;z) := \frac{1}{2\pi i} \oint_{|\zeta|=r'} f(\psi(\zeta)) \left(\frac{1}{2\pi i} \oint_{|t|=r} \frac{\psi'(\zeta)}{\psi(\zeta) - \psi(t)} K_{m,n,j}(w,\zeta,t) dt \right) d\zeta,
$$
$$
j = 1, 2, 3, 4, 5, 6.
$$

and the kernels $K_{m,n,j}(w,\zeta,t)$ $(j=1,2,3,4)$ are given explicitly as follows:

$$
K_{m,n,1}(w,\zeta,t) := \frac{\zeta^{s(m+1)}}{\zeta^{s(m+1)}} \sum_{k=1}^{p(n+1)-1} \sum_{\nu=1}^{\infty} \alpha_{k,\nu} w^{-\nu} t^{-(k+1)}
$$

$$
- \frac{\zeta}{t} \sum_{k=1}^{p(n+1)-1} \sum_{\nu=1}^{\infty} \alpha_{k,\nu} w^{-\nu} \zeta^{-(k+1)}
$$

$$
K_{m,n,2}(w,\zeta,t) := - \frac{t^{s(m+1)}}{\zeta^{s(m+1)}} \sum_{k=1}^{p(n+1)-1} \sum_{\nu=1}^{\infty} \alpha_{k,\nu} w^{-\nu} t^{-(k+1)}
$$

$$
K_{m,n,3}(w,\zeta,t) := O(1) \frac{\sqrt{m+1}}{|\zeta|^{m+1}} \frac{\zeta^{s(m+1)} - t^{s(m+1)}}{\zeta^{s(m+1)}} \sum_{k=1}^{p(n+1)-1} \sum_{\nu=1}^{\infty} \alpha_{k,\nu} w^{-\nu} t^{-(k+1)}.
$$

$$
K_{m,n,4}(w,\zeta,t) := \frac{t^{p(n+1)} - w^{p(n+1)}}{(t-w)t^{p(n+1)}} - \frac{\zeta}{t} \frac{\zeta^{p(n+1)} - w^{p(n+1)}}{(\zeta-w)\zeta^{p(n+1)}}, \qquad (6.4.4)
$$

$$K_{m,n,5}(w,\zeta,t) := O(1)\frac{\sqrt{m+1}}{|\zeta|^{m+1}}\frac{\zeta^{s(m+1)}-t^{s(m+1)}}{\zeta^{s(m+1)}}\frac{t^{p(n+1)}-w^{p(n+1)}}{(t-w)t^{p(n+1)}}, \quad (6.4.5)$$

and

$$K_{m,n,6}(w,\zeta,t) := -\frac{t^{s(m+1)}}{\zeta^{s(m+1)}}\frac{t^{p(n+1)}-w^{p(n+1)}}{(t-w)t^{p(n+1)}}. \quad (6.4.6)$$

We estimate now $\Delta^{*p,s}_{m,n,\ell,j}(f;z)$ for $1 \leq j \leq 6$.

Estimate of $\Delta^{*p,s}_{m,n,\ell,1}(f;z)$. By (5.7) we have

$$\frac{1}{2\pi i}\oint_{|t|=r}\frac{\psi'(\zeta)}{\psi(\zeta)-\psi(t)}K_{m,n,1}(w,\zeta,t)dt = 0.$$

Hence

$$\limsup_{n\to\infty}\left|\Delta^{*p,s}_{m,n,\ell,1}(f;z)\right|^{1/n} = 0 \quad \text{for } |w| > R.$$

Estimate of $\Delta^{*p,s}_{m,n,\ell,2}(f;z)$. By (3.14) we have

$$K_{m,n,2}(w,\zeta,t) = O(1)\left(\frac{r}{r'}\right)^{s(m+1)}\sum_{k=1}^{p(n+1)-1}\sqrt{k}r^{-k+1} = O(1)\left(\frac{r}{r'}\right)^{s(m+1)}.$$

Hence

$$\limsup_{n\to\infty}|K_{m,n,2}(w,\zeta,t)|^{1/n} \leq \left(\frac{r}{r'}\right)^{sq}.$$

This implies

$$\limsup_{n\to\infty}\left|\Delta^{*p,s}_{m,n,\ell,2}(f;z)\right|^{1/n} \leq \left(\frac{r}{r'}\right)^{sq}.$$

Letting $r \searrow 1$ and $r' \nearrow R$ we get

$$\limsup_{n\to\infty}\left|\Delta^{*p,s}_{m,n,\ell,2}(f;z)\right|^{1/n} \leq r^{-sq} < 1.$$

Estimate of $\Delta^{*p,s}_{m,n,\ell,3}(f;z)$. By (3.14) we have

$$K_{m,n,3}(w,\zeta,t) = O(1)\frac{\sqrt{m+1}}{|\zeta|^{m+1}}.$$

Hence

$$\limsup_{n\to\infty}|K_{m,n,3}(w,\zeta,t)|^{1/n} \leq \frac{1}{(r')^q}.$$

This implies

$$\limsup_{n\to\infty}\left|\Delta^{*p,s}_{m,n,\ell,3}(f;z)\right|^{1/n} \leq \frac{1}{(r')^q}.$$

Letting $r' \nearrow R$ we get

$$\limsup_{n\to\infty}\left|\Delta^{*p,s}_{m,n,\ell,3}(f;z)\right|^{1/n} \leq \frac{1}{R^q} < 1.$$

Estimate of $\Delta_{m,n,\ell,4}^{*p,s}(f;z)$. By (5.7) we have

$$\frac{1}{2\pi i}\oint_{|t|=r}\frac{\psi'(\zeta)}{\psi(\zeta)-\psi(t)}K_{m,n,4}(w,\zeta,t)dt=0.$$

Hence

$$\limsup_{n\to\infty}\left|\Delta_{m,n,\ell,4}^{*p,s}(f;z)\right|^{1/n}=0.$$

Estimate of $\Delta_{m,n,\ell,5}^{*p,s}(f;z)$. We have

$$K_{m,n,5}(w,\zeta,t)=O(1)\frac{\sqrt{m+1}}{|\zeta|^{m+1}}\frac{|w|^{p(n+1)}}{r^{p(n+1)}}.$$

Therefore

$$\limsup_{n\to\infty}|K_{m,n,5}(w,\zeta,t)|^{1/n}\le\frac{1}{(r')^q}\frac{|w|^p}{r^p}.$$

Hence

$$\limsup_{n\to\infty}\left|\Delta_{m,n,\ell,5}^{*p,s}(f;z)\right|^{1/n}\le\frac{1}{(r')^q}\frac{|w|^p}{r^p}.$$

Letting $r'\nearrow R$ and then $r\nearrow R$ we get

$$\limsup_{n\to\infty}\left|\Delta_{m,n,\ell,5}^{*p,s}(f;z)\right|^{1/n}\le\frac{|w|^p}{R^{p+q}}.$$

Estimate of $\Delta_{m,n,\ell,6}^{*p,s}(f;z)$. We have

$$K_{m,n,6}(w,\zeta,t)=O(1)\frac{r^{s(m+1)}}{(r')^{s(m+1)}}\frac{|w|^{p(n+1)}}{r^{p(n+1)}}.$$

Hence

$$\limsup_{n\to\infty}|K_{m,n,6}(w,\zeta,t)|^{1/n}\le\frac{|w|^p}{r^{p-qs}(r')^{qs}}.$$

Therefore

$$\limsup_{n\to\infty}\left|\Delta_{m,n,\ell,6}^{*p,s}(f;z)\right|^{1/n}\le\frac{|w|^p}{r^{p-qs}(r')^{qs}}.$$

Letting $r'\nearrow R$ and $r\nearrow R$ when $p-qs\ge0$ but for $r\searrow1$ when $p-qs<0$ we get

$$\limsup_{n\to\infty}\left|\Delta_{m,n,\ell,6}^{*p,s}(f;z)\right|^{1/n}\le\begin{cases}\dfrac{|w|^p}{R^p}&\text{when }p>qs\\[2mm]\dfrac{|w|^p}{R^{qs}}&\text{when }p\le qs.\end{cases}$$

Combining these estimates we get

$$\limsup_{n\to\infty}\left|\Delta_{m,n}^{*,p,s}\right|^{1/n}\le\begin{cases}\dfrac{|w|^p}{R^p}&\text{when }p\ge qs\\[2mm]|w|^p\cdot\max\left(\dfrac{1}{r^{p+q}},\dfrac{1}{R^{qs}}\right)&\text{when }p<qs.\end{cases}$$

6.5 Proof of Theorem 8

We assume in the proof that $0 < r_0 < \beta < 1 < r < r' < R$ and $|w| > R$. For $w_n(\psi(t))$ is based on Fejér nodes, we have by (5.3) and (5.4)

$$D_{s,p,n}(f;z) = \frac{1}{2\pi i} \oint_{|\zeta|=r'} f(\psi(\zeta)) \frac{1}{2\pi i} \oint_{|t|=r} \frac{\psi'(\zeta)}{\psi(\zeta) - \psi(t)} K_{s,n}(\zeta,t) \times$$

$$\times \sum_{k=1}^{p(n+1)-1} F_k(z) t^{-(k+1)} dt \zeta$$

where

$$K_{s,n}(\zeta,t) = \frac{w_n(\psi(\zeta))^s - w_n(\psi(t))^s}{w_n(\psi(\zeta))^s} - \frac{\omega_{s(n+1)-1}(\psi(\zeta)) - \omega_{s(n+1)-1}(\psi(t))}{\omega_{s(n+1)-1}(\psi(\zeta))}.$$

For $|w| > R$ we have

$$\sum_{k=0}^{p(n+1)-1} \frac{F_k(z)}{t^{k+1}} = \frac{t^{p(n+1)} - w^{p(n+1)}}{(t-w)t^{p(n+1)}} + \sum_{k=1}^{p(n+1)-1} \sum_{\nu=1}^{\infty} \alpha_{k,\nu} w^{-\nu} t^{-k-1}$$

$$=: A_{p,n,1}(w,t) + A_{p,n,2}(w,t). \tag{6.5.1}$$

By (3.12) we have

$$A_{p,n,2}(w,t) = \sum_{k=0}^{p(n+1)-1} \sum_{\nu=1}^{\infty} \alpha_{k,\nu} w^{-\nu} t^{-(k+1)} = O(1) \sum_{k=1}^{p(n+1)-1} \beta^k r^{-(k+1)}$$

$$= O(1) \frac{\beta}{r} \tag{6.5.2}$$

and

$$A_{p,n,1} = O(1) \left(\frac{|w|}{r} \right)^{p(n+1)} \qquad \text{for } |w| > R, 1 < |t| < R. \tag{6.5.3}$$

From (6.5.1),(6.5.2) and (6.5.3) we get

$$\sum_{k=0}^{p(n+1)-1} \frac{F_k(z)}{t^{k+1}} = O(1) \left(\frac{|w|}{r} \right)^{p(n+1)} \qquad \text{for } |w| > R, 1 < |t| < R. \tag{6.5.4}$$

By (4.6) we have, since ∂E is an r_0-analytic curve, for each given β, $1 > \beta > r_0$ and $z = \psi(w)$, $|w| > 1$,

$$K_{s,n}(\zeta,t) = \left(\frac{t^{s(n+1)} - 1}{\zeta^{s(n+1)} - 1}(1 + O(1)\beta^{sn}) - \frac{(t^{n+1} - 1)^s}{(\zeta^{n+1} - 1)^s}(1 + O(1)\beta^n) \right) \times$$

$$\times \sum_{k=1}^{p(n+1)-1} F_k(z) t^{-(k+1)} = \sum_{j=1}^{3} K_{s,n,j}(\zeta,t) \tag{6.5.5}$$

where

$$K_{s,n,1}(\zeta,t) = \left(\frac{t^{s(n+1)} - 1}{\zeta^{s(n+1)} - 1} - \frac{(t^{n+1} - 1)^s}{(\zeta^{n+1} - 1)^s}\right) \sum_{k=0}^{p(n+1)-1} \frac{F_k(z)}{t^{k+1}},$$

$$K_{s,n,2}(\zeta,t) = O(1)\beta^{sn}\frac{t^{s(n+1)} - 1}{\zeta^{s(n+1)} - 1} \sum_{k=0}^{p(n+1)-1} \frac{F_k(z)}{t^{k+1}},$$

$$K_{s,n,3}(\zeta,t) = -O(1)\beta^n\frac{(t^{n+1} - 1)^s}{(\zeta^{n+1} - 1)^s} \sum_{k=0}^{p(n+1)-1} \frac{F_k(z)}{t^{k+1}}.$$

Write for $j = 1, 2, 3$

$$D_{s,p,n,j}(f;z) = \frac{1}{2\pi i} \oint_{|\zeta|=r'} f(\psi(\zeta)) \frac{1}{2\pi i} \oint_{|t|=r} \frac{\psi'(\zeta)}{\psi(\zeta) - \psi(t)} K_{s,n,j}(\zeta,t)\, dt\zeta.$$

We have

$$\frac{t^{s(n+1)} - 1}{\zeta^{s(n+1)} - 1} - \frac{(t^{n+1} - 1)^s}{(\zeta^{n+1} - 1)^s} = \frac{t^{s(n+1)}}{\zeta^{s(n+1)}}\left\{\frac{1 - \frac{1}{t^{s(n+1)}}}{1 - \frac{1}{\zeta^{s(n+1)}}} - \frac{\left(1 - \frac{1}{t^{n+1}}\right)^s}{\left(1 - \frac{1}{\zeta^{n+1}}\right)^s}\right\}$$

$$= \begin{cases} \frac{t^{s(n+1)}}{\zeta^{s(n+1)}}\left(-\frac{s}{t^{n+1}} - \frac{s}{\zeta^{n+1}} - \frac{1}{t^{s(n+1)}} + \frac{1}{\zeta^{s(n+1)}} + \dots\right) & \text{if } s > 1 \\ 0 & \text{if } s = 1. \end{cases}$$

Therefore

$$\frac{t^{s(n+1)} - 1}{\zeta^{s(n+1)} - 1} - \frac{(t^{n+1} - 1)^s}{(\zeta^{n+1} - 1)^s} = O(1)\begin{cases} \frac{r^{(s-1)(n+1)}}{(r')^{s(n+1)}} & \text{if } s > 1, \\ 0 & \text{if } s = 1. \end{cases} \tag{6.5.6}$$

We have

$$O(1)\beta^{sn}\frac{t^{s(n+1)} - 1}{\zeta^{s(n+1)} - 1} = O(1)\frac{r^{s(n+1)}}{(r')^{s(n+1)}}\beta^{sn} \tag{6.5.7}$$

and

$$-O(1)\beta^n\frac{(t^{n+1} - 1)^s}{(\zeta^{n+1} - 1)^s} = O(1)\frac{r^{s(n+1)}}{(r')^{s(n+1)}}\beta^n. \tag{6.5.8}$$

From the above estimates we get

$$\limsup_{n\to\infty} |K_{s,n,1}(w,\zeta,t)|^{1/n} \le \frac{|w|^p}{r^p} \times \begin{cases} \frac{r^{s-1}}{(r')^s} & \text{if } s > 1 \\ 0 & \text{if } s = 1 \end{cases}$$

$$\limsup_{n\to\infty} |K_{s,n,2}(w,\zeta,t)|^{1/n} \le \frac{|w|^p\beta^s}{(r')^s r^{p-s}}$$

$$\limsup_{n\to\infty} |S_{s,n,3}(w,\zeta,t)|^{1/n} \le \frac{|w|^p\beta}{(r')^s r^{p-s}}$$

By the same argument used before for the passage from the estimate of $K_{s,n,j}$ $(w,\zeta,t))$ to that of $\limsup_{n\to\infty}|D_{p,s,n,j}(f;z)|^{1/n}$ we get, after letting $r' \nearrow R$ and $r \nearrow R$ when the exponent of r is negative or $r \searrow 1$ when the exponent of r is positive, and by letting $\beta \searrow r_0$,

$$B_1: \; = \limsup_{n\to\infty}|D_{p,s,n,1}(w,\zeta,t)|^{1/n} \leq \frac{|w|^p}{R^s} \times \begin{cases} \frac{1}{R^{p-s+1}} & \text{if } p+1 \geq s > 1 \\ 1 & \text{if } p+1 < s \\ 0 & \text{if } s = 1. \end{cases}$$

$$B_2 := \limsup_{n\to\infty}|D_{p,s,n,2}(w,\zeta,t)|^{1/n} \leq \frac{|w|^p}{R^s} \times \begin{cases} \frac{r_0^s}{R^{p-s}} & \text{when } p \geq s \geq 1, \\ r_0^s & \text{when } 1 \leq p < s, \end{cases}$$

$$B_3 := \limsup_{n\to\infty}|D_{p,s,n,3}(w,\zeta,t)|^{1/n} \leq \frac{|w|^p}{R^s} \times \begin{cases} \frac{r_0}{R^{p-s}} & \text{when } p \geq s \geq 1 \\ r_0 & \text{when } 1 \leq p < s. \end{cases}$$

Hence

$$\limsup_{n\to\infty}|D_{p,s,n}(f;z)|^{1/n} \leq \max\,(B_1,B_2,B_3) = \max\,(B_1,B_3)\,.$$

Writing out explicitly the values of this max we get Theorem 8.   □

6.6 Proof of Theorem 9

We assume in the proof that $1 < r < r' < R$. It is easy to verify that the following integral representation holds:

$$D^*_{p,s,n}(f;z) =$$

$$= \frac{1}{2\pi i}\oint_{|\zeta|=r'} f(\psi(\zeta))\left(\frac{1}{2\pi i}\oint_{|t|=r}\frac{\psi'(\zeta)}{\psi(\zeta)-\psi(t)}K_{s,n}(\zeta,t)\sum_{k=0}^{p(n+1)-1}\frac{F_k(z)}{t^{k+1}}dt\right)d\zeta,$$

$$(6.6.1)$$

where $K_{s,n}(\zeta,t)$ is the difference between the kernels of Hermite and Lagrange interpolants and where $1 < r < r' < R$. Then by (5.3) and (5.4)

$$K_{s,n}(\zeta,t) := \frac{w_n(\psi(\zeta))^s - w_n(\psi(t))^s}{w_n(\psi(\zeta))^s} - \frac{w_{s(n+1)-1}(\psi(\zeta)) - w_{s(n+1)-1}(\psi(t))}{w_{s(n+1)-1}(\psi(\zeta))}$$

$$= \frac{w_{s(n+1)-1}(\psi(t))}{w_{s(n+1)-1}(\psi)))} - \frac{w_n^s(\psi(t))}{w_n^s(\psi(\zeta))} \qquad (6.6.2)$$

Using (3.2) and (3.14), we then have

$$K_{s,n}(\zeta,t) = \frac{t^{s(n+1)}}{\zeta^{s(n+1)}}O(1)\frac{\sqrt{m+1}}{(r')^{n+1}},$$

uniformly on closed sets of $\{0 < |t| = r < r' = ||\zeta| < R\}$, for each triple $1 < r < r' = |\zeta| < R$. We have, since $z = \psi(w)$ and $|w| > R$, that

$$K_{s,n}(\zeta,t) \sum_{k=0}^{p(n+1)-1} \frac{F_k(z)}{t^{n+1}} = O(1)\frac{1}{(r')^{n+1}} \frac{t^{s(n+1)}}{\zeta^{s(n+1)}} \cdot \frac{t^{p(n+1)} - w^{p(n+1)}}{(t-w)t^{p(n+1)}}$$

$$+ O(1)\frac{1}{(r')^{n+1}} \frac{t^{s(n+1)}}{\zeta^{s(n+1)}} \sum_{k=1}^{p(n+1)-1} \sum_{\nu=1}^{\infty} \alpha_{k,\nu} w^{-\nu} t^{-k-1}$$

$$= J_1 + J_2. \tag{6.6.3}$$

Since $|t| = r < r' = |\zeta|$ and because of (3.14) we have

$$J_2 = O(1)\frac{1}{\rho^{n+1}} \left(\frac{r}{r'}\right)^{s(n+1)} \sum_{k=1}^{p(n+1)-1} \frac{\sqrt{k}}{|w|-1}|t|^{-k-1} = O(1)\left(\frac{1}{r'}\left(\frac{r}{r'}\right)^s\right)^{n+1}.$$

The first term on the right of (6.6.3) can be estimated by

$$J_1 = O(1)\left(\frac{|w|^p}{r'r^p}\right)^{n+1}\left(\frac{r}{r'}\right)^{s(n+1)}.$$

Therefore we have

$$\limsup_{n\to\infty}\left|K_{s,n}(\zeta,t)\cdot\sum_{k=0}^{p(n+1)-1}\frac{F_k(z)}{t^{n+1}}\right|^{1/n} \le \max\left(|J_1|^{1/n},|J_2|^{1/n}\right) \le \frac{|w|^p r^{s-p}}{r'r^p}.$$

When $s \le p$ we get by letting $r' \nearrow R$ and then $r \nearrow R$

$$\limsup_{n\to\infty}\left|K_{s,n}(\zeta,t)\cdot\sum_{k=0}^{p(n+1)-1}\frac{F_k(z)}{t^{n+1}}\right|^{1/n} \le \frac{|w|^p}{R^{2p+1-s}} \quad \text{when } s \le p \quad \text{for } |w| > R.$$

When $s > p$ we get by letting $r' \nearrow R$ and then $r \searrow 1$

$$\limsup_{n\to\infty}\left|K_{s,n}(\zeta,t)\cdot\sum_{k=0}^{p(n+1)-1}\frac{F_k(z)}{t^{n+1}}\right|^{1/n} \le \frac{|w|^p}{R^{p+1}} \quad \text{when } s > p, \text{ for } |w| > R.$$

By (6.6.1) the proof of the theorem is complete. $\qquad\square$

7. Historical Remarks

Equiconvergece results where related to Faber expansion of analytic functions were considered by Brück [19], Brück, Sharma and Varga [22, 23] and for rational interpolats by Saff, Sharma and Varga [92].

Remarks to Section 2 Following Rivlin [88] we obtain equiconvergence theorems for the Faber expansion of a function if the Lagrange or Hermite interpolants are replaced by their sections

$$S_n\left(L_m(f;\cdot);z\right) \qquad \text{or} \quad S_n\left(h_{s,s(m+1)-1}(f;\cdot);z\right).$$

Remarks to Corollary 1.   In the special case of $E = D_1$ (Example 1) where $r_0 = 0$, (2.2) reduces to $\lambda = R^{1+q\ell}$. For $\ell = 1$, this gives a result of Rivlin [88, Theorem 1] and for $q = 1$, this is a result of Cavaretta, Sharma and Varga [28, Theorem 1].

We do not know if λ in (2.3) is best possible. However, it is possible to improve (2.3) of Corollary 1 when $E = E_\delta$ ($\delta > 1$) of Example 3. In this example we have an explicit expression for $w(\psi(w))$ and $(\alpha_{n,k})_{n,k\geq0}$. Using these explicit expressions in the proof of Theorem 3 we get for λ in Corollary 1 the result

$$\lambda = \min\{R^{1+\ell q}, R^{\ell+q}/r_0^{2q}, R^{2q-1}/r_0^{2(q-1)}\}.$$

This is best possible as can be seen by the example $f(z) := \frac{1}{\psi(R)-z}$. If $q = 1$, then $\lambda = R$.

The previous remark also applies when $\delta = 1$, i.e., $E = [-1,1]$ (Example 2). If we use the zeros of the Chebyshev polynomials (the Faber nodes) as interpolation nodes and the identity

$$\omega_m(\psi(w)) = \left(\frac{1}{2}\right)^{m+1} w^{m+1}\left(1 - \frac{1}{w^{2(m+1)}}\right)$$

(Example 3 for $\delta = 1$) we get, by a minor modification of the proof of Theorem 3,

$$\lambda = \begin{cases} R^{2q-1} & \text{for} \quad \ell = 1, \\ R & \text{for} \quad q = 1, \\ R^{q+1} & \text{for} \quad q, \ell > 1. \end{cases}$$

This is a generalization and a new proof of Theorem 2 of Rivlin [88].

An improvement of Theorem 5 may be achieved in the case of an ellipse E_δ (where $\delta > 1$ given in Example 3 for which where ∂E_δ is an r_0-analytic curve with $r_0 = 1/\delta$). An examination of the proof of Theorem 5 shows that in this case

$$\lambda = \begin{cases} r_0(R/r_0)^{2q-1}, & \ell = 1, \\ \min\{R(R/r_0)^{2q-1}; R^{q+1}\}, & \ell \geq 2. \end{cases}$$

The previous remark also applies to the case of the segment $E = [-1,1]$ (where $\delta = 1$) and gives

$$\lambda = \begin{cases} R^{2q-1}, & \ell = 1, \\ R, & q = 1, \\ r^{q+1}, & \ell \geq 2. \end{cases}$$

For $\ell = 1$, this is Theorem 2 of Rivlin [88].

We do not know if λ^* of Theorem 6 is best possible. But for Faber expansions of the sets E of Example 2 and Example 3 the conclusion of Theorem 6 can be improved (see [23]).

We do not know whether λ^* in (2.7) of Theorem 7 is best possible. However (see [23]), if ∂E is an r_0-analytic curve, we can improve the conclusion of Theorem 7. A further improvement of Theorem 7 may be achieved in the case where $E = E_\delta$ is the ellipse treated in Example 3 where this ellipse is an r_0-analytic curve with $r_0 = 1/\delta < 1$.

We do not know if (2.8) of Theorem 8 is the best possible. However (see [23]), it is possible to improve our Theorem 8 when E is the ellipse E_δ considered in Example 3.

We do not know whether λ of (2.9) of Theorem 9 is best possible. However (see [23]) it is possible to improve (2.9) if ∂E is r_0-analytic curve. A further improvement may be achieved in the case of the ellipse E_δ (where $\delta > 1$) considered in Example 3 when the $\partial E\delta$ is an r_0-analytic curve with $r_0 = 1/\delta$.

CHAPTER 10

EQUICONVERGENCE ON LEMNISCATES

1. Equiconvergence and Lemniscates In this chapter we consider the case when the nodes of interpolation are points on a lemniscate and we prove equiconvergence and sharpness results for them.

Let $\lambda \geq 1$ be an integer, and $P_\lambda(z)$ an arbitrary fixed monic polynomial of degree λ. We will assume that the greatest common divisor of the multiplicities of the roots of $P_\lambda(z)$ is 1. The set of points satisfying $|P_\lambda(z)| = \mu^\lambda$ with some $\mu > 0$ is called a *lemniscate* (denoted as $\Gamma(\mu)$). The set of points satisfying $|P_\lambda(z)| < \mu^\lambda$ consists of at most λ disjoint finite regions denoted by $G(\mu)$. Evidently, if $0 < \mu_1 < \mu_2$ then $G(\mu_1) \subset G(\mu_2)$. The class of functions analytic in $G(\rho)$ ($\rho > 1$) but having a singularity on the boundary of $G(\rho)$ will be denoted by $AG(\rho)$. (We do not require that the functions analytic in connected regions of $G(\rho)$ should be analytic continuations of each other.) For an $f(z) \in AG(\rho^\lambda)$, consider the Hermite interpolation polynomial

$$S_n(f;z) = \frac{1}{2\pi i} \oint_{\Gamma(R)} \frac{P_\lambda(t)^n - P_\lambda(z)^n}{P_\lambda(t)^n(t-z)} f(t)dt, \quad 0 < R < \rho \qquad (1.1)$$

of degree at most $\lambda n - 1$, interpolating to $f(z)$ in the roots of $P_\lambda(z)$ with multiplicity n. (Recall that $P_\lambda(z)$ may have multiple roots, thus the actual multiplicity of interpolation may be higher.) In the special case $P_\lambda(z) = z$, the lemniscate is a circle with origin as center and (1.1) is nothing but the Taylor section of $f(z)$ of order $n - 1$ (here $\lambda = 1$).

Evidently, (1.1) can be written in the form

$$S_n(f;z) = \sum_{k=0}^{n-1} q_k(f;z) P_\lambda(z)^k \qquad (1.2)$$

where

$$q_k(f;z) = \frac{1}{2\pi i} \oint_{\Gamma(R)} \frac{P_\lambda(t) - P_\lambda(z)}{t-z} \cdot \frac{f(t)}{P_\lambda(t)^{k+1}} dt, \quad 0 < R < k = 0, 1, \ldots$$

$$(1.3)$$

are polynomials of degree at most $\lambda - 1$. Hence and from (1.1)

$$f(z) - S_n(f; z) = \frac{1}{2\pi i} \oint_{\Gamma(R)} \left(\frac{P_\lambda(z)}{P_\lambda(t)} \right)^n \frac{f(t)}{t - z} dt$$

$$= O\left(\left| \frac{P_\lambda(z)}{P\lambda(t)} \right| \right) \to 0 \quad \text{as} \quad n \to \infty$$

provided $z \in G(\rho)$. Thus

$$f(z) = \sum_{k=0}^{\infty} q_k(f; z) P_\lambda(z)^k, \tag{1.4}$$

the convergence of the series on the right hand side being uniform in every compact subset of $G(\rho)$. Moreover, $f(z) \in AG(\rho)$ is equivalent to

$$\varlimsup_{k=\infty} \max_{z \in \Gamma(\rho)} |q_k(f; z)|^{1/k} = \frac{1}{\rho^\lambda}. \tag{1.5}$$

Let now $r \geq 1$ be an arbitrary fixed integer, and consider

$$H_{\lambda rn-1}(f; z) = \frac{1}{2\pi i} \oint_{\Gamma(R)} \frac{\left(P_\lambda(t)^n - 1 \right)^r - \left(P_\lambda(z)^n - 1 \right)^r}{\left(P_\lambda(t)^n - 1 \right)^r (t - z)} f(t) dt,$$
$$\text{for } 0 < R < \rho. \tag{1.6}$$

This is a polynomial of degree at most $\lambda rn - 1$ interpolating $f^{(j)}(z)$, $j = 0, \ldots, r$ in the λn roots of the polynomial $P_\lambda(z)^n - 1$ which are on $\Gamma(1) \subset G(\rho)$. Notice that the latter polynomial may have at most $\lambda - 1$ multiple roots (namely, the roots of $P'_\lambda(z) = 0$), therefore this interpolation polynomial may be an Hermite interpolation of higher order than r at multiple nodes. Of course, in the special case $P_\lambda(z) = z$, (1.6) is the ordinary Hermite interpolation at the n-th roots of unity.

We want to compare this Hermite interpolation polynomial with the "shifted Taylor sections"

$$S_{rn,\ell}(f; z) := \frac{1}{2\pi i} \oint_{\Gamma(R)} \left[1 - \frac{P_\lambda(z)^{rn}}{P_\lambda(t)^{rn}} + (P_\lambda(t)^n - P_\lambda(z)^n) \sum_{j=1}^{\ell-1} \frac{\beta_{j,r}(P_\lambda(z)^n)}{P_\lambda(t)^{(r+j)n}} \right] \times$$

$$\times \frac{f(t)}{t - z} dt$$

of (1.1), where $\ell \geq 1$ is an arbitrary fixed integer, and

$$\beta_{j,r}(z) = \sum_{k=0}^{r-1} \binom{r+j-1}{k} (z-1)^k, \qquad j = 0, 1, \ldots$$

(see Ch. 2, (1.7)). (Note that for $\ell = 1$ we take the sum as 0, and $S_{rn,1}$ is nothing else but S_{rn} from (1.1).) Using the identity (1.6) from Ch. 2, this can be written as

$$S_{rn,\ell}(f;z) := \frac{1}{2\pi i} \oint_{\Gamma(R)} \left[1 - \frac{(P_\lambda(z)^n - 1)^r}{(P_\lambda(t)^n - 1)^r} + (P_\lambda(t)^n - P_\lambda(z)^n) \times \right.$$

$$\left. \times \sum_{j=\ell}^{\infty} \frac{\beta_{j,r}(P_\lambda(z)^n)}{P_\lambda(t)^{(r+j)n}} \right] \frac{f(t)}{t - z} dt.$$

We now want to compare this with (1.6), therefore we define

$$\Delta_{\lambda rn-1,\ell}(f;z) := H_{\lambda rn-1}(f;z) - S_{n,\ell}(f;z) \qquad (1.7)$$

$$= \frac{1}{2\pi i} \oint_{\Gamma(R)} (P_\lambda(t)^n - P_\lambda(z)^n) \sum_{j=\ell}^{\infty} \frac{\beta_{j,r}(P_\lambda(z)^n)}{P_\lambda(t)^{(r+j)n}} \frac{f(t)}{t - z} dt$$

$$= \sum_{j=\ell}^{\infty} \beta_{j,r}(P_\lambda(z)^n) \sum_{k=0}^{n-1} q_{(r+j-1)n+k}(f;z)P_\lambda(z)^k)).$$

The analogue of Theorem 6, Ch. 4 is the following:

THEOREM 1. *If $f(z) \in AG(\rho)$ $(\rho > 1, \lambda \geq 1)$ and $\ell \geq 1$ is an integer then*

$$\varlimsup_{n \to \infty} \max_{z \in \Gamma(R)} |\Delta_{\lambda rn-1,r,\ell}(f;z)|^{1/(\lambda rn)} = \rho^{-1-\frac{\ell-1}{r}} \max\{1, R^{1-1/r}, R\rho^{-1/r}\}$$

$$:= K_{r,\ell}(\rho, R). \qquad (1.8)$$

This theorem shows that if $R < \rho^{1+\ell/r}$ then we have overconvergence.

PROOF. First we prove the upper estimate. (1.5) implies

$$|q_k(f;z)| = O((\rho - \varepsilon)^{-\lambda k}), \quad k = 0, 1, \ldots, \qquad (1.9)$$

with an arbitrary $\varepsilon > 0$, and we also use

$$|\beta_{j,r}(P_\lambda(z)^n)| \leq cj^{r-1} \max\{1, P_\lambda(z)^{(r-1)n}\}, \qquad j = 1, \ldots$$

(see Lemma 2 in Ch. 2). Thus (1.7) yields

$$|\Delta_{\lambda rn-1,\ell}(f;z)| \leq c \max\{1, R^{(r-1)n}\} \sum_{j=\ell}^{\infty} j^{r-1} \sum_{k=0}^{n-1} (\rho - \varepsilon)^{-\lambda[(r+j-1)n-k}R^k.$$

Choosing $\varepsilon > 0$ properly and distinguishing the cases $0 < R < 1, 1 \leq R < \rho, \rho \leq R,$, we obtain the upper estimates in (1.8).

In order to show the opposite inequality, we assume that for some $\varepsilon > 0$,

$$\max_{z \in \Gamma(R)} |\Delta_{\lambda rn-1,\ell}(f;z)| \leq K_{r,\ell}(\rho,\mu) - \varepsilon. \tag{1.10}$$

Hence and by (1.7)

$$\left| \beta_{\ell,r}(P_\lambda(z)^n) \sum_{k=0}^{n-1} q_{(r+\ell-1)n+k}(f;z) P_\lambda(z)^k \right| = \tag{1.11}$$

$$= |\Delta_{\lambda rn-1,\ell}(f;z) - \Delta_{\lambda rn-1,\ell}(f;z)| \leq c(K_{r,\ell+1}(\rho,\mu) - \varepsilon)^{\lambda rn},$$

since evidently $K_{r,\ell}(\rho,\mu)$ is a monotone decreasing function of ℓ. Taking into account the structure of the polynomials $\beta_{\ell,r}$ (see (1.7) in Ch. 2), the latter relation can be written in the form

$$\sum_{k=0}^{n-1} q_{(r+\ell-1)n+k}(f;z) \sum_{s=0}^{r-1} c_{s,r,\ell} P_\lambda(z)^{sn+k} \leq\leq c(K_{r,\ell}(\rho,\mu) - \varepsilon)^{\lambda rn},$$

where $c_{s,r,\ell} \neq 0$ are some real numbers.

We now make use of the following general remark: if $Q(z)$ is an arbitrary polynomial of degree at most $\lambda n - 1$, then it can be represented in the form

$$Q(z) = \sum_{k=0}^{n-1} q_k(Q;z) P_\lambda(z)^k \tag{1.12}$$

where $q_k(Q;z)$ are polynomials of degree at most $\lambda - 1$ such that

$$\max_{z \in \Gamma(R)} |q_k(Q;z)| \leq \frac{c}{R^{\lambda(k+1)}} \max_{z \in \Gamma(R)} |Q(z)|, \quad R > 0, \ k = 0, 1, \ldots, n-1 \tag{1.13}$$

with $c > 0$ depending only on P_λ. This follows easily from (1.1) - (1.3) and the uniqueness of Hermite interpolation.

Thus (1.11) and (1.13) imply

$$\max_{z \in \Gamma(R)} |q_{k+(r+j-1)n}(f;z)| \leq c \frac{(K_{r,\ell}(\rho,\mu) - \varepsilon)^{\lambda rn}}{R^{\lambda(sn+k+1)}}$$
$$\text{for} \quad k = 0, 1, \ldots, n-1; \ s = 0, \ldots, r-1. \tag{1.14}$$

We now distinguish three cases:

CASE 1: $0 < R \leq 1$. Then $K_{r,\ell}(\rho,\mu) = \rho^{-1-\frac{\ell-1}{r}}$, and using (1.14) with $s = 0$ we get

$$\max_{z \in \Gamma(R)} |q_{(r+\ell-1)n+k}(f;z)| \leq \frac{c}{(\rho+\varepsilon)^{\lambda(r+\ell-1)n} R^{\lambda(k+1)}}, \quad k = 0, \ldots, n-1.$$

Representing any integer $m \geq r + \ell - 1$ in the form $m = (r + \ell - 1)n + k$, $0 \leq k \leq r + \ell - 2$, we get

$$\limsup_{m \to \infty} \max_{z \in \Gamma(R)} |q_m(f;z)|^{1/m} \leq \limsup_{m \to \infty} (\rho + \varepsilon)^{-\frac{\lambda(m-k)}{m}} = (\rho + \varepsilon)^{-\lambda} < \rho^{-\lambda},$$

which contradicts $f \in AG(\rho)$.

CASE 2: $1 < R \leq \rho$. Then $K_{r,\ell}(\rho, \mu) = \rho^{-1 - \frac{\ell-1}{r}} R^{1 - \frac{1}{r}}$, and using (1.14) with $s = r - 1$ we get

$$\max_{z \in \Gamma(R)} |q_{(r+\ell-1)n+k}(f;z)| :\leq \frac{R^{\lambda(r-1)n}}{(\rho + \varepsilon)^{\lambda(r+\ell-1)n} R^{\lambda((r-1)n+k+1)}}$$

$$= \frac{c}{(\rho + \varepsilon)^{\lambda(r+\ell-1)n}}, \qquad k = 0, \ldots, n-1,$$

and this is the same as Case 1.

CASE 3: $\rho < R$. Then $K_{r,\ell}(\rho, \mu) = \rho^{-1 - \frac{\ell}{r}}$, and using (1.14) with $s = r - 1$ again, we get

$$\max_{z \in \Gamma(R)} |q_{(r+\ell-1)n+k}(f;z)| :\leq \frac{R^{\lambda rn}}{(\rho + \varepsilon)^{\lambda(r+\ell)n} R^{\lambda((r-1)n+k+1)}}$$

$$= \frac{c R^{\lambda(n-k)}}{(\rho + \varepsilon)^{\lambda(r+\ell)n}}, \qquad k = 0, \ldots, n-1,$$

Representing any integer $m \geq r + \ell - 1$ in the form $m = (r + \ell - 1)n + k$, $n - r - \ell + 1 \leq k \leq n - 1$, we obtain

$$\limsup_{m \to \infty} \max_{z \in \Gamma(R)} |q_m(f;z)|^{1/m} \leq \limsup_{m \to \infty} (\rho + \varepsilon)^{\frac{\lambda(r+\ell)n}{(r+\ell)n+c}} = (\rho + \varepsilon)^{-\lambda},$$

a contradiction again. The theorem is proved.   □

As in Chapter 4, we will call a point z *distinguished* if there exists an $f \in AG(\rho)$ such that the maximum in (1.9) is not attained, i.e.

$$\overline{\lim_{n \to \infty}} |\Delta_{\lambda rn-1, \ell}(f;z)|^{1/(\lambda n)} < K_{r,\ell}(\rho, |z|).$$

We will show that the number of distinguished points not on the lemniscate $|P_\lambda(z)| = \rho^\lambda$ is finite, and that one can characterize whether a given set of points is distinguished.

To this aim, let $Z = \{z_1, z_2, \ldots, z_m\}$ be a given sequence of pairwise different points and define the $m \times \lambda$ matrices

$$X_i(Z) := \begin{pmatrix} P_\lambda(z_1)^i & z_1 P_\lambda(z_1)^i & \cdots & z_1^{\lambda-1} P_\lambda(z_1)^i \\ \vdots & \vdots & & \vdots \\ P_\lambda(z_m)^i & z_m P_\lambda(z_m)^i & \cdots & z_m^{\lambda-1} P_\lambda(z_m)^i \end{pmatrix}, \qquad i = 0, 1, \ldots.$$

Further let $U := \{u_1, \ldots, u_\mu\}$ and $V := \{v_1, \ldots, v_\nu\}$ be two sets of pairwise different points such that

$$|P_\lambda(u_j)| < \rho^\lambda, \qquad j = 1, \ldots, \mu \tag{1.15}$$

and

$$|P_\lambda(v_j)| > \rho^\lambda, \qquad j = 1, \ldots, \nu. \tag{1.16}$$

We also assume that $|P_\lambda(z)| \neq 1$, $z \in U \cup V$ when $\beta_{\ell,r}(P_\lambda(z)^n)$ has a zero such that $|P_\lambda(z)| = 1$.

Define the $\mu \times \lambda(r + \ell - 1)$ and $\nu \times \lambda(r + \ell)$ matrices

$$X(U) := [X_0(U) X_1(U) \ldots X_{r+\ell-2}(U)]$$

and

$$X(V) := [X_0(V) X_1(V) \ldots X_{r+\ell-1}(V)],$$

respectively. Finally, let

$$M(X(U), X(V)) = \begin{pmatrix} X(U) & & & & & \\ & X(U) & & & & \\ & & \ddots & & & \\ & & & X(U) & & \\ X(V) & & & & & \\ & X(V) & & & & \\ & & \ddots & & & \\ & & & & & X(V) \end{pmatrix}, \tag{1.17}$$

where $X(U)$ and $X(V)$ are repeated $r + \ell$ and $r + \ell - 1$ times, respectively (so that $M(X(U), X(V))$ is an $((r+\ell)\mu + (r+\ell-1)\nu) \times \lambda(r+\ell-1)(r+\ell)$ matrix). (Note that we allow the cases $U = \emptyset$ (then $\mu = 0$), or $V = \emptyset$ (then $\nu = 0$). In the above matrix the $X(V)$'s begin below the last row of the last $X(U)$.

We now state

THEOREM 2. *With the above notations, the points $U \cup V$ form a distinguished set if and only if*

$$\text{rank } M(X(U), X(V)) < \lambda(r + \ell - 1)(r + \ell). \tag{1.18}$$

COROLLARY 1. *There are at most $\lambda(r+\ell-1)-1$ and $\lambda(r+\ell)-1$ distinguished points in $|P_\lambda(z)| < \rho^\lambda$ and $|P_\lambda(z)| > \rho^\lambda$, respectively.*

Namely, assume that (1.18) holds, and $\mu \geq \lambda(r + \ell - 1)$. Then we take that minor of M which consists of the first $L := \lambda(r + \ell - 1)$ rows of each $X(U)$ in M. The determinant of this minor will be the $(r + \ell)^{\text{th}}$ power of the following determinant:

$$
\begin{vmatrix}
1 & u_1 & \dots & u_1^{\lambda-1} & P_\lambda(u_1) & \dots & u_1^{\lambda-1}P_\lambda(u_1) & \dots & P_\lambda(u_1)^{r+\ell-2} & \dots & u_1^{\lambda-1}P_\lambda(u_1)^{r+\ell-2} \\
\vdots & \vdots & & \vdots & \vdots & & \vdots & & \vdots & & \vdots \\
1 & u_L & \dots & u_L^{\lambda-1} & P_\lambda(u_L) & \dots & u_L^{\lambda-1}P_\lambda(u_L) & \dots & P_\lambda(u_L)^{\ell-1} & \dots & u_L^{\lambda-1}P_\lambda(u_L)^{\ell-1}
\end{vmatrix},
$$

In order to calculate this determinant, multiply the first λ columns by the subsequent coefficients of the polynomial $P_\lambda(z)$, and subtract them from the $(\lambda+1)^{\text{st}}$ column. Then this column becomes $(u_1^\lambda\, u_2^\lambda\, \dots\, u_L^\lambda)^{\text{T}}$, because $P_\lambda(z)$ is a monic polynomial. Now we multiply the first $\lambda + 1$ columns of the resulting determinant by subsequent coefficients of the polynomial $zP_\lambda(z)$, and subtract these columns from the $(\lambda + 1)$st column. The new $(\lambda + 1)^{\text{st}}$ column will be $(u_1^{\lambda+1}\, u_2^{\lambda+1} \vdots u_L^{\lambda+1})^{\text{T}}$. Now the pattern is clear: after $\lambda(r+\ell-2)$ steps, we arrive at the Vandermonde determinant $V(u_1, \dots, u_L) \neq 0$. This contradicts (1.18), since we have found a minor of size $\lambda(r+\ell-1)(r+\ell)$ with nonzero determinant. Similar reasoning applies for the points outside the lemniscate $|P_\lambda(z)| = \rho^\lambda$.

COROLLARY 2. *If $(r + \ell)\mu + (r + \ell - 1)\nu < \lambda(r + \ell - 1)(r + \ell)$, then the corresponding points are always distinguished.*

Namely, then the number of rows in M is less than the number of columns, i.e. (1.18) is satisfied.

PROOF OF THEOREM 2. *Sufficiency.* Suppose (1.18) holds. Then there exists a nonzero vector $\mathbf{b} = (b_0, \dots, b_{\lambda(r+\ell-1)(r+\ell)-1})$ such that $M \cdot \mathbf{b} = \mathbf{0}$. By (1.17), this is equivalent to the following:

$$
\sum_{s=0}^{r+\ell-2} \sum_{k=0}^{\lambda-1} b_{\lambda(v(r+\ell-1)+s)+k} z^k P_\lambda(z)^s = 0, \qquad v = 0, \dots, r+\ell-1;\ z \in U \quad (1.19a)
$$

and

$$
\sum_{s=0}^{r+\ell-1} \sum_{k=0}^{\lambda-1} b_{\lambda(v(r+\ell)+s)+k} z^k P_\lambda(z)^s = 0, \qquad v = 0, \dots, r + \ell - 2;\ z \in V. \quad (1.19b)
$$

Define

$$
f(z) := \sum_{s=0}^{(r+\ell-1)(r+\ell)-1} \sum_{k=0}^{\lambda-1} b_{\lambda s+k} z^k P_\lambda(z)^s \left[1 - \left(\frac{P_\lambda(z)}{\rho^\lambda} \right)^{(r+\ell-1)(r+\ell)} \right]^{-1}
$$

$$
= \sum_{u=0}^{\infty} \sum_{s=0}^{(r+\ell-1)(r+\ell)-1} \rho^{-u\lambda(r+\ell-1)(r+\ell)} \sum_{k=0}^{\lambda-1} b_{\lambda s+k} z^k P_\lambda(z)^{u(r+\ell-1)(r+\ell)+s}.
$$

Evidently, $f \in AG(\rho)$, and

$$q_{u(r+\ell-1)(r+\ell)+s}(f;z) = \rho^{-u\lambda(r+\ell-1)(r+\ell)} \sum_{k=0}^{\lambda-1} b_{\lambda s+k} z^k, \qquad (1.20)$$

$u = 0, \ldots;\; s = 0, \ldots, (r+\ell-1)(r+\ell) - 1$, so that (1.4) takes the form

$$f(z) = \sum_{u=0}^{\infty} \sum_{s=0}^{(r+\ell-1)(r+\ell)-1} q_{u(r+\ell-1)(r+\ell)+s}(f;z) P_\lambda(z)^{u(r+\ell-1)(r+\ell)+s}.$$

Replacing s in (1.20) by $v(r+\ell-1)+s$, $v = 0, \ldots, r+\ell-1$ and by $v(r+\ell)+s$, $v = 0, \ldots, r+\ell-2$, respectively, we obtain from (1.19a)–(1.19b)

$$\sum_{s=0}^{r+\ell-2} q_{u(r+\ell-1)(r+\ell)+v(r+\ell-1)+s}(f;z) P_\lambda(z)^s = 0$$

$$\text{for} \quad u = 0, \ldots;\; v = 0, \ldots, r+\ell-1;\; z \in U$$

and

$$\sum_{s=0}^{r+\ell-1} q_{u(r+\ell-1)(r+\ell)+v(r+\ell)+s}(f;z) P_\lambda(z)^s = 0$$

$$\text{for} \quad u = 0, \ldots;\; v = 0, \ldots, r+\ell-2;\; z \in V.$$

On writing v for $u(r+\ell)+v$ and $u(r+\ell-1)+v$ in the above equations, they are equivalent to

$$\sum_{s=0}^{r+\ell-2} q_{v(r+\ell-1)+s}(f;z) P_\lambda(z)^s = 0, \qquad v = 0, 1, \ldots;\; z \in U \qquad (1.21a)$$

and

$$\sum_{s=0}^{r+\ell-1} q_{v(r+\ell)+s}(f;z) P_\lambda(z)^s = 0, \qquad v = 0, 1, \ldots;\; z \in V, \qquad (1.21b)$$

respectively.

(1.20) implies

$$\max_{z \in \Gamma(\rho)} |q_n(f;z)| = O\left(\rho^{-\lambda n}\right).$$

Hence (1.7) yields

$$\Delta_{\lambda rn-1,\ell}(f;z) := \beta_{r,\ell}(P_\lambda(z)^n) \sum_{k=0}^{n-1} q_{(r+\ell-1)n+k}(f;z) P_\lambda(z)^k$$

$$+ O\left(\sum_{j=\ell+1}^{\infty} j^{r-1} \sum_{k=0}^{n-1} \frac{|P_\lambda(z)|^k}{\rho^{\lambda((r+j-1)n+k)}}\right).$$

Thus defining t and s by $(r + \ell - 1)t + s = (r + \ell)n$, $0 \le s \le r + \ell - 1$, we get from (1.21a)

$$
\begin{aligned}
\Delta_{\lambda rn-1,\ell}(f;z) = {}& P_\lambda(z)^{(r+\ell-1)(t-n)} \sum_{k=0}^{s-1} q_{(r+\ell-1)t+k}(f;z)P_\lambda(z)^k \\
& + \sum_{v=0}^{t-1} P_\lambda(z)^{(r+\ell-1)(v-n)} \sum_{k=0}^{r+\ell-2} q_{v(r+\ell-1)+k}(f;z)P_\lambda(z)^k \\
& + O\left(\frac{1}{\rho^{\lambda(r+\ell)n}}\right) \\
= {}& O\left(\frac{|P_\lambda(z)|^n}{\rho^{\lambda(r+\ell)}} + \frac{1}{\rho^{\lambda(r+\ell)}}\right) = O\left(\frac{1}{(\rho+\varepsilon)^{\lambda(r+\ell-1)n}}\right), \qquad z \in U,
\end{aligned}
$$

i.e. U forms indeed a set of distinguished points.

Similarly, defining t and s by $(r + \ell)t = (r + \ell - 1)n + s$, $0 \le s < r + \ell$, we get from (1.21b)

$$
\begin{aligned}
\Delta_{\lambda rn-1,\ell}(f;z) = {}& \sum_{k=0}^{s-1} q_{(r+\ell-1)n+k}(f;z_j)P_\lambda(z)^k \\
& + \sum_{v=r}^{n-1} P_\lambda(z)^{(r+\ell)(v-(r+\ell-1)n)} \sum_{k=0}^{\ell} q_{(r+\ell)v+k}(f;z)P_\lambda(z)^k \\
& + O\left(\frac{|P_\lambda(z)|^{rn}}{\rho^{\lambda(r+\ell+1)n}}\right) \\
= {}& O\left(\frac{1}{\rho^{\lambda(r+\ell-1)n}} + \frac{|P_\lambda(z)|^{rn}}{\rho^{\lambda(r+\ell+1)n}}\right) = O\left(\frac{|P_\lambda(z)|^{rn}}{(\rho+\varepsilon)^{\lambda(r+\ell)n}}\right), \qquad z \in V,
\end{aligned}
$$

i.e. the points in V are also distinguished.

Necessity. Suppose the points in $U \cup V$ are distinguished, i.e.

$$
\limsup_{n\to\infty} \Delta_{\lambda rn-1,\ell}(f;z)^{1/(\lambda n)} < K_{r,\ell}(\rho,|z|) \tag{1.22}
$$

but

$$
\operatorname{rank} M(X(U), X(V)) = \lambda(r + \ell - 1)(r + \ell). \tag{1.23}
$$

We show that this leads to a contradiction. (1.11) yields

$$
\begin{aligned}
\Delta_{\lambda rn-1,\ell}(f;z) = {}& \beta_{\ell,r}(P_\lambda(z)^n) \sum_{k=0}^{n-1} q_{(r+\ell-1)n+k}(f;z)P_\lambda(z)^k \\
& + O\left((K_{r,\ell+1}(\rho,|z|) + \varepsilon)^{\lambda rn}\right),
\end{aligned}
$$

where $\varepsilon > 0$ is arbitrarily small. Thus

$$
\begin{aligned}
h(z) := {} & \beta_{r,\ell}(P_\lambda(z)^{n+1})\Delta_{\lambda rn-1,\ell}(f;z) \\
& - \beta_{r,\ell}(P_\lambda(z)^n)P_\lambda(z)^{r+\ell-1}\Delta_{\lambda(n+1)-1,\ell}(f;z) \\
= {} & \beta_{\ell,r}(P_\lambda(x)^n)\beta_{\ell,r}(P_\lambda(z)^{n+1}) \sum_{k=0}^{r+\ell-2} q_{(r+\ell-1)n+k}(f;z)P_\lambda(z)^k \\
& - \sum_{k=0}^{r+\ell-1} q_{(r+\ell)n+k}(f;z)P_\lambda(z)^{n+k} + \Delta_{\lambda rn-1,\ell+1}(f;z)\beta_{\ell,r}(P_\lambda(z)^{n+1}) \\
& - \Delta_{\lambda r(n+1)-1,\ell+1}(f;z)\beta_{\ell,r}(P_\lambda(z)^n)P_\lambda(z)^{r+\ell-1}.
\end{aligned}
$$

On the other hand, by (1.22)

$$
h(z) = O\left((K_{r,\ell}(\rho,|z|) - \varepsilon)^{\lambda rn}\right)\left(|\beta_{\ell,r}(P_\lambda(z)^n)| + |\beta_{\ell,r}(P_\lambda(z)^{n+1})|\right), \quad z \in U \cup V,
$$

and

$$
\Delta_{\lambda rn-1,\ell+1}(f;z) = O\left((K_{r,\ell+1}(\rho,|z|) + \varepsilon)^{\lambda rn}\right)
$$

by Theorem 1. Thus

$$
\begin{aligned}
\sum_{k=0}^{r+\ell-2} q_{(r+\ell-1)n+k}(f;z)P_\lambda(z)^k := {} & O\left(\left(\frac{1}{\rho^{\lambda(r+\ell)}} + \varepsilon\right)^n\right) + \\
& + O\left((K_{r,\ell+1}(\rho,|z|) + \varepsilon)^{\lambda rn}\right) \\
& + O\left((K_{r,\ell}(\rho,|z|) - \varepsilon)^{\lambda rn}\right)\left(|\beta_{\ell,r}(P_\lambda(z)^n)|^{-1} + |\beta_{\ell,r}(P_\lambda(z)^{n+1})|^{-1}\right), \quad z \in U.
\end{aligned}
$$

On using Lemma 2 from Ch. 4 we get

$$
\sum_{k=0}^{r+\ell-2} q_{(r+\ell-1)n+k}(f;z)P_\lambda(z)^k = O\left(\left(\frac{1}{\rho^{\lambda(r+\ell-1)}} - \varepsilon\right)^n\right), \quad z \in U.
$$

Similarly,

$$
\sum_{k=0}^{r+\ell-1} q_{(r+\ell)n+k}(f;z)P_\lambda(z)^k = O\left(\left(\frac{1}{\rho^{\lambda(r+\ell)}} - \varepsilon\right)^n\right), \quad z \in V.
$$

With the notation

$$
q_n(f;z) := \sum_{s=0}^{\lambda-1} b_{\lambda n+s}z^s \tag{1.24}
$$

these relations take the form

$$
\sum_{k=0}^{r+\ell-2}\sum_{s=0}^{\lambda-1} b_{\lambda(r+\ell-1)n+\lambda k+s}z^s P_\lambda(z)^k = O\left(\left(\frac{1}{\rho^{\lambda(r+\ell-1)}} - \varepsilon\right)^n\right), \quad z \in U \tag{1.25a}
$$

and

$$\sum_{k=0}^{r+\ell-1}\sum_{s=0}^{\lambda-1}b_{\lambda(r+\ell)n+\lambda k+s}z^s P_\lambda(z)^k = O\left(\left(\frac{1}{\rho^{\lambda(r+\ell)}}-\varepsilon\right)^n\right), \qquad z \in V. \quad (1.25b)$$

For an arbitrary positive integer v, put $n = (r+\ell)v + t$, $0 \le t \le r+\ell-1$ in (1.25a), and $n = (r+\ell)v + t$, $0 \le t \le r+\ell-2$ in (1.25b). Then we obtain

$$\sum_{k=0}^{r+\ell-2}\sum_{s=0}^{\lambda-1}b_{\lambda(r+\ell-1)[(r+\ell)v+t]+\lambda k+s}z^s P_\lambda(z)^k = O\left(\left(\frac{1}{\rho^{\lambda(r+\ell-1)}}-\varepsilon\right)^{(r+\ell)v}\right),$$

$$z \in U; \ t = 0,\ldots,r+\ell-1,$$

and

$$\sum_{k=0}^{r+\ell-1}\sum_{s=0}^{\lambda-1}b_{\lambda(r+\ell)[(r+\ell)v+t]+\lambda k+s}z^s P_\lambda(z)^k = O\left(\left(\frac{1}{\rho^{\lambda(r+\ell)}}-\varepsilon\right)^{(r+\ell-1)v}\right),$$

$$z \in V; \ t = 0,\ldots,r+\ell-2.$$

Because of (1.23), we can choose $\lambda(r+\ell-1)(r+\ell)$ points out of $U \cup V$ such that the corresponding system of linear equations is solvable for the unknowns $b_{\lambda(r+\ell-1)(r+\ell)v},\ldots,b_{\lambda(r+\ell-1)(r+\ell)(v+1)-1}$. Due to the structure of the right hand sides, for the solution we obtain

$$\left|b_{\lambda(r+\ell-1)(r+\ell)v+k}\right| = O\left(\left(\frac{1}{\rho^\lambda}-\varepsilon\right)^{(r+\ell-1)(r+\ell)v}\right),$$

$$\text{for } k = 0,\ldots,\lambda(r+\ell-1)(r+\ell)-1; \text{ and } v = 0,1,\ldots,$$

whence by (1.24)

$$\limsup_{n\to\infty} \max_{z\in\Gamma(\rho)} |q_n(f;z)|^{1/(\lambda n)} < \frac{1}{\rho},$$

a contradiction to $f \in AG(\rho)$. $\qquad\square$

AN EXAMPLE. Let $\lambda = 2$ and $P_2(z) = z^2 - a^2$, $a > 0$ (this leads us to the so-called Bernoulli lemniscate; cf. also Example 4 in Ch.9). For an arbitrary $f \in AG(\rho^2)$, (1.3) yields

$$q_k(f;z) = \frac{1}{2\pi i}\oint_{\Gamma(R)}(t+z)\frac{f(t)}{(t^2-a^2)^{k+1}}dt, \qquad k = 0,1,\ldots.$$

Expanding into partial fractions we obtain

$$\frac{t+z}{(t^2-a^2)^{k+1}} = \sum_{j=1}^{k+1}\left(\frac{a_j}{(t-a)^j} + \frac{b_j}{(t+a)^j}\right)$$

with

$$a_j = \frac{(-1)^{k+1-j}}{(2a)^{2k+2-j}}\left[(z+a)\binom{2k+1-j}{k} - 2a\binom{2k-j}{k}\right], \qquad j = 1,\ldots,k+1$$

and

$$b_j = \frac{(-1)^{k+1}}{(2a)^{2k+2-j}}\left[(z-a)\binom{2k+1-j}{k} + 2a\binom{2k-j}{k}\right], \qquad j = 1,\ldots,k+1.$$

Hence using

$$\frac{1}{2\pi i}\oint_{\Gamma(R)}\frac{f(t)}{(t \pm a)^j}\,dt = \frac{f^{(j-1)}(\pm a)}{(j-1)!}, \qquad j = 1,\ldots$$

we get

$$q_k(f;z) = (-1)^{k+1}\sum_{j=1}^{k+1}\frac{1}{(2a)^{2k+2-j}(j-1)!}\times$$

$$\times\left[\binom{2k+1-j}{k}\left[(-1)^j(z+a)f^{(j-1)}(-a) + (z-a)f^{(j-1)}(a)\right]\right.$$

$$\left. - 2a\binom{2k-j}{k}\left[f^{(j-1)}(a) - (-1)^{j-1}f^{(j-1)}(-a)\right]\right],$$

for $k = 0, 1,\ldots$.

Choosing $r = \ell = 1$, let

(a) $U = \{u_1\}$, $V = \emptyset$,

(b) $U = \{u_1\}$, $V = \{v_1\}$,

(c) $U = \emptyset$, $V = \{v_1\}$,

(d) $U = \emptyset$, $V = \{v_1, v_2\}$,

(e) $U = \emptyset$, $V = \{v_1, v_2, v_3\}$.

According to Corollary 2, the set $U \cup V$ given by the examples (a)–(e) above are always distinguished.

Now let $U = \{u_1\}$, $V = \{v_1, v_2\}$. Then

$$X(U) = X_0(U) = (1\ u_1), \quad X_0(V) = \begin{pmatrix} 1 & v_1 \\ 1 & v_2 \end{pmatrix}, \quad X_1(V) = \begin{pmatrix} v_1^2 - a^2 & v_1^3 - a^2 v_1 \\ v_2^2 - a^2 & v_2^3 - a^2 v_2 \end{pmatrix},$$

$$X(V) = \begin{pmatrix} 1 & v_1 & v_1^2 - a^2 & v_1^3 - a^2 v_1 \\ 1 & v_2 & v_2^2 - a^2 & v_2^3 - a^2 v_2 \end{pmatrix}, \quad M = \begin{pmatrix} 1 & u_1 & 0 & 0 \\ 0 & 0 & 1 & u_1 \\ 1 & v_1 & v_1^2 - a^2 & v_1^3 - a^2 v_1 \\ 1 & v_2 & v_2^2 - a^2 & v_2^3 - a^2 v_2 \end{pmatrix}.$$

According to Theorem 2, in order that $U \cup V$ be distinguished, the determinant of the matrix M must vanish, i.e.

$$\det M = (v_1 - v_2)(v_1 + v_2)(u_1 - v_1)(u_1 - v_2) = 0. \qquad (1.26)$$

Since u_1 is inside and $v_1 \neq v_2$ are outside the lemniscate $|z^2 - a^2| = \rho^2$, this condition can be satisfied only if $v_1 = -v_2$. For example, the points $u_1 = a > 0$, $v_1 = -v_2 = a + \rho$ will be distinguished.

By Corollary 1, the maximal number of distinguished points in and out of the lemniscate $|z^2 - a^2| = \rho^2$ is at most 1 and 3, respectively. However, with

$$U = \{u_1\}, \quad V = \{v_1, v_2, v_3\}$$

the matrix M (with similar calculations as above) takes the form

$$M = \begin{pmatrix} 1 & u_1 & 0 & 0 \\ 0 & 0 & 1 & u_1 \\ 1 & v_1 & v_1^2 - a^2 & v_1^3 - a^2 v_1 \\ 1 & v_2 & v_2^2 - a^2 & v_2^3 - a^2 v_2 \\ 1 & v_3 & v_3^2 - a^2 & v_3^3 - a^2 v_3 \end{pmatrix}.$$

In order that $U \cup V$ be distinguished, all fourth order minors of this 5×4 matrix must vanish. Taking the first four rows, we arrive at the same condition as in (1.26), which means $v_1 = -v_2$. Now taking the first three rows and the last row from M we get $v_1 = -v_3$, i.e. $v_2 = v_3$, a contradiction. So in this case $U \cup V$ is never a distinguished set.

2. Historical Remarks

Theorem 1 was proved by Lou Yuanren [73]. Theorem 2 is new; the special case $\lambda = 1$ (i.e. the circle) was considered by Ivanov and Sharma [51]; cf. also Chapter 4.

CHAPTER 11

WALSH EQUICONVERGENCE AND EQUISUMMABILITY

1. Introduction

In the equiconvergence theorems the convergence to 0 or the order of the difference of two interpolating operators acting on an analytic function from some class are considered. In this chapter we consider the regular summability, instead of convergence, of the difference of the above two operators. Since convergence is also a regular summability method we get at the same time also equiconvergence theorems.

In this chapter the equisummability of a sequence of operators $(\Lambda_n(z, f))_{n \geq 0}$ that operate on elements of the class A_R, i.e., the class of functions regular in the disk $D_R \equiv \{z : |z| < R\}$ but not in $\overline{D}_R$, $R > 1$, are considered. The above operators are defined in Sec. 2.

A star-shaped domain that includes 0 is defined as a domain that is intersected by any linear ray originating at 0 by a single linear segment.

A star-shaped Jordan curve is defined as a closed Jordan curve that contains 0 in its interior and is intersected by any linear ray originating at 0 by a single point.

The unit disk $\{z : |z| < 1\}$ is denoted by D.

DEFINITION 1.1. The A-summability considered here is defined in the following way. Suppose $X \subset \mathbb{R}$. Let $x^* \in \mathbb{R}$ be an accumulation point of X. Let $(a_n(x))_{n \geq 0}$ be a given sequence of functions on X. For a sequence $(s_n)_{n \geq 0}$ of complex numbers, denote, formally, $\sigma(x) := \sum_{n=0}^{\infty} a_n(x)s_n$ ($x \in X$). The sequence $(s_n)_{n \geq 0}$ is A-summable to the value $s \in \mathbb{C}$, which is written as $A - \lim_{n \to \infty} s_n = s$, if the series $\sigma(x)$ is convergent for all $x \in X$, and $\sigma(x) \to s$ as $x \to x^*$. In the definition of A-summability the following additional assumptions are made: For each $x \in X$ the power series of z, $\phi(x, z) := \sum_{n=0}^{\infty} a_n(x)z^n$ is an entire function. A star-shaped domain $G_A \subset \mathbb{C}$, $\infty \notin G_A$, is given and for this domain we have $D \subset G_A$, $1 \notin G_A$. and $\lim_{x \to x^*} \phi(x, z) = 0$, $\forall z \in G_A$, the convergence being uniform on compact subsets of G_A. Finally, we require that

257

$\lim_{x \to x^*} \phi(x, 1) = 1$. By a compact subset of $\mathbb{C}$ we mean a closed and bounded set.

The Mittag-Leffler A-summability (see G.H. Hardy [45,p.77] is given by $X \equiv \{x : x > 0\}$, $x^* := 0$ and $a_n(x) := \exp(n \log n - \exp(n+1) \log(n+1))$, $n \geq 0$. For the Mittag-Leffler A-summability the associated domain G_A is the the finite complex plane less the real ray $x \geq 1$.

The function $(1 - z^2)^{-1}$ has only two singular points, (± 1) in the complex plane. Therefore the set S_f generated by the Mittag-Leffler A-summability for this function is the finite complex less the two real rays $x \geq 1$ and $x \leq -1$.

DEFINITION 1.2. Suppose $f(z) \in A_R$, $R > 1$. The function $f(z)$ is continued analytically along a ray beginning at 0 and up to the first singular point ζ on this ray. This is done along each ray beginning at 0. The union of all the segments $[0, \zeta)$, that correspond to all the rays beginning at 0 is denoted by $S \equiv S_f$ and is called the Mittag-Leffler star-domain of the given function $f(z)$ (see E.Hille [47,p.38]).

For A-summability we have the following simple result.

LEMMA 1.3. Suppose an A-summability is given. Let p, r be non-negative integers. Then: (i) For each $x \in X$ and $z \in \mathbb{C} \backslash \{\infty\}$ we have

$$\frac{d^p}{dz^p} \phi(x, z) = p! \sum_{n=p}^{\infty} \binom{n}{p} a_n(x) z^{n-p};$$

(ii) For each $z \in G_A$ we have

$$\lim_{x \to x^*} \sum_{n=r}^{\infty} \binom{n}{p} a_n(x) z^{n-r} = 0$$

where the convergence is uniform in z on any compact and bounded subset of G_A.

PROOF. By our assumptions $\phi(x, z) := \sum_{n=0}^{\infty} a_n(x) z^n$ is an entire function for each $x \in X$ and $\lim_{x \to x^*} \phi(x, z) = 0$ where the convergence is uniform in z on compact and bounded subsets of G_A. Therefore by the Weierstrass double-series theorem (i) holds and also $\lim_{x \to x^*} a_n(x) = 0$ for $n = 0, 1, 2, \cdots$. Also, for each $n = p, p+1, \cdots$ $\lim_{x \to x^*} \binom{n}{p} a_n(x) z^{n-p} = 0$ where the convergence is uniform in z, on bounded subsets of G_A (because z^{n-p} (for a fixed n) is uniformly bounded on bounded sets). (a) Assume first that $0 \leq r \leq p$. Then

$$\sum_{n=r}^{\infty} \binom{n}{p} a_n(x) z^{n-r} = z^{p-r} \sum_{n=p}^{\infty} \binom{n}{p} a_n(x) z^{n-p} \to 0 \text{ as } x \to x^*$$

and the convergence is uniform on closed and bounded subsets of G_A because z^{p-r} is bounded on such sets. This proves (ii) when $0 \leq r \leq p$. (b) Assume $r > p$. Since $\lim_{x \to x^*} \binom{n}{p} a_n(x) z^{n-p} = 0$ for $n = 0, 1, \cdots$, uniformly in z on bounded sets, we get from (i) that $\lim_{x \to x^*} \sum_{n=r}^{\infty} \binom{n}{p} a_n(x) z^{n-p} = 0$, where the convergence is uniform on compact and bounded subsets of G_A. Given a compact and bounded subset K of G_A, let γ be a rectifiable closed Jordan curve which is included in G_A and includes 0 and K in its interior. Then $d \equiv d(0, \gamma) > 0$, i.e., $|z| \geq d \quad \forall z \in \gamma$. Therefore we have (since $\gamma \subset G$ is a compact and bounded set)

$$\max_{z \in \gamma} \left| \sum_{n=r}^{\infty} \binom{n}{p} a_n(x) z^{n-r} \right| = \max_{z \in \gamma} \left\{ |z|^{-(r-p)} \left| \sum_{n=r}^{\infty} \binom{n}{p} a_n(x) z^{n-p} \right| \right\} \leq$$

$$\leq d^{-(r-p)} \max_{z \in \gamma} \left| \sum_{n=r}^{\infty} \binom{n}{p} a_n(x) z^{n-p} \right| \to 0 \text{ as } x \to x^* .$$

By the maximum principle it follows that $\lim_{x \to x^*} \sum_{n=r}^{\infty} \binom{n}{p} a_n(x) z^{n-r} = 0$, where the convergence is uniform for $z \in K$. □

2. Definition of the kernels $(\Lambda_n^{\alpha,\beta}(z,f))_{n \geq 1}$.

2.1 Notations and assumptions used in the definition of the kernels $\Lambda_n^{\alpha,\beta}(z,f)$.

In the definition of the operators $(\Lambda_n^{\alpha,\beta}(z,f))_{n \geq 1}$ we use the following notations and assumptions. D_ρ will denote the open disk of radius ρ and center 0. For a set $U \subset \mathbb{C}$ which includes a disk D_η, $\eta > 0$, we write $\infty * U := \mathbb{C}$ and $0 * U := \{0\}$. G and G_A will be open star-shaped sets that include the unit disk, $1 \notin G$ and $1 \notin G_A$.

Let R be a real number satisfying $R > 1$. Let α, β be two distinct complex numbers satisfying $\alpha\beta \neq 0$, and $|\alpha| \neq |\beta|$. Set $\gamma := \max(|\alpha|, |\beta|)$ and $\delta := \max(1, |\alpha|, |\beta|)$. Assume $\gamma < R$. For a positive integer ν and any given set $A \subset \mathbb{C}$, let $\phi_\nu^{-1}(A) := \{\omega : \omega \in \mathbb{C}, \omega^\nu \in A\}$.

Let i_0 be a fixed positive integer. For each i, $1 \leq i \leq i_0$, let n_i be a given positive integer and let N_i be a finite set of positive integers. When $i_0 > 1$ let $M_2^{(i)}, \ldots, M_{n_i}^{(i)}$ be non-empty sets of non-negative integers and suppose that to each i, $1 \leq i \leq i_0$, and each choice of $\nu_i \in N_i$, $k_j^{(i)} \in M_j^i$, $2 \leq j \leq i_0$ there corresponds a set $M_1^{(i)} := M_1^{(i)}(\nu_i, k_{n_i}^{(i)}, \ldots, k_2^{(i)})$ of non-negative integers with the property that

(1) $\min M_1^{(i)}(\nu_i \in N_i, k_j^{(i)} \in M_j^i, 2 \leq j \leq i_0) = a_0^{(i)} + \sum_{j=2}^{n_i} a_j^{(i)} k_j^{(i)} + a_{n+1}^{(i)} \nu_i \geq 0$,

and either

(2) $\max M_1^{(i)}(\nu_i \in N_i,\ k_j^{(i)} \in M_j^i,\ 2 \le j \le i_0) = b_0^{(i)} + \sum_{j=2}^{n_i} b_j^{(i)} k_j^{(i)} + b_{n+1}^{(i)} \nu_i < \infty$

or

(3) $\max M_1^{(i)}(\nu_i \in N_i,\ k_j^{(i)} \in M_j^i,\ 2 \le j \le i_0) = +\infty,$

where $(a_j^{(i)})_{j=0}^{n_i}$, $(b_j^{(i)})_{j=0}^{n_i}$ are fixed integers which depend only on i. When $i_0 = 1$ then $M_1^{(i_0)}$ depends only on the elements of N_1 and satisfies (1), (2) and (3).

We use the notation $\boldsymbol{k}^{(i)} \in \boldsymbol{M}^{(i)}$ to mean that $k_{n_i}^{(i)} \in M_{n_i}^{(i)}, \ldots, k_1^{(i)} \in M_1^{(i)}$ for $1 \le i \le i_0$,

Let $\boldsymbol{\lambda}^{(i)} := (\lambda_j^{(i)})_{1 \le j \le n_i}$, $\boldsymbol{\psi}^{(i)} := (\psi_j^{(i)})_{1 \le j \le n_i}$, be two given sequences of integers.

We use the abbreviation $(\boldsymbol{\lambda}^{(i)}, \boldsymbol{k}^{(i)})_{r,s} := \sum_{j=r}^{s} \lambda_j^{(i)} k_j^{(i)}$, and $\boldsymbol{\lambda}^{(i)} \cdot \boldsymbol{k}^{(i)} := (\boldsymbol{\lambda}^{(i)}, \boldsymbol{k}^{(i)})_{1,n_i}$. $(\boldsymbol{\psi}^{(i)} \cdot \boldsymbol{\lambda}^{(i)})_{r,s})$ and $\boldsymbol{\psi}^{(i)} \cdot \boldsymbol{\lambda}^{(i)}$ are defined similarly.

We assume that for $1 \le i \le i_0$ and for each choice of $\boldsymbol{k}^{(i)} \in \boldsymbol{M}^{(i)}$ we have

$$\boldsymbol{\lambda}^{(i)} \cdot \boldsymbol{k}^{(i)} \ge 0, \quad \boldsymbol{\psi}^{(i)} \cdot \boldsymbol{k}^{(i)} \ge 0 \quad \text{and} \quad \boldsymbol{\lambda}^{(i)} \cdot \boldsymbol{k}^{(i)} + \boldsymbol{\psi}^{(i)} \cdot \boldsymbol{k}^{(i)} \ge 1. \qquad (2.1.1)$$

We further suppose that for each i, $1 \le i \le i_0$, and for each positive integer V the set of indices $\boldsymbol{k}^{(i)}$ for which $\boldsymbol{\lambda}^{(i)} \cdot \boldsymbol{k}^{(i)} + \boldsymbol{\psi}^{(i)} \cdot \boldsymbol{k}^{(i)} < V$ holds is finite.

Let i_1 be a fixed positive integer. For each i, $1 \le i \le i_1$, let m_i be a given positive integer. When $i_1 > 1$ let $P_2^{(i)}, \ldots, P_{m_i}^{(i)}$ be non-empty sets of non-negative integers and suppose that to each i, $1 \le i \le i_1$, and each choice of $\ell_j^{(i)} \in P_j^i$, $2 \le j \le i_1$ there corresponds a set $P_1^{(i)} := P_1^{(i)}(\nu_i, \ell_{n_i}^{(i)}, \ldots, \ell_2^{(i)})$ of non-negative integers with the property that

(1) $\min_{\ell_j^{(i)} \in P_j^i,\ 2 \le j \le i_1} P_1^{(i)} = c_0^{(i)} + \sum_{j=2}^{n_i} c_j^{(i)} \ell_j^{(i)} \ge 0,$

and either

(2) $\max_{\ell_j^{(i)} \in P_j^i,\ 2 \le j \le i_1} P_1^{(i)} = d_0^{(i)} + \sum_{j=2}^{n_i} d_j^{(i)} \ell_j^{(i)} < \infty$

or

(3) $\sup_{\ell_j^{(i)} \in P_j^i,\ 2 \le j \le i_1} P_1^{(i)} = +\infty$, where $(c_j^{(i)})_{j=1}^{n_i}$, $(d_j^{(i)})_{j=1}^{n_i}$ are fixed integers which depend only on i. When $i_1 = 1$ then $P_1^{(i_1)}$ is a given set of non-negative integers satisfying (1), (2) and (3).

Moreover let $\boldsymbol{\ell}^{(i)} := (\ell_j^{(i)})_{j=1}^{m_i}$, $\boldsymbol{\mu}^{(i)} := (\mu_j^{(i)})_{j=1}^{m_i}$, $\boldsymbol{\chi}^{(i)} := (\chi_j^{(i)})_{j=1}^{m_i}$ be given sequences of integers. We use the abbreviation $(\boldsymbol{\mu}^{(i)}, \boldsymbol{\ell}^{(i)})_{r,s} := \sum_{j=r}^{s} \mu_j^{(i)} \ell_j^{(i)}$, and $\boldsymbol{\mu}^{(i)} \cdot \boldsymbol{\ell}^{(i)} := (\boldsymbol{\mu}^{(i)}, \boldsymbol{\ell}^{(i)})_{1,m_i}$.

We also assume that for all $i \in [0, i_1]$ and for any $\ell_{m_i}^{(i)} \in P_{m_i}^{(i)}, \ldots, \ell_1^{(i)} \in \ldots, P^{(i)}$, we have

$$\boldsymbol{\mu}^{(i)} \cdot \boldsymbol{\ell}^{(i)} \ge 0, \quad \boldsymbol{\chi}^{(i)} \cdot \boldsymbol{\ell}^{(i)} \ge 0 \quad \text{and} \quad \boldsymbol{\mu}^{(i)} \cdot \boldsymbol{\ell}^{(i)} + \boldsymbol{\chi}^{(i)} \cdot \boldsymbol{\ell}^{(i)} \ge 1. \qquad (2.1.2)$$

2.2 Definition of the kernels $(\Lambda_n^{(\alpha,\beta)}(z,f))_{n\geq 1}$

Let $(\Lambda_n^{(\alpha,\beta)}(z,f))_{n\geq 1}$ be a sequence of operators that act on functions $f \in A_R$ and having the following representation

$$\Lambda_n^{(\alpha,\beta)}(z,f) = \frac{1}{2\pi i} \oint_\Gamma \frac{f(t)}{t-z} K_n(z,t)dt \qquad (2.1.3)$$

where $\Gamma := \{t \in \mathbb{C} : |t| = \tau\}$, $\delta < \tau < R$, $|z| > \tau$, and

$$K_n(z,t) = \sum_{1\leq i\leq i_0} A_n^{(i)}(z,t) + \sum_{1\leq i\leq i_1} B_n^{(i)}(t). \qquad (2.1.4)$$

Using the notation $\boldsymbol{k}^{(i)} \in \boldsymbol{M}^{(i)}$ to mean that $k_{n_i}^{(i)} \in M_{n_i}^{(i)}, \ldots, k_1^{(i)} \in M_1^{(i)}$ for $1 \leq i \leq i_0$, the abbreviations $\eta_i(t) := \left(\frac{\alpha}{t}\right)^{\boldsymbol{\lambda}^{(i)}.\boldsymbol{k}^{(i)}} \left(\frac{\beta}{t}\right)^{\boldsymbol{\psi}^{(i)}.\boldsymbol{k}^{(i)}}$, and $\xi_i(t) := \left(\frac{\alpha}{t}\right)^{\boldsymbol{\mu}^{(i)}.\boldsymbol{\ell}^{(i)}} \left(\frac{\beta}{t}\right)^{\boldsymbol{\chi}^{(i)}.\boldsymbol{\ell}^{(i)}}$, we assume that

$$A_n^{(i)}(z,t) := \sum_{\nu_i \in N_i} \sum_{\boldsymbol{k}^{(i)} \in \boldsymbol{M}^{(i)}} b_i(\boldsymbol{k}^{(i)}) \left(\left(\frac{\alpha}{t}\right)^{\boldsymbol{\lambda}^{(i)}.\boldsymbol{k}^{(i)}} \left(\frac{\beta}{t}\right)^{\boldsymbol{\psi}^{(i)}.\boldsymbol{k}^{(i)}} \left(\frac{z}{t}\right)^{\nu_i} \right)^n$$
$$\text{for } 1 \leq i \leq i_0, \qquad (2.1.5)$$

$$B_n^{(i)}(t) := \sum_{\boldsymbol{\ell}^{(i)} \in \boldsymbol{P}^{(i)}} c_i(\boldsymbol{\ell}^{(i)}) \left(\left(\frac{\alpha}{t}\right)^{\boldsymbol{\mu}^{(i)}.\boldsymbol{\ell}^{(i)}} \left(\frac{\beta}{t}\right)^{\boldsymbol{\chi}^{(i)}.\boldsymbol{\ell}^{(i)}} \right)^n, \quad \text{for } 1 \leq i \leq i_1. \qquad (2.1.6)$$

where $\boldsymbol{\ell}^{(i)} \in \boldsymbol{P}^{(i)}$ means $\ell_{m_{i_0}}^{(i)} \in P_{m_{i_0}}^{(i)}, \cdots, \ell_1^{(i)} \in P_1^{(i)}$. The coefficients $b_i(\boldsymbol{k}^{(i)})$ and $c_i(\boldsymbol{\ell}^{(i)})$ do not depend on n. We assume that for each x, $0 \leq x < 1$, and for any positive number d the series

$$\sum_{\nu_i \in N_i} \sum_{\boldsymbol{k}^{(i)} \in \boldsymbol{M}^{(i)}} |b_i(\boldsymbol{k}^{(i)})| d^{\nu_i} x^{\boldsymbol{\lambda}^{(i)}.\boldsymbol{k}^{(i)} + \boldsymbol{\psi}^{(i)}.\boldsymbol{k}^{(i)}} \qquad (2.1.7)$$

is convergent for each i, $1 \leq i \leq i_0$. We assume also that

$$\sum_{\boldsymbol{\ell}^{(i)} \in \boldsymbol{P}^{(i)}} |c_i(\boldsymbol{\ell}^{(i)})| x^{\boldsymbol{\mu}^{(i)}.\boldsymbol{\ell}^{(i)} + \boldsymbol{\chi}^{(i)}.\boldsymbol{\ell}^{(i)}}. \qquad (2.1.8)$$

is convergent for each i, $1 \leq i \leq i_1$.

Remark. In the definition of the operators $\Lambda_n^{(\alpha,\beta)}(z,t)$ two parameters α, β were used. The results, except for obvious changes, remain the same if only one parameter α is used, i.e., the powers of $\frac{\beta}{t}$ and the parameters associated with β do not appear in the definition of the operators $\Lambda_n^{(\alpha)}(z,f)$.

3. Equiconvergence and equisummability of the operators $\Lambda_n^{(\alpha,\beta)}(z,f)$

DEFINITION 3.1 OF THE WALSH RADIUS. For $R \geq \gamma$ the Walsh radius $\widetilde{R}(r)$ associated with an operator $(\Lambda_n^{(\alpha,\beta)}(z,f))_{n\geq1}$ is defined by

$$\widetilde{R}(r) \equiv \widetilde{R}(r; i_0, N_i, \boldsymbol{M}^i, \boldsymbol{\lambda}^i, \boldsymbol{\psi}) \tag{3.1}$$

$$:= \inf_{1\leq i\leq i_0} \inf_{\nu_i \in N_i, \, \boldsymbol{k}^{(i)} \in \boldsymbol{M}^{(i)}} \left\{ r\left(\frac{r}{|\alpha|}\right)^{\boldsymbol{\lambda}^{(i)}\cdot\boldsymbol{k}^{(i)}/\nu_i} \left(\frac{r}{|\beta|}\right)^{\boldsymbol{\psi}^{(i)}\cdot\boldsymbol{k}^{(i)}/\nu_i} \right\}.$$

REMARK 3.2. From the definition of $\widetilde{R}(r)$ it follows that $\widetilde{R}(r) \geq r\left(\frac{r}{\gamma}\right)^{1/\nu^*}$ and $\widetilde{R}(R) > R$, since (2.1.1) is satisfied.

We have the following equiconvergence theorem.

THEOREM 3.3. *Assume conditions (2.1.1) and (2.1.2) are satisfied and that* $f \in A_R$, *where* $R > \delta := \max(1, |\alpha|, |\beta|)$. *Then for each* r_1, $0 < r_1 < \widetilde{R}(R)$

$$\limsup_{n\to\infty} \max_{|z|\leq r_1} \left|\Lambda_n^{(\alpha,\beta)}(z,f)\right|^{1/n} \leq \max\left(\frac{\gamma}{R}, \left(\frac{r_1}{\widetilde{R}(R)}\right)^{\min_{1\leq i\leq i_0} N_i}\right). \tag{3.2}$$

The proof of Theorem 3.3 is given in Sec. 4.

In the statement of the next equisummability theorem for operators $\Lambda_n^{(\alpha,\beta)}(z,f)$ recall the following assumptions made in defining the operators: $\alpha\beta \neq 0$, $|\beta| \neq |\alpha|$, $R > \delta := \max(1, |\alpha|, |\beta|$. Let $S \equiv S_f$ be the Mittag-Leffler star-domain of a given function $f \in A_R$ and let $G \equiv G_A$ denote the A-summability domain associated with the A-summability which is given by $(a_n(x))_{n\geq1}$. We write for $1 \leq i \leq i_0$,

$$\boldsymbol{A}_i := \bigcap_{\nu_i \in N_i} \bigcap_{\boldsymbol{k}^{(i)} \in \boldsymbol{M}^{(i)}} \bigcap_{c \notin S} \phi_{\nu_i}^{-1}\left(c^{\mathcal{V}_i}\left(\frac{c}{\alpha}\right)^{\boldsymbol{\lambda}^{(i)}\cdot\boldsymbol{k}^{(i)}} \left(\frac{c}{\beta}\right)^{\boldsymbol{\psi}^{(i)}\cdot\boldsymbol{k}^{(i)}} G_A\right);$$

$$\mathcal{E} := \bigcap_{1\leq i\leq i_0} \boldsymbol{A}_i. \tag{3.3}$$

We have the following equisummability theorem for operators $(\Lambda_n^{(\alpha,\beta)}(z,f))_{n\geq1}$.

THEOREM 3.4. *Suppose we are given an A-summability method A (see Definition 1.1). Then for each* $f \in A_R$ *we have* $A-\lim_{n\to\infty} \Lambda_{n+1}^{(\alpha,\beta)}(z,f) = 0$, $\forall z \in \mathcal{E}$. *More precisely, if* $\Psi(x,z) := \sum_{n=0}^{\infty} a_n(x)\Lambda_{n+1}^{(\alpha,\beta)}(z,f)$ $(x \in X)$ *then* $\Psi(x,.)$ *is an entire function for every fixed* $x \in X$, *and on any compact and bounded subset* K *of* $\mathcal{E}$ *there holds* $|\Psi(x,z)| \leq M.\max_{w\in B}|\phi(x,w)|$, $\forall z \in K$, $\forall x \in X$, *where M is a constant, and B is a compact subset of G_A. Both M and B depend on $f, \alpha, \beta, \mathcal{E}$ and K but not on x.*

In Sec. 5 some properties of particular sets are stated and proved. These results are used in the proof of Theorem 3.4. The proof of Theorem 3.4 is given in Sec. 6.

4. Proof of Theorem 3.3

First we obtain some properties of the Walsh radius and then prove Theorem 3.3.

LEMMA 4.1. *For given numbers α, β, δ satisfying $\alpha\beta \neq 0$ and $|\alpha| \neq |\beta| =$ and $\delta = \max(1, |\alpha|, |\beta|)$ define for any positive number $c \geq \delta$ the number H by:*

$$H \equiv H(c) := \inf_{1 \leq i \leq i_0} \inf_{\nu_i \in N_i, \, \mathbf{k}^{(i)} \in \mathbf{M}^{(i)}} \left| \phi_{\nu_i}^{-1} \left(c^{\nu_i} \left(\frac{c}{\alpha} \right)^{\boldsymbol{\lambda}^{(i)} \cdot \mathbf{k}^{(i)}} \left(\frac{c}{\beta} \right)^{\boldsymbol{\psi}^{(i)} \cdot \mathbf{k}^{(i)}} \right) \right|.$$
$$(4.1)$$

Then either $H = 0$ or H is a positive minimum. In both cases $H = \tilde{R}(c)$. H vanishes only in the following case: For some i, $1 \leq i \leq i_0$, and for some j, $1 \leq j \leq n_i$, the set $M_j^{(i)}$ is unbounded and either: (1) $\lambda_j^{(i)} < 0$, $\psi_j^{(i)} \leq 0$; or (2) $\lambda_j^{(i)} \leq 0$, $\psi_j^{(i)} < 0$; or (3) $\lambda_j^{(i)} < 0$, $\psi_j^{(i)} > 0$ and $\left| \frac{c}{\alpha} \right|^{\lambda_j^{(i)}} \left| \frac{c}{\beta} \right|^{\psi_j^{(i)}} < 1$. In all other cases the minimum of H is attained for a $k_j^{(i)}$ which is either $\min_i M_j^{(i)}$ or $\max_i M_j^{(i)}$ and for ν_i which is either $\max N_i$ or $\min N_i$.

RAMARK. The conditions when $k_j^{(i)}$ is $\min_i M_j^{(i)}$ or $\max_i M_j^{(i)}$ and when ν_i is $\max N_i$ or $\min N_i$ are given explicitly in the proof.

The last statement is not exact when the set $M_1^{(i)}$ depends on the sets $M_2^{(i)}, \ldots, M_{n_i}^{(i)}$. The statement is made precise in the proof when the case where $M_1^{(i)}$ depends on the sets $M_2^{(i)}, \ldots, M_{n_i}^{(i)}$ is considered.

If $S \equiv S_f$ is the Mittag-Leffler star-domain of some $f \in A_R$, $R > \delta$, then from Lemma 4.1, we have

COROLLARY 4.2. *Let $f \in A_R$, $R > 1$. Suppose the assumptions of Lemma 4.1 hold. Then*

$$\bigcap_{1 \leq i \leq i_0} \bigcap_{\nu_i \in N_i} \bigcap_{\mathbf{k}^{(i)} \in \mathbf{M}^{(i)}} \bigcap_{c \notin S} \phi_{\nu_i}^{-1} \left(c^{\nu_i} \left(\frac{c}{\alpha} \right)^{\boldsymbol{\lambda}^{(i)} \cdot \mathbf{k}^{(i)}} \left(\frac{c}{\beta} \right)^{\boldsymbol{\psi}^{(i)} \cdot \mathbf{k}^{(i)}} D_1 \right) = D_{\tilde{R}(R)}.$$

PROOF OF COROLLARY 4.2. We have the obvious relation

$$\phi_{\nu_i}^{-1}\left(r^{\nu_i}\left(\frac{r}{\alpha}\right)^{\boldsymbol{\lambda}^{(i)}\cdot\boldsymbol{k}^{(i)}}\left(\frac{r}{\beta}\right)^{\boldsymbol{\psi}^{(i)}\cdot\boldsymbol{k}^{(i)}}D_1\right)=D_a,$$

where $a=r\left(\frac{r}{|\alpha|}\right)^{\boldsymbol{\lambda}^{(i)}\cdot\boldsymbol{k}^{(i)}/\nu_i}\left(\frac{r}{|\beta|}\right)^{\boldsymbol{\psi}^{(i)}\cdot\boldsymbol{k}^{(i)}/\nu_i}$, Since S^c has at least one point common with $\partial\overline{D}_R$, the proof follows from Lemma 4.1. $\qquad\square$

PROOF OF LEMMA 4.1. Since $H=\min_{1\le i\le i_0}H_i$ where

$$H_i\equiv H_i(c)$$
$$:=\inf\left\{\left|\phi_{\nu_i}^{-1}\left(c^{\nu_i}\left(\frac{c}{\alpha}\right)^{\boldsymbol{\lambda}^{(i)}\cdot\boldsymbol{k}^{(i)}}\left(\frac{c}{\beta}\right)^{\boldsymbol{\psi}^{(i)}\cdot\boldsymbol{k}^{(i)}}\right)\right|\ :\ \nu_i\in N_i,\ \boldsymbol{k}^{(i)}\in\boldsymbol{M}^{(i)}\right\}$$

it is enough to prove Lemma 4.1 for each H_i $(1\le i\le i_0)$ separately; hence without loss of generality we may in the rest of the proof omit the index i. In this case we write for $c\ge\delta$

$$H(c)\equiv H\left(r;N,(M_j)_{1\le j\le n},(\lambda_j)_{1\le j\le n},(\psi_j)_{1\le j\le n}\right)$$
$$=\inf_{\nu\in N,\boldsymbol{k}\in\boldsymbol{M}}c\left|\frac{c}{\alpha}\right|^{\boldsymbol{\lambda}\cdot\boldsymbol{k}/\nu}\left|\frac{c}{\beta}\right|^{\boldsymbol{\psi}\cdot\boldsymbol{k}/\nu}.$$

where $\boldsymbol{k}\in\boldsymbol{M}$ means $k_{n_0}\in M_{n_0},\cdots,k_1\in M_1$. The rest of the proof is divided into two parts. The first part of the proof considers the case when the set M_1 is independent of $k_2,\cdots,k_n,\nu$. The second part of the proof considers the case when M_1 is dependent on $k_2,\cdots,k_n$ and ν.

Proof of the case when when M_1 is independent of M_n, $\cdots$ M_2. We use here the notation $\boldsymbol{k}':=(k_2,\ldots,k_n)$, $\boldsymbol{M}':=(M_2,\ldots,M_n)$ and $\boldsymbol{k}'\in\boldsymbol{M}'$ means that $k_n\in M_n,\ldots,k_2\in M_2$. With respect to k_1 the following cases can occur: *(A) when the set M_1 is bounded* or *(B) when the set M_1 is unbounded.*

The case (A) where M_1 is bounded. In this case we have to consider several subcases. (i) Assume $\lambda_1=\psi_1=0$. Then

$$H=\inf_{\nu\in N,\ \boldsymbol{k}'\in\boldsymbol{M}'}c\left|\frac{c}{\alpha}\right|^{((\boldsymbol{\lambda}\cdot\boldsymbol{k})_{2,n})/\nu}\left|\frac{c}{\beta}\right|^{((\boldsymbol{\psi}\cdot\boldsymbol{k})_{2,n})/\nu}$$
$$=\inf_{\nu\in N,\ \boldsymbol{k}'\in\boldsymbol{M}'}c\left|\frac{c}{\alpha}\right|^{(\lambda_1k_1^*+(\boldsymbol{\lambda}\cdot\boldsymbol{k})_{2,n})/\nu}\left|\frac{c}{\beta}\right|^{(\psi_1k_1^*+(\boldsymbol{\psi}\cdot\boldsymbol{k})_{2,n})/\nu}$$

where $k_1^*=\min M_1$.

(ii) Assume $\lambda_1\ne0$, $\psi_1=0$. In this case, since $\nu\in N\subset\text{IN}$ and $M_1\subset\text{IN}^+$ is bounded, we have

$$H = \inf_{\nu \in N, \, \boldsymbol{k'} \in M'} c \left| \frac{c}{\alpha} \right|^{((\boldsymbol{\lambda} \cdot \boldsymbol{k})_{2,n})/\nu} \times$$

$$\times \left| \frac{c}{\beta} \right|^{((\boldsymbol{\psi} \cdot \boldsymbol{k})_{2,n})/\nu} \times \begin{cases} \left| \frac{c}{\alpha} \right|^{\lambda_1 \cdot (\min M_1)/\nu} & \text{if} \quad \lambda_1 > 0 \\[2mm] \left| \frac{c}{\alpha} \right|^{\lambda_1 \cdot (\max M_1)/\nu} & \text{if} \quad \lambda_1 < 0 \end{cases}$$

$$= \inf_{\nu \in N, \, \boldsymbol{k'} \in M'} c \left| \frac{c}{\alpha} \right|^{(\lambda_1 k_1^* + (\boldsymbol{\lambda} \cdot \boldsymbol{k})_{2,n})/\nu} \left| \frac{c}{\beta} \right|^{(\psi_1 k_1^* + (\boldsymbol{\psi} \cdot \boldsymbol{k})_{2,n})/\nu}$$

where $k_1 = k_1^*$ and $k_1^* = \min M_1$ if $\lambda_1 > 0$ and $k_1^* = \max M_1$ if $\lambda_1 < 0$.

(iii) Assume $\lambda_1 = 0$, $\quad \psi_1 \neq 0$. As in the previous case we have

$$H = \inf_{\nu \in N, \, \boldsymbol{k'} \in M'} c \left| \frac{c}{\alpha} \right|^{((\boldsymbol{\lambda} \cdot \boldsymbol{k})_{2,n})/\nu} \times$$

$$\times \left| \frac{c}{\beta} \right|^{((\boldsymbol{\psi} \cdot \boldsymbol{k})_{2,n})/\nu} \times \begin{cases} \left| \frac{c}{\beta} \right|^{\psi_1 \cdot (\min M_1)/\nu} & \text{if} \quad \psi_1 > 0 \\[2mm] \left| \frac{c}{\beta} \right|^{\psi_1 \cdot (\max M_1)/\nu} & \text{if} \quad \psi_1 < 0. \end{cases}$$

$$= \inf_{\nu \in N, \, \boldsymbol{k'} \in M'} c \left| \frac{c}{\alpha} \right|^{(\lambda_1 k_1^* + (\boldsymbol{\lambda} \cdot \boldsymbol{k})_{2,n}/\nu} \left| \frac{c}{\beta} \right|^{(\psi_1 k_1^* + (\boldsymbol{\psi} \cdot \boldsymbol{k})_{2,n})/\nu}$$

where $k_1 = k_1^*$ and $k_1^* = \min M_1$ if $\psi_1 > 0$, and $k_1^* = \max M_1$ if $\psi_1 < 0$.

(iv) $\lambda_1 \neq 0$, $\psi_1 \neq 0$. In this case the following four cases can occur:

a) $\lambda_1 > 0, \psi_1 > 0$. In this case

$$H = \inf_{\nu \in N, \, \boldsymbol{k'} \in M'} c \left| \frac{c}{\alpha} \right|^{(\lambda_1 k_1^* + (\boldsymbol{\lambda} \cdot \boldsymbol{k})_{2,n})/\nu} \left| \frac{c}{\beta} \right|^{(\psi_1 k_1^* + (\boldsymbol{\psi} \cdot \boldsymbol{k})_{2,n})/\nu}.$$

where $k_1 = k_1^*$ and $k_1^* = \min M_1$.

b) Assume $\lambda_1 < 0$, $\psi_1 < 0$. In this case

$$H = \inf_{\nu \in N, \, \boldsymbol{k'} \in M'} c \left| \frac{c}{\alpha} \right|^{(\lambda_1 k_1^* + (\boldsymbol{\lambda}^{(i)} \cdot \boldsymbol{k}^{(i)})_{2,n})/\nu} \left| \frac{c}{\beta} \right|^{(\psi_1 k_1^* + (\boldsymbol{\psi} \cdot \boldsymbol{k})_{2,n})/\nu}.$$

where $k_1 = k_1^*$ and $k_1^* = \max M_1$.

c) Assume $\lambda_1 > 0$, $\psi_1 < 0$. In this case

$$H = \inf_{\nu \in N, \, \boldsymbol{k'} \in M'} c \left| \frac{c}{\alpha} \right|^{(\lambda_1 k_1^* + (\boldsymbol{\lambda} \cdot \boldsymbol{k})_{2,n})/\nu} \left| \frac{c}{\beta} \right|^{(\psi_1 k_1^* + (\boldsymbol{\psi} \cdot \boldsymbol{k})_{2,n})/\nu}$$

where

$$k_1^* = \min M_1 \quad \text{if} \quad \left|\frac{c}{\alpha}\right|^{\lambda_1} \geq \left|\frac{c}{\beta}\right|^{-\psi_1}$$

and

$$k_1^* = \max M_1 \quad \text{if} \quad \left(\frac{R}{|\alpha|}\right)^{\lambda_1} < \left(\frac{R}{|\beta|}\right)^{-\psi_1}.$$

d) Assume $\lambda_1 < 0$, $\psi_1 > 0$. As in case c) we get

$$H = \inf_{\nu \in N,\ \boldsymbol{k}' \in \boldsymbol{M}'} c \left|\frac{c}{\alpha}\right|^{(\lambda_1 k_1^* + (\boldsymbol{\lambda}\cdot\boldsymbol{k})_{2,n})/\nu} \left|\frac{c}{\beta}\right|^{(\psi_1 k_1^* + (\boldsymbol{\psi}\cdot\boldsymbol{k})_{2,n})/\nu}$$

where $k_1^* = \min M_1$ if $\left|\frac{c}{\beta}\right|^{\psi_1} \geq \left|\frac{c}{\alpha}\right|^{-\lambda_1}$ and $k_1^* = \max M_1$ if $\left|\frac{c}{\beta}\right|^{\psi_1} < \left|\frac{c}{\alpha}\right|^{-\lambda_1}$. In conclusion we get that

$$H = \inf_{\nu \in N,\ \boldsymbol{k}' \in \boldsymbol{M}'} c \left|\frac{c}{\alpha}\right|^{(\lambda_1 k_1^* + (\boldsymbol{\lambda}\cdot\boldsymbol{k})_{2,n})/\nu} \left|\frac{c}{\beta}\right|^{(\psi_1 k_1^* + (\boldsymbol{\psi}\cdot\boldsymbol{k})_{2,n})/\nu}$$

where k_1^* is either equal to $\min M_1$ or else, equal to $\max M_1$.

The case (B) when the set M_1 is not bounded. In this case we have to consider separately the following five cases:

(a) Assume $\lambda_1 > 0$, $\psi_1 \geq 0$ or $\lambda_1 \geq 0$ and $\psi_1 > 0$. If $k_1^{(1)} \in M_1$, $k_1^{(2)} \in M_1$ and $k_1^{(1)} < k_1^{(2)}$ we see that

$$c\left|\frac{c}{\alpha}\right|^{(\lambda_1 k_1^{(1)} + (\boldsymbol{\lambda}\cdot\boldsymbol{k})_{2,n})/\nu} \left|\frac{c}{\beta}\right|^{(\psi_1 k_1^{(1)} + (\boldsymbol{\psi}\cdot\boldsymbol{k})_{2,n})/\nu}$$
$$< c\left|\frac{c}{\alpha}\right|^{(\lambda_1 k_1^{(2)} + (\boldsymbol{\lambda}\cdot\boldsymbol{k})_{2,n})/\nu} \left|\frac{c}{\beta}\right|^{(\psi_1 k_1^{(2)} + (\boldsymbol{\psi}\cdot\boldsymbol{k})_{2,n})/\nu},$$

so that

$$H(c) = \min_{\nu \in N,\ \boldsymbol{k}^{(i)} \in \boldsymbol{M}^{(i)}} c \left|\frac{c}{\alpha}\right|^{(\lambda_1 k_1^* + (\boldsymbol{\lambda}\cdot\boldsymbol{k})_{2,n})/\nu} \left|\frac{c}{\beta}\right|^{(\psi_1 k_1^* + (\boldsymbol{\psi}\cdot\boldsymbol{k})_{2,n})/\nu}$$

where $k_1^* = \min M_1$.

(b) Assume $\lambda_1 < 0$, $\psi_1 \leq 0$ or $\lambda_1 \leq 0$ and $\psi_1 < 0$. Then by an argument similar to that of part (a) of the first part of the proof when the set $M - 1$ is bounded it follows, since now the set M_1 is not bounded from above, that $H = 0$ which corresponds to $k^* = \infty = \sup M_1$. In this case it is not necessary to check the other indices $k_2, \ldots, k_{n+1}$.

(c) Assume $\lambda_1 > 0$, $\psi_1 < 0$ or $\lambda_1 < 0$, $\psi_1 > 0$. Then

$$H = \inf_{\nu \in N,\ \boldsymbol{k}^{(i)} \in \boldsymbol{M}^{(i)}} c \left|\frac{c}{\alpha}\right|^{((\boldsymbol{\lambda}\cdot\boldsymbol{k})_{2,n})/\nu} \left|\frac{c}{\beta}\right|^{((\boldsymbol{\psi}\cdot\boldsymbol{k})_{2,n})/\nu} \min_{k_1 \in M_1} \cdot \left(\left|\frac{c}{\alpha}\right|^{\lambda_1} \left|\frac{c}{\beta}\right|^{\psi_1}\right)^{k_1}.$$

We have to consider in this case three possibilities:

(1) If $\left|\frac{c}{\alpha}\right|^{\lambda_1}\left|\frac{c}{\beta}\right|^{\psi_1} > 1$ then

$$H = \inf_{\nu \in N,\, \mathbf{k}^{(i)} \in \mathbf{M}^{(i)}} c \left|\frac{c}{\alpha}\right|^{(\lambda_1 k_1^* + (\boldsymbol{\lambda}\cdot\mathbf{k})_{2,n})/\nu} \left|\frac{c}{\beta}\right|^{(\psi_1 k_1^* + (\boldsymbol{\psi}\cdot\mathbf{k})_{2,n})/\nu}$$

where $k_1^* = \min M_1$.

(2) If $\left|\frac{c}{\alpha}\right|^{\lambda_1}\left|\frac{c}{\beta}\right|^{\psi_1} = 1$ then

$$H = \inf_{\nu \in N,\, \mathbf{k}^{(i)} \in \mathbf{M}^{(i)}} c \left|\frac{c}{\alpha}\right|^{((\boldsymbol{\lambda}\cdot\mathbf{k})_{2,n})/\nu} \left|\frac{c}{\beta}\right|^{(\boldsymbol{\psi}\cdot\mathbf{k})_{2,n}},$$

which corresponds to $k_1^* = \min M_1$.

(3) If $\left|\frac{c}{\alpha}\right|^{\lambda_1}\left|\frac{c}{\beta}\right|^{\psi_1} < 1$ then since M_1 is not bounded from above, $H = 0$ which corresponds to $k^* = \infty = \sup M_1$. In this case it is not necessary to consider all other indices $k_2, \ldots, k_n$.

(d) Assume $\lambda_1 = 0$, $\psi_1 = 0$. In this case H is independent of k_1, and we can choose $k_i = k_1^*$ where $k_1^* = \min M_1$.

(e) The same argument is applied now to $k_2, \ldots, k_n$. The final result is that either $H = 0$ (and it is known when this happens) or there exist $k_1^*, \ldots, k_n^* \in M_1, \ldots, M_n$, respectively such that

$$H = \inf_{\nu \in N} c \left(\left|\frac{c}{\alpha}\right|^{\lambda_1 k_1^* + \cdots + \lambda_n k_n^*} \left|\frac{c}{\beta}\right|^{\psi_1 k_1^* + \cdots + \psi_n k_n^*} \right)^{\frac{1}{\nu}}$$

(it is important to remember that the set N is bounded and $\nu \geq 1$)

$$= c \left|\frac{c}{\alpha}\right|^{(\lambda_1 k_1^* + \cdots + \lambda_n k_n^*)/\nu^*} \left|\frac{c}{\beta}\right|^{(\psi_1 k_1^* + \cdots + \psi_n k_n^*)/\nu^*}$$

where $\nu^* = \max N$ or $\min N$ according as

$$\left|\frac{c}{\alpha}\right|^{\lambda_1 k_1^* + \cdots + \lambda_n k_n^*} \left|\frac{c}{\beta}\right|^{\psi_1 k_1^* + \cdots + \psi_n k_n^*} \geq 1 \quad \text{or} \quad < 1, \quad \text{respectively.}$$

This completes the proof of the first part of the proof where it is assumed that M_1 is independent of $k_2, \ldots, k_n, \nu$.

  The second part of the proof when M_1 is dependent on $k_2, \cdots, k_n$, ν. Two cases can happen for M_1. Either **(a)** $\min M_1 = a_0 + \sum_{j=2}^{n} a_j k_j + a_{n+1}\nu$, and $\max M_1 = b_0 + \sum_{j=2}^{n} b_j k_j + b_{n+1}\nu$; or **(b)** $\min M_1 = a_0 + \sum_{j=2}^{n} a_j k_j + a_{n+1}\nu$ and $\max M_1 = +\infty$.

Using the same notation as in part **(a)** of the proof and remembering that i is omitted in the suffixes, we see that $H \equiv H_i$ is obtained when k_1 is either equal to $\min M_1$ or equal to $\max M_1$. Assume first that k_1 is equal to $\min M_1$. In this case we get

$$H := \inf_{\nu \in N, \mathbf{k}' \in \mathbf{M}'} \left| c \frac{c}{\alpha} \right|^{(a_0 + \sum_{j=2}^{n}(\lambda_j + a_j) + a_{n+1}\nu)/\nu} \times$$

$$\times \left| \frac{c}{\beta} \right|^{(a_0 + \sum_{j=2}^{n}(\psi_j + a_j) + a_{n+1}\nu)/\nu} .$$

If the minimum is obtained for $k_1 = \max M_1$ the result will be similar but with b's instead of a's. But the expression obtained for H is exactly the same as before but with different sequences of constants λ, ψ. Therefore the proof is completed as in the first part. If we have the case (b) then as in the proof of the first main part either $H = 0$ in which case the proof is finished, or H is obtained for $k_1 = \min M_1$. The rest of the proof follows now again from the first part of the above proof.   □

LEMMA 4.3. *If* $\boldsymbol{\lambda}^{(i)}$, $\mathbf{k}^{(i)}$, $\boldsymbol{\psi}^{(i)}$ *have the property (2.1.1) then* $\widetilde{R}(r)$ *satisfies:*

$$\begin{cases} \text{(i) } \widetilde{R}(r) \geq r > 0 \quad \text{for} \quad r \geq \gamma; & \text{(ii) } \widetilde{R}(r) > r \quad \text{for} \quad r > \gamma; \\ \text{(iii) } \widetilde{R}(r) \text{ is non-decreasing for } r \geq \gamma; & \text{(iv) } \widetilde{R}(r) \text{is continuous for } r \geq \gamma. \end{cases}$$

$$(4.2)$$

PROOF. From the definition of $\widetilde{R}(r)$ for $r \geq \gamma$ we get that (i) and (ii) hold. Now (2.1.1) implies that for given $1 \leq i \leq i_0$, $\nu_i \in N_i$, $\mathbf{k}^{(i)} \in \mathbf{M}^{(i)}$ the function of r in the curly braces in (3.1) is non-decreasing for $r \geq \gamma$ so that (iii) holds. It remains to prove that $\widetilde{R}(r)$ is continuous on the left and also continuous on the right for each $r \geq \gamma$.

For the function

$$H \equiv H(c) := \inf_{1 \leq i \leq i_0} \inf_{\nu_i \in N_i, \mathbf{k}^{(i)} \in \mathbf{M}^{(i)}} \left| \phi_{\nu_i}^{-1} \left(c^{\nu_i} \left(\frac{c}{\alpha}\right)^{\boldsymbol{\lambda}^{(i)} \cdot \mathbf{k}^{(i)}} \left(\frac{c}{\beta}\right)^{\boldsymbol{\psi}^{(i)} \cdot \mathbf{k}^{(i)}} \right) \right|$$

where $c \geq \gamma$, we have $H = \widetilde{R}(c)$. By Lemma 4.1 H is either $= 0$ or is a positive minimum. The lemma gives also necessary and sufficient conditions under which H can vanish.

Proof of the continuity on the left for $r > \gamma$. For a fixed r, $r > \gamma$, let $\gamma < r_1 < r_2 < \ldots < r_j < r_{j+1} \nearrow r$. By Lemma 4.3(i) $\widetilde{R}(r_j) \geq r_j > 0$ for $j \geq 1$, and by Lemma 4.1

$$\widetilde{R}(r_j) = r_j \left(\frac{r_j}{|\alpha|}\right)^{\boldsymbol{\lambda}^{(i_j)} \cdot \mathbf{k}_{j,1}^{(i_j)}/\nu_j^{(i_j)}} \left(\frac{r_j}{|\beta|}\right)^{\boldsymbol{\psi}^{(i_j)} \cdot \mathbf{k}_{j,1}^{(i_j)}/\nu_j^{(i_j)}}$$

where $k_{j,1}^{(i_j)}$ is equal either to $\min M_1^{(i_j)}$ or to $\max M_1^{(i_j)}$, $k_{j,2}^{(i_j)}$ is equal either to $\min M_2^{(i_j)}$ or to $\max P_2^{(i_j)}$; $k_{j,n^{(i_j)}}^{(i_j)}$ is equal either to $\min P_{n^{(i_j)}}^{(i_j)}$ or to $\max P_{n^{(i_j)}}^{(i_j)}$, and $\nu^{(i_j)}$ is equal either to $\min N^{(i_j)}$ or to $\max N^{(i_j)}$. Since $1 \leq i_j \leq i_0$, the number of possible different values of $(i_j)_{j \geq 1}$ is finite, hence there is an infinite subsequence of $(r_j)_{j \geq 1}$ for which all the values i_j's are equal. We may assume without loss of generality that $i_j = \tilde{i}$ for $j \geq 1$. Again, since the number of values of $(\nu_j^{(\tilde{i})})_{j \geq 1}$ is finite, we may assume that $\nu_j^{(\tilde{i})} = \tilde{\nu} \in N^{(\tilde{i})}$. Since $k_{j,n^{\tilde{i}}}^{\tilde{i}}$ can take only one of two values (either $\max M^{(j)}$ or $\min M^{(j)}$) we can again assume that $\forall j \geq 1$ $k_{j,n^{\tilde{i}}}^{\tilde{i}} = \tilde{k}_{n^{(\tilde{i})}}^{(\tilde{i})}$. Repeating the same argument a finite number of times we see that

$$\tilde{R}(r_j) = r_j \left(\frac{r_j}{|\alpha|} \right)^{\boldsymbol{\lambda}^{(\tilde{i})} \cdot \boldsymbol{k}^{(\tilde{i})}/\tilde{\nu}} \left(\frac{r_j}{|\beta|} \right)^{\boldsymbol{\psi}^{(\tilde{i})} \cdot \boldsymbol{k}^{(\tilde{i})}/\tilde{\nu}} , \quad \forall j \geq 1,$$

where $\boldsymbol{k}_j^{(i)} = (k_{j,1}^{(i)}, \dots, k_{j,n_{i_j}}^{(i)})$, $\boldsymbol{\lambda}_j^{(i)} = (\lambda_1^{(i)}, \dots, \lambda_{n_{i_j}}^{(i)})$. Letting $j \to \infty$ we get

$$\lim_{j \to \infty} \tilde{R}(r_j) = r_0 \left(\frac{r_0}{|\alpha|} \right)^{\boldsymbol{\lambda}^{(\tilde{i})} \cdot \boldsymbol{k}^{(\tilde{i})}/\tilde{\nu}} \left(\frac{r_0}{|\beta|} \right)^{\boldsymbol{\psi}^{(\tilde{i})} \cdot \boldsymbol{k}^{(\tilde{i})}/\tilde{\nu}} \geq \tilde{R}(r_0).$$

Since $\tilde{R}(r)$ is an increasing function for $r \geq \gamma$ we have $\lim_{r_j \nearrow r_0} \tilde{R}(r_j) \leq \tilde{R}(r_0)$. Combining these two inequalities we get $\lim_{j \to \infty} \tilde{R}(r_j) = \tilde{R}(r_0)$. Therefore $\tilde{R}(r)$ is continuous on the left for $r > r_0$.

Proof of the continuity on the right for $r \geq \gamma$ Choose $r_0 > 0$. $\tilde{R}(r)$ is non-decreasing for $r > 0$. Given a strictly decreasing sequence $r_j \searrow r_0$ we have $w := \lim_{j \to \infty} \tilde{R}(r_j) \geq \tilde{R}(r_0)$. Assume $w > \tilde{R}(r_0)$. Choose a number η, $w > \eta > \tilde{R}(r_0)$. Hence there are numbers $1 \leq i \leq i_0$, $\boldsymbol{k}^{(i)} \subset \boldsymbol{M}^{(i)}$ such that

$$\tilde{R}(r_0) \leq r_0 \left(\frac{r_0}{|\alpha|} \right)^{\boldsymbol{\lambda}^{(i)} \cdot \boldsymbol{k}^{(i)}/\nu_i} \left(\frac{r_0}{|\beta|} \right)^{\boldsymbol{\psi}^{(i)} \cdot \boldsymbol{k}^{(i)}/\nu_i} < \eta < w.$$

For each $j > 0$ we have

$$r_j \left(\frac{r_j}{|\alpha|} \right)^{\boldsymbol{\lambda}^{(i)} \cdot \boldsymbol{k}^{(i)}/\nu_i} \left(\frac{r_j}{|\beta|} \right)^{\boldsymbol{\psi}^{(i)} \cdot \boldsymbol{k}^{(i)}/\nu_i} \geq \tilde{R}(r_j).$$

Letting $j \to \infty$ we get (because $r_j \searrow r_0$)

$$w > \eta > r_0 \left(\frac{r_0}{|\alpha|} \right)^{\boldsymbol{\lambda}^{(i)} \cdot \boldsymbol{k}^{(i)}/\nu_i} \left(\frac{r_0}{|\beta|} \right)^{\boldsymbol{\psi}^{(i)} \cdot \boldsymbol{k}^{(i)}/\nu_i}$$

$$= \lim_{j \to \infty} r_j \left(\frac{r_j}{|\alpha|} \right)^{\boldsymbol{\lambda}^{(i)} \cdot \boldsymbol{k}^{(i)}/\nu_i} \left(\frac{r_j}{|\beta|} \right)^{\boldsymbol{\psi}^{(i)} \cdot \boldsymbol{k}^{(i)}/\nu_i} \geq \lim_{j \to \infty} \tilde{R}(r_j) = w.$$

But this is a contradiction. Hence $w \equiv \lim_{j \to \infty} \widetilde{R}(r_j) = \widetilde{R}(r_0)$.   □

PROOF OF THEOREM 3.3. By Lemma 4.3, $\widetilde{R}(R) > R$. Since $\Lambda_n^{(\alpha,\beta)}(z,f)$ is a holomorphic function, it is enough, by the maximum principle, to prove only the case $|z| = r_1$, $R < r_1 < \widetilde{R}(R)$. Since $\widetilde{R}(r)$ is an increasing and continuous function for $r \geq \gamma$ there exists a number r_0^*, $\delta < r_0^* < R$ such that for each r_0, $r_0^* < r_0 < R$ we have $r_1 < \widetilde{R}(r_0)$. Then

$$\Lambda_n^{(\alpha,\beta)}(z,f) = \frac{1}{2\pi i} \oint_\Gamma \frac{f(t)}{t-z} K_n(z,t)dt$$

where $\Gamma := \{t \in \mathbb{C} : |t| = r_0\}$, $\delta < r_0 < R$, and

$$K_n(z,t) = \sum_{1 \leq i \leq i_0} A_n^{(i)}(z,t) + \sum_{1 \leq i \leq i_1} B_n^{(i)}(t).$$

From the definition of $\widetilde{R}(r_0)$ it follows that $\forall\, i \in [0, i_0]$, and $\forall\, \boldsymbol{k}^{(i)} \in \boldsymbol{M}^{(i)}$ we have

$$\widetilde{R}(r_0)^{\nu_i} \leq r_0^{\nu_i} \left(\frac{r_0}{|\alpha|}\right)^{\boldsymbol{\lambda}^{(i)}\cdot\boldsymbol{k}^{(i)}} \left(\frac{r_0}{|\beta|}\right)^{\boldsymbol{\psi}^{(i)}\cdot\boldsymbol{k}^{(i)}} ;$$

or

$$\widetilde{R}(r_0)^{-\nu_i} \geq \left(\frac{1}{r_0}\right)^{\nu_i} \left(\frac{|\alpha|}{r_0}\right)^{\boldsymbol{\lambda}^{(i)}\cdot\boldsymbol{k}^{(i)}} \left(\frac{|\beta|}{r_0}\right)^{\boldsymbol{\psi}^{(i)}\cdot\boldsymbol{k}^{(i)}} . \tag{4.3}$$

By using (4.3), we have (see (2.1.5) for the definition of $A_n^{(i)}(z,t)$)

$$\left| \frac{1}{2\pi i} \oint_\Gamma \frac{f(t)}{t-z} A_n^{(i)}(z,t)dt \right| \leq \frac{1}{2\pi} 2\pi r_0 (\max_{|t|=r_0} |f(t)|) \cdot \frac{1}{r_1 - r_0} \times$$

$$\times \sum_{\nu_i \in N_i} \sum_{\boldsymbol{k}^{(i)} \in \boldsymbol{M}^{(i)}} |b(\boldsymbol{k}^{(i)})| \left(\left(\frac{|\alpha|}{r_0}\right)^{\boldsymbol{\lambda}^{(i)}\cdot\boldsymbol{k}^{(i)}} \left(\frac{|\beta|}{r_0}\right)^{\boldsymbol{\psi}^{(i)}\cdot\boldsymbol{k}^{(i)}} \left(\frac{r_1}{r_0}\right)^{\nu_i} \right)^n$$

$$\leq c_1^{(i)} \left(\left(\frac{r_1}{\widetilde{R}(r_0)}\right)^{\nu_i} \right)^{n-1} \sum_{i \in N_i} \sum_{\boldsymbol{k}^{(i)} \in \boldsymbol{M}^{(i)}} |b(\boldsymbol{k}^{(i)})| \left(\frac{\gamma}{r_0}\right)^{\boldsymbol{\lambda}^{(i)}\cdot\boldsymbol{k}^{(i)}+\boldsymbol{\psi}^{(i)}\cdot\boldsymbol{k}^{(i)}} \left(\frac{r_1}{r_0}\right)^{\nu_i}$$

$$\leq c \left(\left(\frac{r_1}{\widetilde{R}(r_0)}\right)^{\nu_i} \right)^n \leq c_2 \left(\left(\frac{r_1}{\widetilde{R}(r_0)}\right)^{\min_{1 \leq i \leq i_0} N_j} \right)^n \qquad n = 1, 2, \cdots,$$

where c_2 is a suitable constant. Hence

$$\limsup_{n \to \infty} \max_{|z|=r_1} \left| \frac{1}{2\pi i} \oint_\Gamma \frac{f(t)}{t-z} A_n^{(i)}(z,t)dt \right|^{1/n} \leq \left(\frac{r_1}{\widetilde{R}(r_0)}\right)^{\min_{1 \leq j \leq i_0} N_i} .$$

Since r_0 can be chosen arbitrarily between r_0^* and R by letting r_0 tend to R we also have

$$\limsup_{n \to \infty} \max_{|z|=r_1} \left| \frac{1}{2\pi i} \oint_\Gamma \frac{f(t)}{t-z} A_n^{(i)}(z,t) dt \right|^{1/n} \leq \left(\frac{r_1}{\widetilde{R}(R)} \right)^{\min_{1 \leq j \leq i_0} N_i}. \qquad (4.4)$$

In a similar way we have using (2.1.2) (see (2.1.6) for the definition of $B_n^{(i)}(t)$)

$$\left| \frac{1}{2\pi i} \oint_{|t|=r_0} \frac{f(t)}{t-z} B_n^{(i)}(t) dt \right| \leq \frac{1}{2\pi} 2\pi r_0 \max_{|t|=r_0} |f(t)| \frac{1}{r_1 - r_0} \sum_{\ell^{(i)}} |c(\ell^{(i)})|$$

$$\times \left(\frac{\gamma}{r_0} \right)^{(\boldsymbol{\mu}^{(i)} \cdot \boldsymbol{\ell}^{(i)} + \boldsymbol{\chi}^{(i)} \cdot \boldsymbol{\ell}^{(i)})n}$$

$$\leq d_1 \left(\frac{\gamma}{r_0} \right)^{n-1} \sum_{\ell^{(i)} \in P^{(i)}} |c(\ell^{(i)})| \left(\frac{\gamma}{t} \right)^{\boldsymbol{\mu}^{(i)} \cdot \boldsymbol{\ell}^{(i)} + \boldsymbol{\chi}^{(i)} \cdot \boldsymbol{\ell}^{(i)}}$$

$$\leq d \left(\frac{\gamma}{r_0} \right)^n \qquad n = 1, 2, \ldots$$

where d is a suitable constant. From the last inequality we obtain, by first taking the n'th root and then letting r_0 tend to R,

$$\limsup_{n \to \infty} \max_{|z|=r_1} \left| \frac{1}{2\pi i} \oint_{|t|=r_0} \frac{f(t)}{t-z} B_n^{(i)}(t) dt \right|^{1/n} \leq \frac{\gamma}{r_0}. \qquad (4.5)$$

Now (3.2) follows from (4.4) and (4.5). □

5. Some Topological Results

5.1. Properties of some compact sets.

LEMMA 5.1.1. *Assume $K \subset \mathbb{C}$ is a bounded compact set and $0 \notin K$. Then the set $K^{-1} := \{\omega : \omega^{-1} \in K\}$ is compact.*

PROOF. Since the map $z \longrightarrow 1/z$ of $\mathbb{C} \backslash \{0, \infty\}$ onto itself is one-to-one and bi-continuous it follows that K^{-1} is compact. Also $0 \notin K$ implies the existence of a disk D_ζ, $\zeta > 0$, which is disjoint with K. Hence $K^{-1} \subset D_{1/\zeta}$. □

Assume G is a star-shaped domain (and therefore simply connected). Let $0 < \rho < 1$. By the Riemann mapping theorem $\exists\, g : D_1 \to G$ which is one-to-one and onto with $g(0) = 0$. Let $H \equiv H_\rho := g(D_\rho) \bigcup D$. Obviously $D \subset H_\rho \subset G$. Since G is star-shaped (see Z.Nehari [80, p.220]), therefore $g(D_\rho)$ and $H_\rho = g(D_\rho) \bigcup D$ are star-shaped. Hence the boundary of $g(D_\rho)$ is a star-shaped and rectifiable Jordan curve. Since the boundary is a star-shaped and rectifiable Jordan curve, it follows that ∂H_ρ is also a star-shaped rectifiable Jordan curve.

LEMMA 5.1.2. *For each ρ, $0 < \rho < 1$, the set H_ρ is bounded.*

PROOF. The function g is continuous and $g(\overline{D}_\rho)$ is compact, and therefore bounded.

$\square$

DEFINITION 5.1.3. *Assume $f \in A_R$, $R > 1$. Let $0 \neq |\alpha|, |\beta| < R$ and for $1 \leq i \leq i_0$ let $\mathbf{k}^{(i)} \in \mathbf{M}^{(i)}$ means $k_{n_i}^{(i)} \in M_{n_i}^{(i)}, \cdots, k_1^{(i)} \in M_1^{(i)}$. Define for each $0 < \rho < 1$ (similar to the definition of A_i) and $\mathcal{E}$ in (3.3)*

$$A_i(\rho) := \bigcap_{\nu_i \in N_i} \bigcap_{\mathbf{k}^{(i)} \in \mathbf{M}^{(i)}} \bigcap_{c \notin S} \phi_{\nu_i}^{-1} \left(c^{\nu_i} \left(\frac{c}{\alpha}\right)^{\boldsymbol{\lambda}^{(i)} \cdot \mathbf{k}^{(i)}} \left(\frac{c}{\beta}\right)^{\boldsymbol{\psi}^{(i)} \cdot \mathbf{k}^{(i)}} H_\rho \right)$$

$$\text{and} \quad \mathcal{E}(\rho) := \bigcap_{1 \leq i \leq i_0} A_i(\rho).$$

In particular if G is replaced by H_ρ then the same result remains true with $\mathcal{E}$ replaced by $\mathcal{E}(\rho)$.

LEMMA 5.1.4. *Suppose the assumptions of Definition 5.1.3 are satisfied. Then for each i, $1 \leq i \leq i_0$ and each ρ, $0 < \rho < 1$, the sets $A_i(\rho)$ and $\mathcal{E}$ are star-shaped domains.*

PROOF. The relation $z_0 \in A_i(\rho)$ is equivalent to the statement: $\forall i (1 \leq i \leq i_0)$, $\forall \nu_i \in N_i$, $\forall \mathbf{k}^{(i)} \in \mathbf{M}^{(i)}$ and $\forall c \notin S$ $\exists \zeta_0 \in H_\rho$ such that

$$z_0^{\nu_i} = c^{\nu_i} \left(\frac{c}{\alpha}\right)^{\boldsymbol{\lambda}^{(i)} \cdot \mathbf{k}^{(i)}} \left(\frac{c}{\beta}\right)^{\boldsymbol{\psi}^{(i)} \cdot \mathbf{k}^{(i)}} \zeta_0 .$$

For each $0 < \lambda < 1$, we have $0 < \lambda^{\nu_i} < 1$. Hence $\lambda^{\nu_i} \in H_\rho$. We have

$$(\lambda z_0)^{\nu_i} = c^{\nu_i} \left(\frac{c}{\alpha}\right)^{\boldsymbol{\lambda}^{(i)} \cdot \mathbf{k}^{(i)}} \left(\frac{c}{\beta}\right)^{\boldsymbol{\psi}^{(i)} \cdot \mathbf{k}^{(i)}} (\lambda^{\nu_i} \zeta_0).$$

Therefore $\lambda z_0 \in A_i(\rho)$. So that $A_i(\rho)$ is star-shaped, which implies that $A(\rho)$ is star-shaped. Similarly if we choose now $\lambda = 1 + \delta e^{i\phi}$, $0 \leq \phi < 2\pi$, where $\delta > 0$ is sufficiently small we get again

$$(\lambda z_0)^{\nu_i} = c^{\nu_i} \left(\frac{c}{\alpha}\right)^{\boldsymbol{\lambda}^{(i)} \cdot \mathbf{k}^{(i)}} \left(\frac{c}{\beta}\right)^{\boldsymbol{\psi}^{(i)} \cdot \mathbf{k}^{(i)}} (\lambda^{\nu_i} \zeta_0).$$

Since H_ρ is open it follows that for a sufficiently small $\delta_0 > 0$, $(1 + \delta e^{i\phi}) z_0 \in A_i(\rho)$. Hence $A_i(\rho)$ is an open star-shaped domain. Now, since $\mathcal{E}$ is a finite intersection of open star-shaped domains it is also an open star-shaped domain.

$\square$

LEMMA 5.1.5. *Assume the assumptions of Definition 5.1.3 are satisfied. Let K be a non-empty set of complex numbers. Then $K \subset \mathcal{E}$, if and only if,*

$$\mathcal{L} := \bigcup_{1 \leq i \leq i_0} \bigcup_{\nu_i \in N_i} \bigcup_{\boldsymbol{k}^{(i)} \in \boldsymbol{M}^{(i)}} \bigcup_{c \notin S} \{ \frac{1}{c^{\nu_i}} \left(\frac{\alpha}{c}\right)^{\boldsymbol{\lambda}^{(i)} \cdot \boldsymbol{k}^{(i)}} \left(\frac{\beta}{c}\right)^{\boldsymbol{\psi}^{(i)} \cdot \boldsymbol{k}^{(i)}} K^{\nu_i} \} \subset G.$$

PROOF. We have

$$K \subset \mathcal{E} \iff \forall \zeta \in K, \ \forall i \in [1, i_0], \ \forall \nu_i \in N_i, \ \forall \boldsymbol{k}^{(i)} \in \boldsymbol{M}^{(i)}, \quad \forall c \notin S$$

$$\zeta \in \bigcap_{c \notin S} \phi_{\nu_i}^{-1} \left(c^{\nu_i} \left(\frac{c}{\alpha}\right)^{\boldsymbol{\lambda}^{(i)} \cdot \boldsymbol{k}^{(i)}} \left(\frac{c}{\beta}\right)^{\boldsymbol{\psi}^{(i)} \cdot \boldsymbol{k}^{(i)}} G \right).$$

This is equivalent to the statement that $\forall i \in [1, i_0], \ \forall \nu_i \in N_i, \ \forall \boldsymbol{k}^{(i)} \in \boldsymbol{M}^{(i)}$, $\forall c \notin S$

$$\frac{1}{c^{\nu_i}} \left(\frac{\alpha}{c}\right)^{\boldsymbol{\lambda}^{(i)} \cdot \boldsymbol{k}^{(i)}} \left(\frac{\beta}{c}\right)^{\boldsymbol{\psi}^{(i)} \cdot \boldsymbol{k}^{(i)}} K^{\nu_i} \subset G.$$

And the last statement is equivalent to

$$\bigcup_{1 \leq i \leq i_0} \bigcup_{\nu_i \in N_i} \bigcup_{\boldsymbol{k}^{(i)} \in \boldsymbol{M}^{(i)}} \bigcup_{c \notin S} \left(\frac{1}{c^{\nu_i}} \left(\frac{\alpha}{c}\right)^{\boldsymbol{\lambda}^{(i)} \cdot \boldsymbol{k}^{(i)}} \left(\frac{\beta}{c}\right)^{\boldsymbol{\psi}^{(i)} \cdot \boldsymbol{k}^{(i)}} K^{\nu_i} \right) \subset G. \qquad \square$$

LEMMA 5.1.6. *If the set K is bounded, then the set $\mathcal{L}$ is bounded.*

PROOF. When the set K is bounded, each of the sets K^{ν} is bounded. Now for each $c \notin S$, we have $|c| \geq R$ and $|\frac{\alpha}{R}| < 1$, $|\frac{\beta}{R}| < 1$. Therefore for each $c \notin S$

$$\left| \frac{1}{c^{\nu}} \left(\frac{\alpha}{c}\right)^k \left(\frac{\beta}{c}\right)^j \right| \leq \frac{1}{|c|^{\nu}} \leq \frac{1}{R^{\nu}} \leq \frac{1}{R}, \quad \text{if} \quad \nu > 0 \quad \text{and} \quad R > 1.$$

The proof follows now from the definition of $\mathcal{L}$ (in Lemma 5.1.5). $\qquad \square$

LEMMA 5.1.7. *Let K be a bounded and compact set satisfying $0 \in K$. Then the set $\mathcal{L}$ is bounded, compact and $0 \in \mathcal{L}$.*

PROOF. By Lemma 5.1.6 the set $\mathcal{L}$ is bounded. From $0 \in K$, it follows that $0 \in \mathcal{L}$. Hence it only remains to show that $\mathcal{L}$ is closed. Assume $z_r \in \mathcal{L}$ for $r \geq 1$ and that $z_r \longrightarrow z_0$. We will prove that $z_0 \in \mathcal{L}$. To each $z_r \in \mathcal{L}, \quad r = 1, 2, \ldots$, there correspond numbers $i_r \in [1, i_0], \ \nu_r \in N_{i_r}, \ k_{n_{(i_r)}}^{(i_r)} \in M_{n_{(i_r)}}^{(i_r)}, \ \ldots, \ k_1^{(i_r)} \in$

$M_1^{(ir)}$, $c_r \notin S$ and $\zeta_r \in K$, such that $z_r = \frac{1}{c_r^{\nu_r}} \left(\frac{\alpha}{c_r}\right)^{k_r} \left(\frac{\beta}{c_r}\right)^{j_r} \zeta_r^{\nu_r}$ where $k_r :=$ $\lambda^{(ir)} \cdot \boldsymbol{k}^{(ir)} \geq 0$, and $j_r := \boldsymbol{\psi}^{(ir)} \cdot \boldsymbol{k}^{(ir)} \geq 0$ (see 2.1.1). Since $1 \leq i_r \leq i_0$ for $r \geq 1$, there is an integer $\widetilde{i}$, $1 \leq \widetilde{i} \leq i_0$ (at least one) such that $i_r = \widetilde{i}$ for an infinite subsequence $r_1 < r_2 < r_3 < \ldots \nearrow \infty$. Taking a subsequence of the original sequence $(z_r)_{r \geq 1}$, which we may assume without loss of generality to be the original sequence $(z_r)_{r \geq 1}$ itself, we may assume $i_r = \widetilde{i}$ for $r = 1, 2, \ldots$. Similarly by passing to subsequences we may assume without loss of generality that $\nu_r = \widetilde{\nu} \in N_{\widetilde{i}}$ for $r = 1, 2, \ldots$.

Assume first that $(k_r)_{r \geq 1}$ is not bounded. Again by passing to subsequences we may assume $k_r \to \infty$. Hence $(\frac{\alpha}{c_r})^{k_r} \to 0$, because $|c_r| > R > |\alpha|$ and we see in this case that $z_r \longrightarrow 0 = z_0 \in \mathcal{L}$ (since $0 \in K$). Similarly if $j_r \to \infty$ then again $z_0 \in K$. Assume now that the sequences $(k_r)_{r \geq 1}$ and $(j_r)_{r \geq 1}$ are bounded. Then we may again assume that the sequences $(k_r)_{r \geq 1}$, and $(j_r)_{r \geq 1}$ are constant, i.e., $k_r = \widetilde{k}$, and $j_r = \widetilde{j}$, $r = 1, 2, \ldots$. Now $\zeta_r^{\nu_r} = \zeta_r^{\widetilde{\nu}}$ are bounded because K is bounded. Taking further subsequences we see that we may assume $\zeta_r \to \zeta \in K$ and either for some subsequence $c_{r_p} \longrightarrow \infty$, in which case we see again that $z_{r_p} \longrightarrow 0 = z_0 \in \mathcal{L}$, or $c_{r_p} \longrightarrow c \notin S$ (because S^c is closed). In this case we get $z_r = \frac{1}{c_r^{\widetilde{\nu}}}(\frac{\alpha}{c_r})^{\widetilde{k}}(\frac{\beta}{c_r})^{\widetilde{j}}\zeta_r^{\widetilde{\nu}} \to \frac{1}{c^{\widetilde{\nu}}}(\frac{\alpha}{c})^{\widetilde{k}}(\frac{\beta}{c})^{\widetilde{j}}\zeta^{\widetilde{\nu}} = z_0 \in \mathcal{L}$, from the definition of the set $\mathcal{L}$ in Lemma 5.1.5. $\qquad \square$

LEMMA 5.1.8. *Let K be a compact set satisfying $0 \in K$ and $K \subset \mathcal{E}$. Then the set $\mathcal{L}$ is compact, $0 \in \mathcal{L}$ and $\mathcal{L} \subset G$.*

PROOF. By Lemma 5.1.5 we have $\mathcal{L} \subset G$. By Lemma 5.1.7, $\mathcal{L}$ is compact. $\square$

LEMMA 5.1.9. *Let K be a compact and bounded set satisfying $0 \in K$ and $K \subset \mathcal{E}$. Then there exists a number ρ_0, $0 < \rho_0 < 1$, such that for all ρ, $\rho_0 < \rho < 1$, $\mathcal{L} \subset H_\rho$.*

PROOF. The map $g : D \to G$ is bi-continuous. By Lemma 5.1.8, $\mathcal{L} \subset G$ and therefore $g^{-1}(\mathcal{L}) \subset D$ and $g^{-1}(\mathcal{L})$ is compact. Therefore there exists a number ρ_0, $0 < \rho_0 < 1$, such that $g^{-1}(\mathcal{L}) \subset D_{\rho_0}$. This implies that for all ρ, $(\rho_0 < \rho < 1)$, we have $\mathcal{L} \subset g(D_\rho) \cup D = H_\rho$. $\qquad \square$

5.2. Some paticular Jordan curves.

By the Riemann mapping theorem $\exists h \equiv h_\rho : D \to \mathcal{E}(\rho)$ with $h(0) = 0$. For any ρ', $0 < \rho' < 1$, we set $C_{\rho'} := h_\rho(\overline{D}_{\rho'}) = \overline{h_\rho(D_{\rho'})} \subset \mathcal{E}(\rho)$. $C_{\rho'}$ is star-shaped and $\partial C_{\rho'}$ is a star-shaped, rectifiable and closed Jordan curve.

LEMMA 5.2.1. *Let K be a compact and bounded set satisfying $0 \in K$ and $K \subset \mathcal{E}$. Then $\forall \rho$, $\rho_0 < \rho < 1$ (ρ_0 is defined in Lemma 5.1.9) $K \subset \mathcal{E}(\rho)$. And to each ρ, $\rho_0 < \rho < 1$ there corresponds a number $\rho_1 \equiv \rho_1(\rho)$, $0 < \rho_1 < 1$ such that for all $\rho', \rho_1(\rho) < \rho' < 1$, $K \subset h_\rho(D_{\rho'}) \underset{\neq}{\subset} \overline{h_\rho(D_{\rho'})} \equiv C_{\rho'} \subset \mathcal{E}(\rho), K \subset \mathrm{Int}\, C_{\rho'}$, and K is in the interior of $\partial C_{\rho'}$.*

PROOF. **a)** By Lemma 5.1.9 for all ρ, $\rho_0 < \rho < 1$, $\mathcal{L} \subset H_\rho$. By Lemma 5.1.5, with G replaced by H_ρ, we get $K \subset \mathcal{E}(\rho)$. **b)** $h_\rho : D \to \mathcal{E}(\rho)$, $h_\rho(0) = 0$, is univalent and onto. Since K is compact and bounded and $K \subset \mathcal{E}(\rho)$ the set $h_\rho^{-1}(K)$ is compact and $h_\rho^{-1}(K) \subset D$. Hence to each ρ, $\rho_0 < \rho < 1$, there corresponds a number $\rho_1 \equiv \rho_1(\rho)$, $0 < \rho_1 < 1$ such that for all ρ', $\rho_1(\rho) < \rho' < 1$, $h_\rho^{-1}(K) \subset D_{\rho'}$, which implies $K \subset h_\rho(D_{\rho'}) \underset{\neq}{\subset} \overline{h_\rho(D_{\rho'})} = C_{\rho'} \subset \mathcal{E}_\rho$, and in particular $K \subset \mathrm{Int}\, C_{\rho'}$. Also, since the domains are star-shaped K is in the interior of $\partial C_{\rho'}$.   □

LEMMA 5.2.2. *For each ρ, $0 < \rho < 1$, we have $\mathcal{E}(\rho) \supset D_{\widetilde{R}(r)}$ for $0 < r < R$.*

PROOF. Since for each ρ, $0 < \rho < 1$, $D \subset H_\rho$, we obtain from Corollary 4.2 that $D_{\widetilde{R}(r)} \subset D_{\widetilde{R}(R)} \subset \mathcal{E}(\rho)$ for $0 \leq r \leq R$.   □

LEMMA 5.2.3. *Assume that r satisfies $\max(1, |\alpha|, |\beta|) \equiv \delta < r < R$. Then to each ρ, $0 < \rho < 1$ there corresponds a number $\rho_2 \equiv \rho_2(\rho)$, $0 < \rho_2 < 1$ such that for each ρ', $\rho_2(\rho) < \rho' < 1$, we have $\overline{D}_{\widetilde{R}(r)} \subset h_\rho(D_{\rho'}) \underset{\neq}{\subset} \overline{h_\rho(D_{\rho'})} \equiv C_{\rho'}$.*

PROOF. From the definition of $\widetilde{R}(r)$ we see, since $0 < \delta < r < R$, that $\widetilde{R}(R) > \widetilde{R}(r)$. By Lemma 5.2.2 we have now for each ρ, $0 < \rho < 1$, $\mathcal{E}(\rho) \supset D_{\widetilde{R}(R)} \supset D_{\widetilde{R}(r)}$. Since $h_\rho : D \to \mathcal{E}(\rho)$ is onto, one-to-one and bi-continuous, we see that the preimage $h_\rho^{-1}(\overline{D}_{\widetilde{R}(R)}) \subset D$ is compact and bounded. Therefore there exists some $\rho_2 \equiv \rho_2(\rho)$, $0 < \rho_2 < 1$, such that $h_\rho^{-1}(D_{\widetilde{R}(R)}) \subset D_{\rho_2}$. Hence for each ρ', $\rho_2(\rho) < \rho' < 1$, $C_{\rho'} := \overline{h_\rho(D_{\rho'})} \underset{\neq}{\supset} h_\rho(D_{\rho'}) \supset D_{\widetilde{R}(R)} \supset D_{\widetilde{R}(r)}$.   □

LEMMA 5.2.4. *Assume that r satisfies $\max(1, |\alpha|, |\beta|) \equiv \delta < r < R$. Set $\gamma := \max\{|\alpha|, |\beta|\}$. Suppose ρ, $0 < \rho < 1$, is given. Let ρ' satisfy $\rho_2(\rho) < \rho' < 1$ (see Lemma 5.2.3). Then for each $z \in \partial C_{\rho'}$ we have $|z| > r \left(\frac{r}{\gamma}\right)^{1/\nu^*} > r$, where $\nu^* = \max \bigcup_{1 \leq i \leq i_0} N_i < \infty$.*

PROOF. By Lemma 5.2.3 we have for ρ', $\rho_2 < \rho' < 1$,

$$\overline{D}_{\widetilde{R}(r)} \subset h_\rho(D_{\rho'}) \underset{\neq}{\subset} \overline{h_\rho(D_{\rho'})} \equiv C_{\rho'}.$$

Therefore for each $z \in \partial C_{\rho'}$, $z \notin \overline{D}_{\tilde{R}(r)}$ which, by (2.1.1) and Remark 3.2, implies that $|z| > \tilde{R}(r) \geq r \left(\frac{r}{\gamma}\right)^{\frac{1}{\nu^*}} > r$.   □

5.3. The sets $K_{\nu jk}(z)$ and their properties.

Let $\tilde{i}$, $\tilde{\nu}$, $\tilde{k}_1$, $\cdots$, $\tilde{k}_{n^{\tilde{i}}}$ be given non-negative integers satisfying $1 \leq \tilde{i} \leq i_0$, $\tilde{\nu} \in N_{\tilde{i}}$, $\tilde{k}_{n_i} \in M_{n^{\tilde{i}}}^{\tilde{i}}$, $\cdots$, $\tilde{k}_1 \in M_1^{\tilde{i}}$. For the sake of simplicity write $\nu := \tilde{\nu}_j$, $k := \boldsymbol{\lambda}^{\tilde{i}}.\tilde{\boldsymbol{k}}$ and $j := \boldsymbol{\psi}^{\tilde{i}}.\tilde{\boldsymbol{k}}$. Then k and j satisfy $k \geq 0$, $j \geq 0$ and $k + j \geq 1$. For $0 < \rho < 1$, set $H_\rho^c := \mathbb{C}\backslash H_\rho$ and $(H_\rho^c)^{-1} := \{\omega \in \mathbb{C} : \frac{1}{\omega} \in H_\rho^c\}$,

REMARK. Some of the following definitions and lemmas remain valid when it is assumed only that the integers ν k, j satisfy $\nu \geq 1$, $k \geq 0$, $j \geq 0$ and $k + j \geq 1$.

DEFINITION 5.3.1. *For a positive integer ν and two non-negative integers ν, k, j, satisfying $k \geq 0$, $j \geq 0$ and $k + j \geq 1$ write*

$$K_{\nu jk}(z) := \left\{\omega \in \mathbb{C} : \omega^\nu \left(\frac{\omega}{\alpha}\right)^k \left(\frac{\omega}{\beta}\right)^j \in z^\nu (H_\rho^c)^{-1}\right\}.$$

which is equivalent to

$$K_{\nu jk}(z) = \phi_{\nu+k+j}^{-1}\left(z^\nu \alpha^k \beta^j (H_\rho^c)^{-1}\right). \tag{5.3.1}$$

We now give a sequence of lemmas which lead to the construction of the star-shaped closed Jordan curves $\Gamma_{\nu jk}(z)$ that are used later in the proof of Theorem 3.4.

LEMMA 5.3.2. *For each ρ, $0 < \rho < 1$, and three integers ν, k, j satifying the assumptions of Definition 5.3.1 the following assertions hold:*

(i) $(H_\rho^c)^{-1}$ and $\phi_\nu^{-1}((H_\rho^c)^{-1})$ are star-shaped, compact and bounded,

$$0 \in (H_\rho^c)^{-1}, \ 0 \in \phi_\nu^{-1}((H_\rho^c)^{-1}), \ (H_\rho^c)^{-1} \supset D_\eta,$$

and

$$\phi_\nu^{-1}((H_\rho^c)^{-1}) \supset D_{\eta'} \quad for \ \eta = (2 \max |H_\rho|)^{-1} \ and \ \eta' = \eta^{1/\nu}.$$

(ii) $(H_\rho^c)^{-1} \subset \overline{D}$, and $\phi_\nu^{-1}((H_\rho^c)^{-1}) \subset \overline{D}$.

(iii) For any finite complex z, $z(H_\rho^c)^{-1}$ and $z\phi_\nu^{-1}((H_\rho^c)^{-1})$ are star-shaped, compact and bounded.

(iv) For any finite complex z the sets $\alpha^k \beta^j z^\nu (H_\rho^c)^{-1}$ and $\phi_{\nu+k+j}^{-1} \alpha^k \beta^j z^\nu (H_\rho^c)^{-1})$ are star-shaped and compact,

$$0 \in \alpha^k \beta^j z^\nu (H_\rho^c)^{-1} \quad \text{and} \quad 0 \in \phi_{\nu+k+j}^{-1} (\alpha^k \beta^j z^\nu (H_\rho^c)^{-1}).$$

PROOF. Proof of (i) and (ii). By Lemma 5.1.1 and the remark made after its proof and by Lemma 5.1.2 the set $(H_\rho^c)^{-1}$ is compact, open, bounded, star-shaped domain and $0 \in (H_\rho^c)^{-1}$. From $H_\rho \supset D$ it follows that $(H_\rho^c)^{-1} \subset \overline{D}$. Since H_ρ is compact and bounded the rest of the proof of (i) follows. The proof for $\phi_\nu^{-1}((H_\rho^c)^{-1})$ is immediate. (iii) and (iv) are immediate too.   □

LEMMA 5.3.3. *Assume the assumptions of Definition 5.3.1 are satisfied. Then for each finite complex z, the set $K_{\nu jk}(z)$ is star-shaped, compact and bounded, $0 \in K_{\nu jk}(z)$, $K_{\nu jk}(z)$ includes a disk with center 0 and radius*

$$\left(\frac{|\alpha|^k |\beta|^j |z|^\nu}{2 \max |H_\rho|} \right)^{1/(\nu+k+j)},$$

and $\partial K_{\nu jk}(z)$ is a rectifiable star-shaped Jordan curve.

PROOF. From the definition of $K_{\nu jk}(z)$ we get

$$K_{\nu jk}(z) = \phi_{\nu+j+k}^{-1}(\alpha^k \beta^j z^\nu (H_\rho^c)^{-1}).$$

By Lemma 5.3.2(i), $(H_\rho^c)^{-1}$ includes a disk with center 0. Therefore $K_{\nu jk}(z)$ contains a disk with center 0. By Lemma 3.3. 2(iv), the set $K_{\nu jk}(z)$ is star-shaped, compact and bounded. By the remark after the proof of Lemma 5.1.1 $\partial K_{\nu jk}(z)$ is a rectifiable and star-shaped Jordan curve. The rest of the proof follows from the definition of $(H_\rho^c)^{-1}$ and Lemma 5.3.1(i).   □

LEMMA 5.3.4. *Assume that r satisfies $\max(1, |\alpha|, |\beta|) \equiv \delta < r < R$ and $0 < \rho < 1$. Let ρ' satisfy $\rho_2(\rho) < \rho' < 1$ (where $\rho_2(\rho)$ is defined in Lemma 5.2.3). Then $\forall z \in C_{\rho'}$, we have $K_{\nu jk}(z) \subset S$.*

PROOF. We have (from the definition of $C_{\rho'}$ in Lemma 5.2.1 $C_{\rho'} \subset \mathcal{E}(\rho)$. Now $z \in \mathcal{E}(\rho)$ is equivalent to the assertion that for all $c \notin S$, for all i, $1 \leq i \leq i_0$ and for all i, ν, $k_1^{(i)}, \ldots, k_n^{(i)}$ belonging to $N_i, M_1^{(i)}, \ldots, M_n^{(i)}$ respectively, and $\forall \lambda_0^{(i)}, \ldots, \lambda_n^{(i)}, \psi_0^{(i)}, \ldots, \psi_n^{(0)}$ we have the following chain of equivalent statements:

$$(i) \qquad z \in \phi_{\nu_i}^{-1}\left(c^{\nu_i} \left(\frac{c}{\alpha}\right)^{\boldsymbol{\lambda}^{(i)} \cdot \boldsymbol{k}^{(i)}} \left(\frac{c}{\beta}\right)^{\boldsymbol{\psi}^{(i)} \cdot \boldsymbol{k}^{(i)}} H_\rho \right)$$

$$(ii) \qquad z \notin \phi_{\nu_i}^{-1}\left(c^{\nu_i}\left(\frac{c}{\alpha}\right)^{\boldsymbol{\lambda}^{(i)}\cdot\boldsymbol{k}^{(i)}}\left(\frac{c}{\beta}\right)^{\boldsymbol{\psi}^{(i)}\cdot\boldsymbol{k}^{(i)}} H_\rho^c\right)$$

$$(iii) \qquad z \notin \phi_{\nu_i}^{-1}\left(c^{\nu_i}\left(\frac{c}{\alpha}\right)^{\boldsymbol{\lambda}^{(i)}\cdot\boldsymbol{k}^{(i)}}\left(\frac{c}{\beta}\right)^{\boldsymbol{\psi}^{(i)}\cdot\boldsymbol{k}^{(i)}} \zeta\right) \qquad \text{for each} \quad \zeta \in H_\rho^c$$

$$(iv) \qquad \frac{z^{\nu_i}}{\zeta} \neq c^{\nu_i}\left(\frac{c}{\alpha}\right)^{\boldsymbol{\lambda}^{(i)}\cdot\boldsymbol{k}^{(i)}}\left(\frac{c}{\beta}\right)^{\boldsymbol{\psi}^{(i)}\cdot\boldsymbol{k}^{(i)}} \qquad \qquad \text{for each} \quad \zeta \in H_\rho^c$$

$$(v) \qquad c^{\nu_i}\left(\frac{c}{\alpha}\right)^{\boldsymbol{\lambda}^{(i)}\cdot\boldsymbol{k}^{(i)}}\left(\frac{c}{\beta}\right)^{\boldsymbol{\psi}^{(i)}\cdot\boldsymbol{k}^{(i)}} \notin z^{\nu_i}(H_\rho^c)^{-1}.$$

From the definition of $K_{\nu jk}(z)$ it follows now that for each $z \in C_{\rho'}$, $K_{\nu jk}(z) := \{\omega \in \mathbb{C} : \omega^\nu \left(\frac{\omega}{\alpha}\right)^k \left(\frac{\omega}{\beta}\right)^j \in z^\nu(H_\rho^c)^{-1}\} \subset S$. $\qquad\qquad\square$

LEMMA 5.3.5. *Assume that r satisfies $\max(1, |\alpha|, |\beta|) \equiv \delta < r < R$ and that ρ, $0 < \rho < 1$. Choose θ_0 so that $1 > \theta_0 > \frac{\gamma}{r}$. Let ρ' satisfy $\rho_2(\rho) < \rho' < 1$ (where $\rho_2(\rho)$ is defined in Lemma 5.2.3). Then for each $z \in \partial C_{\rho'}$ and for each $\omega \in K_{\nu jk}(z)$ we have $|\omega| < \theta_0|z|$, $\bigcup_{z\in\partial C_{\rho'}} K_{\nu jk}(z) \subset \text{Int } \theta_0\partial C_{\rho'}$ and $d(\partial C_{\rho'}, \bigcup_{z\in\partial C_{\rho'}} K_{\nu jk}(z) \geq d > 0$.*

PROOF. By Lemma 5.2.4 for each $z \in \partial C_{\rho'}$, where $\rho_2(\rho) < \rho' < 1$, $|z| \geq r|\frac{r}{\gamma}|^{1/\nu^*} > r > 1$. Assume that for some $\omega \in K_{\nu jk}$, we have $|\omega| \geq \theta_0|z|$. By Definition 5.3.1 we have $\omega^\nu\left(\frac{\omega}{\alpha}\right)^k\left(\frac{\omega}{\beta}\right)^j \in z^\nu(H_\rho^c)^{-1}$ which is equivalent to $\frac{1}{z^\nu}\left(\omega^\nu\left(\frac{\omega}{\alpha}\right)^k\left(\frac{\omega}{\beta}\right)^j\right) \in (H_\rho^c)^{-1} \subset \overline{D}$ (see Lemma 5.3.2(ii)). This implies that $1 \geq \left|\frac{1}{z^\nu}(\omega^\nu\left(\frac{\omega}{\alpha}\right)^k\left(\frac{\omega}{\beta}\right)^j)\right|$. Hence

$$1 \geq \left|\frac{\omega}{z}\right|^\nu\left|\frac{\omega}{\alpha}\right|^k\left|\frac{\omega}{\beta}\right|^j \geq \left|\frac{\omega}{z}\right|^\nu\left(\frac{r}{\gamma}\right)^{k+j} \geq \theta_0\left(\frac{r}{\gamma}\right)^{k+j} \geq \theta_0\frac{r}{\gamma} > 1. \qquad (5.3.2)$$

But (5.3.2) is a contradiction. This contradiction implies that $|\omega| < \theta_0|z|$. Since $\theta_0\partial C_{\rho'}$ and $K_{\nu jk}(z)$ are compact, bounded, star-shaped, $0 \in K_{\nu jk}(z)$ and 0 is in the interior of $\partial C_{\rho'}$ we get $\bigcup_{z\in\partial C_{\rho'}} K_{\nu jk}(z) \subset \text{Int } \theta_0\partial C_{\rho'}$. The set $\partial C_{\rho'}$ is a compact, bounded and star-shaped closed Jordan curve. Therefore $d := d(\partial C_{\rho'}, \theta_0\partial C_{\rho'}) > 0$. Combining the last two results we get

$$d(\partial C_{\rho'}, \bigcup_{z\in\partial C_{\rho'}} K_{\nu jk}) \geq d > 0.$$

$\qquad\qquad\square$

LEMMA 5.3.6. *Suppose the assumptions of Lemma 5.3.5 are satisfied. Then the set $\bigcup_{z\in\partial C_{\rho'}} K_{\nu jk}(z)$ is compact, bounded and $\bigcup_{z\in\partial C_{\rho'}} K_{\nu jk}(z) \subset S$. Also there is a number $\epsilon_0 > 0$ such that*

$$\forall \epsilon, \; 0 < \epsilon < \epsilon_0 , \quad \bigcup_{z\in\partial C_{\rho'}} (1+\epsilon)K_{\nu jk}(z) \subset S$$

and

$$d(\partial C_{\rho'}, (1 + \varepsilon^*) K_{\nu jk}(z)) \geq \frac{1}{2} d$$

for

$$\varepsilon^* := \left(2 \max \left| \bigcup_{z \in C_{\rho'}} K_{\nu jk}(z) \right| \right)^{-1} > 0.$$

PROOF. By Lemma 5.3.4, $z \in \partial C_{\rho'}$ implies $K_{\nu jk}(z) \subset S$ so that $\forall z \in \partial C_{\rho'}$, $\bigcup_{z \in \partial C_{\rho'}} K_{\nu jk}(z) \subset S$. By (5.3.3) we see that

$$K_{\nu jk}(z) = \phi_{\nu+k+j}^{-1} \left(\alpha^k \beta^j z^\nu (H_\rho^c)^{-1} \right). \tag{5.3.3}$$

and $S \supset \bigcup_{z \in \partial C_{\rho'}} K_{\nu jk}(z) = \phi_{\nu+k+j}^{-1} \left(\alpha^k \beta^j (\partial C_{\rho'})^\nu (H_\rho^c)^{-1} \right)$. By a standard argument, (since $(H_\rho^c)^{-1}$ and $\partial C_{\rho'}$ are compact and bounded) we see that the set $\bigcup_{z \in C_{\rho'}} K_{\nu jk}(z)$ is compact and bounded, so since S^c is closed we get

$$d^* := d \left(\bigcup_{z \in \partial C_{\rho'}} K_{\nu jk}(z), \; S^c \right) > 0. \tag{5.3.4}$$

The boundedness of $\bigcup_{z \in \partial C_{\rho'}} K_{\nu jk}(z)$ shows that there exists a sufficiently small $\epsilon_0 > 0$ such that for all $0 < \epsilon \leq \epsilon_0$, $\bigcup_{z \in C_{\rho'}} (1 + \epsilon) K_{\nu jk}(z) \subset S$. From Lemma 5.3.5 it follows (since $\bigcup_{z \in C_{\rho'}}$ is compact and bounded) that $d(C_{\rho'}, (1 + \varepsilon^*) \bigcup_{z \in \partial C_{\rho'}} K_{\nu jk}(z)) \geq \frac{1}{2} d > 0$. $\qquad \square$

DEFINITION 5.3.7. Assume the assumptions of Lemma 5.3.6 are satisfied. For each $z \in \partial C_{\rho'}$ and and each $\varepsilon \equiv \varepsilon(z) > 0$, define $\widetilde{K}_{\nu jk}(z, \epsilon) := (1+\epsilon) K_{\nu jk}(z)$.

LEMMA 5.3.8. *Suppose the assumptions of Lemma 5.3.6 are satisfied. Write* $\epsilon_3 := min \; (\varepsilon_0, \; \theta_0, \; \varepsilon^*) > 0$. *Then for each* ϵ, $0 < \epsilon < \epsilon_3$ *the set* $\widetilde{K}_{\nu jk}(z, \epsilon)$ *is star-shaped, compact and*

$$\begin{cases} 0 \in \widetilde{K}_{\nu jk}(z, \epsilon), & z \notin \widetilde{K}_{\nu jk}(z, \epsilon), \quad \widetilde{K}_{\nu jk}(z, \epsilon) \subset S \\ \widetilde{K}_{\nu jk}(z, \epsilon) \supset D_{\eta'} & \text{for some } \eta' > 0 \end{cases} \tag{5.3.5}$$

$$\bigcup_{z \in \partial C_{\rho'}} \widetilde{K}_{\nu jk}(z, \epsilon) \subset \bigcup_{z \in \partial C_{\rho'}} (1 + \epsilon_0) K_{\nu jk} \subset S \tag{5.3.6}$$

and

$$d(\partial C_{\rho'}, \bigcup_{z \in C_{\rho'}} \widetilde{K}_{\nu jk}(z, \varepsilon)) \geq \frac{1}{2} d > 0. \tag{5.3.7}$$

PROOF. Suppose $0 < \varepsilon < \varepsilon_3$. By Lemma 5.3.3 $K_{\nu jk}(z)$ contains a disk with center 0. Therefore for each $\varepsilon > 0$, $\widetilde{K}_{\nu jk}(z, \epsilon)$ also contains a disk with center

0. Again by Lemma 5.3.3 for each $\epsilon > 0$ the set $(1+\epsilon)K_{\nu jk}(z)$ is star-shaped, compact and bounded.

By Lemma 5.3.6 we have $z \notin \widetilde{K}_{\nu jk}(z,\varepsilon)$, and $\widetilde{K}_{\nu jk}(z,\varepsilon) \subset S$. Also (5.3.6) and (5.3.7) follow by Lemma 5.3.6. $\square$

5.4. The curves $\Gamma_{\nu jk}(z)$ and their properties.

LEMMA 5.4.1. *Assume the assumptions of Lemma 5.3.8 are satisfied. Let ρ' satisfy $\rho_2(\rho) < \rho' < 1$. Write $\varepsilon := \frac{1}{2}\varepsilon_3$, Then $\Gamma_{\nu jk}(z) := \partial \widetilde{K}_{\nu jk}(z,\epsilon)$ is a star-shaped and rectifiable closed Jordan curve such that*

$$\forall z \in C_{\rho'} \qquad 0 \in Int \ \Gamma_{\nu jk}(z), \quad z \in Ext \ \Gamma_{\nu jk}(z), \quad \Gamma_{\nu jk}(z) \subset S \qquad (5.4.1)$$

and

$$\forall z \in \partial C_{\rho'}, \ \forall t \in \Gamma_{\nu jk}(z), \qquad \zeta(z,t) := \left(\frac{z}{t}\right)^{\nu}\left(\frac{\alpha}{t}\right)^{k}\left(\frac{\beta}{t}\right)^{j} \in H_{\rho}. \qquad (5.4.2)$$

PROOF. By Lemma 5.3.8 for $z \in \partial C_{\rho'}$ and $\varepsilon = \frac{1}{2}\varepsilon_3$, $\widetilde{K}_{\nu jk}(z,\epsilon)$ is star-shaped, compact and bounded and by Lemma 5.3.3 $\widetilde{K}_{\nu jk}(z,\epsilon)$ includes a disk with center 0. It follows now from Lemma 5.3.8 that $0 \in Int \ \Gamma_{\nu jk}(z)$, $z \in Ext \ \Gamma_{\nu jk}(z)$, $\Gamma_{\nu jk}(z) \subset S$ (since $\Gamma_{\nu jk}(z) = \partial \widetilde{K}_{\nu jk}(z) \subset \widetilde{K}_{\nu jk}(z) \subset S$). Since $\Gamma_{\nu jk}(z) = \partial\{(1+\epsilon)K_{\nu jk}(z)\}$, and $K_{\nu jk}(z)$ is star-shaped, we get $\Gamma_{\nu jk}(z) \cap K_{\nu jk}(z) = \emptyset$. So for each $t \in \Gamma_{\nu jk}(z)$, we have $t \notin K_{\nu jk}(z)$, which means that $t^{\nu}\left(\frac{t}{\alpha}\right)^{k}\left(\frac{t}{\beta}\right)^{j} \notin z^{\nu}(H_{\rho}^{c})^{-1}$. In other words

$$\left(\frac{t}{z}\right)^{\nu}\left(\frac{t}{\alpha}\right)^{k}\left(\frac{t}{\beta}\right)^{j} =: \zeta(z,t)^{-1} \notin (H_{\rho}^{c})^{-1} \Leftrightarrow \zeta(z,t) \notin H_{\rho}^{c} \Leftrightarrow \zeta(z,t) \in H_{\rho}$$

so that (5.4.2) holds. Since

$$\Gamma_{\nu jk}(z) := \partial \widetilde{K}_{\nu jk}(z,\epsilon) = \partial\{(1+\epsilon)K_{\nu jk}(z)\} = (1+\epsilon)\partial K_{\nu jk}(z). \qquad (5.4.3)$$

and the boundary of $K_{\nu jk}(z)$ is a star-shaped and rectifiable Jordan curve (by Lemma 5.3.3), it follows that so is $\Gamma_{\nu jk}(z)$. $\square$

LEMMA 5.4.2. *Suppose the assumptions of Lemma 5.4.1 are satisfied, in particular $0 < \rho < 1$ and $\rho_2(\rho) < \rho' < 1$ (see Lemma 5.2.4). Let k, j, ν be integers satisfying $k \geq 0, j \geq 0, k+j \geq 1$ and $\nu > 0$. Then there exist a compact and bounded set Γ and a star-shaped Jordan rectifiable and closed curve B_2*

satisfying

(i) $\displaystyle\bigcup_{z\in\partial C_{\rho'}} \Gamma_{\nu jk}(z) \subset \Gamma \subset S$;

(ii) $\{\zeta(z,t) : z \in \partial C_{\rho'}, \ t \in \Gamma_{\nu jk}(z)\} \subset B_2 \subset H_\rho \subset G.$

(iii) *The length* $L_{\nu kj}(z)$ *of* $\Gamma_{\nu kj}(z)$ *is a continuous function for* $z \in \partial C_{\rho'}.$

(iv) $\displaystyle\inf_{z\in\partial C_{\rho'}} d\left(\Gamma_{\nu jk}(z),0\right) \geq \sigma > 0$ *where* $\sigma = (1+\varepsilon)\left(\dfrac{|\alpha|^k|\beta|^j\widetilde{R}(r)^\nu}{2 \max H_{\rho'}}\right)^{1/(\nu+j+k)}.$

PROOF. **(a)** By Lemma 5.4.1, Lemma 5.3.8 and   Lemma 5.3.6   the   set $\bigcup_{z\in\partial C_{\rho'}} \Gamma_{\nu jk}(z) \subset \Gamma \subset S$ is star-shaped, compact and bounded. therefore there exists a compact and bounded set Γ satisfying (i). **(b)** Now we prove (ii). By Lemma 5.4.1 to each $z \in \partial C_{\rho'}$ there exists a rectifiable Jordan curve $\Gamma_{\nu jk}(z)$ satisfying $0 \in$ Int $(\Gamma_{\nu jk}(z))$, $z \in$ Ext $(\Gamma_{\nu jk}(z),\ \Gamma_{\nu jk}(z) \subset S,$ and $\forall t \in \Gamma_{\nu jk}(z),\ \zeta(z,t) \in H_\rho \subset G$. By Lemma 5.1.1 $\forall z \in \partial C_{\rho'}$, the set $\{t^{-(\nu+k+j)} : t \in \Gamma_{\nu jk}(z)\}$ is compact and bounded. Hence $\forall z \in \partial C_{\rho'}\ B(z) \equiv B_{\nu jk}(z) := \{\zeta(z,t) : t \in \Gamma_{\nu jk}(z)\}$ is compact and bounded too. In particular $\forall z \in \partial C_{\rho'},\ B(z) \subset H_\rho \subset G.$ For a positive r and a set E of complex numbers we have $\partial\varphi^{-1}(E) = \varphi^{-1}(\partial E)$ and $\partial(E^{-1}) = (\partial E)^{-1}$. By definition 5.3.1 and (5.3.2) we have

$$\Gamma_{\nu jk}(z) = (1+\epsilon)\varphi^{-1}_{\nu+k+j}(\alpha^k\beta^j z^\nu \partial((H_\rho^c)^{-1})). \tag{5.4.4}$$

From the representation (5.4.4) of $\Gamma_{\nu jk}(z)$ we see that $L_{\nu jk}(z)$ is a continuous function of $z \in \partial C_{\rho'}$. From (5.4.4) $\forall z \in \partial C_{\rho'}$ we have

$$B(z) = \{\zeta(z,t) : t \in \Gamma_{\nu jk}(z)\} = \frac{\alpha^k\beta^j z^\nu}{(\Gamma_{\nu jk}(z))^{\nu+k+j}}$$

$$= \frac{\alpha^k\beta^j z^\nu}{(1+\epsilon(z))^{\nu+j+k}\alpha^k\beta^j z^\nu \partial(H_\rho^c)^{-1}}$$

$$= \frac{1}{(1+\epsilon(z))^{\nu+k+j}}\partial(H_\rho^c) = (1+\epsilon(z))^{-(\nu+j+k)}\partial H_\rho.$$

Thus $B(z) \equiv B_{\nu jk}(z) = (1+\epsilon)^{-(\nu+j+k)}\partial H_\rho.$

For each λ, $0 < \lambda < 1$, we have $\lambda\partial H_\rho \subset H_\rho$ since H_ρ and hence also ∂H_ρ are star-shaped. Now from the definition of ϵ in Lemma 5.3.8 it follows that

$$\bigcup_{z\in\partial C_{\rho'}} B_{\nu jk}(z) = \bigcup_{z\in\partial C_{\rho'}} (1+\epsilon)^{-(\nu+j+k)}\partial H_\rho \subset \frac{1}{1+\varepsilon}H_\rho \subset H_\rho \subset G.$$

Because H_ρ and hence also ∂H_ρ is star-shaped we get

$$\bigcup_{z\in\partial C_{\rho'}} B_{\nu jk}(z) \subset B_2 := \frac{1}{1+\varepsilon}\overline{H}_\rho \subset H_\rho \subset G,$$

where the set B_2 is compact, bounded and $B_2 \subset H_\rho \subset G$.

Proof of (iv). By Lemma 5.2.3 we have $d\left(\partial C_{\rho'}|, 0\right) \geq \widetilde{R}(r))$. From the relation

$$\Gamma_{\nu jk}(z) = (1+\varepsilon)\phi_{\nu+j+k}^{-1}\left(\alpha^k \beta^j z^\nu \partial((H_\rho^c)^{-1})\right)$$

we have, by applying Lemma 5.3.2(i) at the last stage,

$$
\begin{aligned}
\min_{z \in \partial C_{\rho'}} |\Gamma_{\nu jk}(z)| &= \min_{z \in \partial C_{\rho'}} (1+\varepsilon)\left|\phi_{\nu+j+k}^{-1}(\alpha^k \beta^j z^\nu \partial (H_{\rho'})^c)^{-1})\right| \\
&= (1+\varepsilon) \min_{z \in \partial C_{\rho'}} \left(|\alpha|^k |\beta|^j |z|^\nu |\partial((H_{\rho'}^c)^{-1})|\right)^{1/(\nu+j+k)} \\
&\geq (1+\varepsilon)\left(\frac{|\alpha|^k |\beta|^j (\widetilde{R}(r))^\nu}{2\max|H_{\rho'}|}\right)^{1/(\nu+j+k)}
\end{aligned}
$$

$\square$

6. Proof of Theorem 3.4

We observe that by the maximum principle, instead of proving Theorem 3.4 for arbitrary compact subsets K of S_f, it is enough to prove the theorem for a family of closed Jordan curves having two properties: (i) each member of the family is included in S_f and (ii) to each compact and bounded subset K of S_f there exists at least one element in the family of curves which includes K in its interior. Lemma 5.2.1 shows the family of curves $\{\partial C_{\rho'}\}$ has these properties. Therefore it is enough to prove the theorem for $z \in \partial C_{\rho'}$ only.

In the lemmas quoted and applied in this proof for the domain G we choose the domain G_A, where G_A is the A-summability domain of the function $f \in A_R$.

We have to estimate the integrals in the representation of $\Lambda_{n+1}^{(\alpha,\beta)}(z, f)$ for z on compact subsets of $\mathcal{E}$. By the above remark it is enough to prove the theorem only for the Jordan curves $\partial C_{\rho'}$. We have

$$\Lambda_{n+1}^{(\alpha,\beta)}(z, f) = \sum_{i=0}^{i_0} \frac{1}{2\pi i}\oint_{|t|=r} \frac{f(t)}{t-z} A_{n+1}^{(i)}(z;t)dt + \sum_{i=1}^{i_0} \frac{1}{2\pi i}\oint_{|t|=r} \frac{f(t)}{z-t} B_{n+1}^{(i)}(t)dt,$$

when $\delta < r < R$ and $z \in \partial C_{\rho'}$. Choose as in Lemma 5.2.3 ρ, $0 < \rho < 1$ and $\rho_2(\rho) < \rho' < 1$. Since $C_{\rho'}$ is compact, we can choose μ, $0 < \mu < +\infty$, large enough so that $\forall z \in C_{\rho'}$, $|z| \leq \mu$. By Lemma 5.2.4, $\forall z \in \partial C_{\rho'}$ we have $|z| > r\left(\frac{r}{\gamma}\right)^{\frac{1}{\nu^*}} > r$. For $|t| = r$ and $z \in C_{\rho'}$ we have (again by Lemma 5.2.4) $|z-t| \geq |z| - |t| \geq r\left|\frac{r}{\gamma}\right|^{\frac{1}{\nu^*}} - r > 0.$

Write $\quad \phi(x,z) := \sum_{n=0}^{\infty} a_n(x)z^n, \quad \epsilon(x,\theta) := \max_{|\varsigma| \le \theta} |\phi(x,\varsigma)|, \quad M :=$
$\max_{|t|=\tau} |f(t)|$ and $\eta_i(t) := \left(\frac{\alpha}{t}\right)^{\boldsymbol{\lambda}^{(i)} \cdot \boldsymbol{k}^{(i)}} \left(\frac{\beta}{t}\right)^{\boldsymbol{\psi}^{(i)} \cdot \boldsymbol{k}^{(i)}}$. For all terms $A_{n+1}^{(i)}(z;t)$,
$1 \le i \le i_0$, we have, by (2.1.5),

$$A^{(i)}(z) :\equiv \sum_{n=0}^{\infty} a_n(x) \frac{1}{2\pi i} \oint_{|t|=r} \frac{f(t)}{t-z} A_{n+1}^{(i)}(z;t) dt$$

$$= \frac{1}{2\pi i} \oint_{|t|=r} \frac{f(t)}{t-z} \sum_{n=0}^{\infty} a_n(x) A_{n+1}^{(i)}(z,t) dt$$

$$= \frac{1}{2\pi i} \oint_{|t|=r} \frac{f(t)}{t-z} \sum_{n=0}^{\infty} a_n(x) \sum_{\nu_i \in N_i} \sum_{\boldsymbol{k}^{(i)} \in \boldsymbol{M}^{(i)}} b_i(\boldsymbol{k}^{(i)}) \, \eta_i(t)^{n+1} \left(\frac{z}{t}\right)^{(n+1)\nu_i}$$

where we recall that $\boldsymbol{k}^{(i)} \in \boldsymbol{M}^{(i)}$ means that $k_{n_i}^{(i)} \in M_{n_i}^{(i)}, \ldots, k_1^{(i)} \in M_1^{(i)}$. So

$$A^{(i)}(z) = \frac{1}{2\pi i} \oint_{|t|=r} \frac{f(t)}{t-z} \sum_{\nu_i \in N_i} \sum_{\boldsymbol{k}^{(i)} \in \boldsymbol{M}^{(i)}} b_i(\boldsymbol{k}^{(i)}) \, \eta_i(t) \left(\frac{z}{t}\right)^{\nu_i} \sum_{n=0}^{\infty} a_n(x) \eta_i(t)^n \times$$

$$\times \left(\frac{z}{t}\right)^{n\nu_i}$$

$$= \frac{1}{2\pi i} \oint_{|t|=r} \frac{f(t)}{t-z} \sum_{\nu_i \in N_i} \sum_{\boldsymbol{k}^{(i)} \in \boldsymbol{M}^{(i)}} b_i(\boldsymbol{k}^{(i)}) \eta_i(t) \left(\frac{z}{t}\right)^{\nu_i} \phi\left(x, \eta_i(t) \left(\frac{z}{t}\right)^{\nu_i}\right).$$

Writing $\xi_i(t) := \left(\frac{\alpha}{t}\right)^{\boldsymbol{\mu}^{(i)} \cdot \boldsymbol{\ell}^{(i)}} \left(\frac{\beta}{t}\right)^{\boldsymbol{\chi}^{(i)} \cdot \boldsymbol{\ell}^{(i)}}$ and recalling the notation $\boldsymbol{\ell}^{(i)} \in \boldsymbol{P}^{(i)}$
$\Longleftrightarrow \ell_{m_i}^{(i)} \in P_{m_i}^{(i)}, \ldots, \ell_1^{(i)} \in P_1^{(i)}$, we get by (2.1.6) for all terms $B_{n+1}^{(i)}(t)$, $\quad 1 \le$
$i \le i_1$,

$$B^{(i)}(x) \equiv \sum_{n=0}^{\infty} a_n(x) \frac{1}{2\pi i} \oint_{|t|=r} \frac{f(t)}{t-z} B_{n+1}^{(i)}(t) dt$$

$$= \frac{1}{2\pi i} \oint_{|t|=r} \frac{f(t)}{t-z} \sum_{\boldsymbol{\ell}^{(i)} \in \boldsymbol{P}^{(i)}} c_i(\boldsymbol{\ell}^{(i)}) \, \xi_i(t) \phi(x, \xi_i(t)).$$

First we estimate the terms $B^{(i)}(x)$. We have

$$|B^{(i)}(x)| \le \frac{M}{2\pi} \frac{2\pi r}{r\left(\frac{r}{\gamma}\right)^{\frac{1}{\nu^*}} - r} \sum_{\boldsymbol{\ell}^{(i)} \in \boldsymbol{P}^{(i)}} |c_i(\boldsymbol{\ell}^{(i)})| \left(\frac{\gamma}{\tau}\right)^{\boldsymbol{\mu}^{(i)} \cdot \boldsymbol{\ell}^{(i)} + \boldsymbol{\chi}^{(i)} \cdot \boldsymbol{\ell}^{(i)}} \times$$

$$\times \max_{|t|=r} |\phi(x, \xi_i(t))|,$$

where $M := \max_{|t|=r} |f(t)|$. Now since $\boldsymbol{\mu}^{(i)} \cdot \boldsymbol{\ell}^{(i)} \ge 0$, $\quad \boldsymbol{\chi}^{(i)} \cdot \boldsymbol{\ell}^{(i)} \ge 0, \boldsymbol{\mu}^{(i)} \cdot \boldsymbol{\ell}^{(i)} +$
$\boldsymbol{\chi}^{(i)} \cdot \boldsymbol{\ell}^{(i)} \ge 1$ and $\gamma < r$, we get from (2.1.2) $|\xi_i(t)| \le \left(\frac{\gamma}{r}\right)^{\boldsymbol{\mu}^{(i)} \cdot \boldsymbol{\ell}^{(i)} + \boldsymbol{\chi}^{(i)} \cdot \boldsymbol{\ell}^{(i)}} \le \frac{\gamma}{r}$.
Hence

$$|B^{(i)}(x)| \le \frac{Mr}{r\left(\frac{r}{\gamma}\right)^{\frac{1}{\nu^*}} - r} \epsilon\left(x, \frac{\gamma}{r}\right) \sum_{\boldsymbol{\ell}^{(i)} \in \boldsymbol{P}^{(i)}} |c_i(\boldsymbol{\ell}^{(i)})| \left(\frac{\gamma}{r}\right)^{\boldsymbol{\mu}^{(i)} \cdot \boldsymbol{\ell}^{(i)} + \boldsymbol{\chi}^{(i)} \cdot \boldsymbol{\ell}^{(i)}}.$$

By (2.1.7) and (2.1.8) we see that for $1 \leq i \leq i_0$ the expression $D^{(i)} := \sum_{\ell^{(i)} \in P^{(i)}}$ $|c_i(\ell^{(i)})| \cdot \left(\frac{\gamma}{r}\right)^{\mu^{(i)} \cdot \ell^{(i)} + \chi^{(i)} \cdot \ell^{(i)}}$ is bounded by a finite number depending on $\frac{\gamma}{r}$. Therefore $\sum_{\ell^{(i)} \in P^{(i)}} D^{(i)}$ is bounded by a finite number depending on $\frac{\gamma}{r}$, say $h_i(\frac{\gamma}{r})$. Thus

$$|B^{(i)}| \leq \frac{Mr}{r\left(\frac{r}{\gamma}\right)^{\frac{1}{\nu^*}} - r} \, h_i\left(\frac{\gamma}{r}\right) \, \epsilon\left(x, \frac{\gamma}{r}\right) \quad \text{for } 1 \leq i \leq i_1.$$

Now we estimate the terms $A^{(i)}(z)$. Choose an integer V so large that for each i, $1 \leq i \leq i_0$, $C^{(i)} := \max_{\nu \in N_i} \left(\frac{\gamma}{r}\right)^V \left(\frac{\mu}{r}\right)^\nu < 1$ (this is possible because each set N_i is finite and $1 \leq i \leq i_0$). As we saw before we have

$$A^{(i)}(z) = \frac{1}{2\pi i} \oint_{|t|=r} \frac{f(t)}{t-z} \sum_{\nu_i \in N_i} \sum_{k^{(i)} \in M^{(i)}} b_i(k^{(i)}) \eta_i(t) \left(\frac{z}{t}\right)^{\nu_i} \phi\left(x, \eta_i(t)\left(\frac{z}{t}\right)^{\nu_i}\right)$$

$$\equiv \frac{1}{2\pi i} \oint_{|t|=r} \frac{f(t)}{t-z} \left(I_1^{(i)} + I_2^{(2)}\right) \phi\left(x, \eta_i(t)\left(\frac{z}{t}\right)^{\nu_i}\right)$$

$$\equiv J_1^{(i)} + J_2^{(i)}$$

where

$$I_1^{(i)} = \underbrace{\sum_{\nu_i \in N_i} \sum_{k^{(i)} \in M^{(i)}}}_{\lambda^{(i)} \cdot k^{(i)} + \psi^{(i)} \cdot k^{(i)} < V} \quad , \quad I_2^{(i)} = \underbrace{\sum_{\nu_i \in N_i} \sum_{k^{(i)} \in M^{(i)}}}_{\lambda^{(i)} \cdot k^{(i)} + \psi^{(i)} \cdot k^{(i)} \geq V} .$$

By the remark made after (2.1.8) the set of indices $k^{(i)}$ for which the numbers $\lambda^{(i)} \cdot k^{(i)} + \psi^{(i)} \cdot k^{(i)}$ satisfy $\lambda^{(i)} \cdot k^{(i)} + \psi^{(i)} \cdot k^{(i)} < V$ is finite. First we estimate $J_2^{(i)}$. We have

$$|J_2^{(i)}| \leq \frac{M}{2\pi} \frac{2\pi r}{r\left(\frac{r}{\gamma}\right)^{\frac{1}{\nu^*}} - r} D^{(i)} \, \varepsilon(x, C^{(i)}),$$

where

$$D^{(i)} = \underbrace{\sum_{\nu_i \in N_i} \sum_{k^{(i)} \in M^{(i)}}}_{\lambda^{(i)} \cdot k^{(i)} + \psi^{(i)} \cdot k^{(i)} \geq V} |b_i(k^{(i)})| \left(\frac{\gamma}{r}\right)^{\lambda^{(i)} \cdot k^{(i)} + \psi^{(i)} \cdot k^{(i)}} \left(\frac{\mu}{r}\right)^{\nu_i}.$$

By the same argument as that used to estimate $B^{(i)}(x)$, we see that

$$D^{(i)} := \underbrace{\sum_{\nu_i \in N_i} \sum_{k^{(i)} \in M^{(i)}}}_{\lambda^{(i)} \cdot k^{(i)} + \psi^{(i)} \cdot k^{(i)} \geq V} |b_i(k^{(i)})| \left(\frac{\gamma}{r}\right)^{\lambda^{(i)} \cdot k^{(i)} + \psi^{(i)} \cdot k^{(i)}} \left(\frac{\mu}{r}\right)^{\nu_i}$$

$$\leq \sum_{\nu_i \in N_i} \sum_{k^{(i)} \in M^{(i)}} |b_i(k^{(i)})| \left(\frac{\gamma}{r}\right)^{\lambda^{(i)} \cdot k^{(i)} + \psi^{(i)} \cdot k^{(i)}} \left(\frac{\mu}{r}\right)^{\nu_i}.$$

Hence $\left| J_2^{(i)} \right| \leq \frac{M}{\left(\frac{r}{\gamma}\right)^{\frac{1}{\nu^*}} - 1} h_i(\frac{\gamma}{r}) \epsilon(x, C^{(i)})$ where $h_i(\frac{\gamma}{r})$ denotes a finite number depending on $\frac{\gamma}{r}$ (see (2.1.7) and (2.1.8) and the estimate of $B^{(i)}$).

Now we estimate the terms of $J_1^{(i)}$. Let $\nu_i \in N_i$, $k_1^{(i)} \in M_1^{(i)}, \ldots, k_{n_i}^{(i)} \in M_{n_i}^{(i)}$ be a given set of indices satisfying $\boldsymbol{\lambda}^{(i)} \cdot \boldsymbol{k}^{(i)} + \boldsymbol{\psi}^{(i)} \cdot \boldsymbol{k}^{(i)} < V$. For such a given set of indices denote, for the sake of brevity, $\nu := \nu_i$, $k := \boldsymbol{\lambda}^{(i)} \cdot \boldsymbol{k}^{(i)}$, $j := \boldsymbol{\psi}^{(i)} \cdot \boldsymbol{k}^{(i)}$. We have $k \geq 0$, $j \geq 0$ and $k + j \geq 1$. We consider each term in the sum defining $J_1^{(i)}$ separately. Let $E_{\nu jk}^{(i)}$ be one of these terms. Then

$$E^{(i)} \equiv E_{\nu jk}^{(i)} := \frac{1}{2\pi i} \oint_\Gamma \frac{f(t)}{t - z} b_i(\boldsymbol{k}^{(i)}) \left(\zeta(z, t)\right) \phi\left(x, \zeta(z, t)\right) dt,$$

where $\zeta(z, t) = \left(\frac{z}{t}\right)^\nu \left(\frac{\alpha}{t}\right)^k \left(\frac{\beta}{t}\right)^j$, $\Gamma = \partial D_r$ (also for $z \in \partial C_\rho$, $z \in$ Ext $(\overline{D_r})$). By Lemma 5.4.1 for each $z \in \partial C_{\rho'}$ there exists a rectifiable Jordan curve $\Gamma_{\nu jk}(z)$ satisfying $0 \in$ Int $(\Gamma_{\nu jk}(z))$, $z \in$ Ext $(\Gamma_{\nu jk}(z)$, $\Gamma_{\nu jk}(z) \subset S$, and $\forall t \in \Gamma_{\nu jk}(z)$, $\zeta(z, t) \in H_\rho \subset G$. Denote by $L_{\nu jk}^{(i)}(z)$ the length of $\Gamma_{\nu jk}(z)$. Using Cauchy's integral theorem we get

$$E^{(i)} = E_{\nu jk}^{(i)} := \frac{1}{2\pi i} \oint_{\Gamma_{\nu jk}(z)} \frac{f(t)}{t - z} b_i(\boldsymbol{k}^{(i)}) \left(\zeta(z, t)\right) \phi\left(x, \zeta(z, t)\right) dt$$

By Lemma 5.4.2 the set $\bigcup_{z \in \partial C_{\rho'}} \Gamma_{\nu jk}(z)$ is included in a compact subset Γ of S, and there exists a compact and bounded set $B_2 \equiv B_2(\nu, j, k)$ such that

$$\{\zeta(z, t) : z \in \partial C_{\rho'}, \ t \in \Gamma_{\nu jk}(z)\} \subset B_2(\nu, j, k).$$

Write

$$M_1 := \sup_t \{|f(t)| \ : \ z \in \partial C_{\rho'}, \ t \in \Gamma_{\nu jk}(z)\} \leq \max_{t \in \Gamma} |f(t)| < +\infty,$$

$$\epsilon(x, B_2(\nu, j, k)) := \max_{\zeta \in B_2(\nu, j, k)} |\phi(x, \zeta)|, \quad B_2(\nu, j, k) \subset H_\rho \subset G_A.$$

Applying now Lemma 5.3.8, Lemma 5.3.6 and (5.4.2) we get

$$|E_{\nu jk}^{(i)}| \leq \frac{M_1}{2\pi} \frac{L}{D} \left(\frac{\gamma}{\sigma}\right)^{k+j} |b_i(\boldsymbol{k}^{(i)})| \epsilon(x, B_2(\nu, j, k)) \left(\frac{\mu}{\sigma}\right)^\nu.$$

Since there is only a finite number of terms such that $k + j < V$ we get

$$|J_1^{(i)}| \leq \text{const}_i \cdot \epsilon\left(x, \bigcup_{k+j<V} B_2(\nu, j, k)\right).$$

Combining the estimates of $J_1^{(i)}$ and $J_2^{(i)}$ the proof follows. $\qquad \square$

7. Applications of Theorems 3.3 and 3.4.

EXAMPLE 7.1. For a function $f \in A_\rho$, $\rho > 1$, $f(z) = \sum_{k=0}^{\infty} a_k z^k$, write as in Chapter 1,Sec. 1.3, $p_{n-1}(f; z) = \sum_{k=0}^{n-1} a_k z^k$. Let $L_{n-1}(f; z)$ denote the Lagrange interpolant to f on the zeros of $z^n - 1$. We have, see Chapter 1,(3.2),

$$L_{n-1}(f; z) - p_{n-1}(f; z) = \frac{1}{2\pi i} \int_{\Gamma_R} \frac{f(t)}{t - z} K_n(t, z)\, dt. \tag{7.1}$$

where

$$K_n(t, z) := \frac{t^n - z^n}{t^n(t^n - 1)} = \left(1 - \left(\frac{z}{t}\right)^n\right) \frac{1}{t^n} \sum_{k=0}^{\infty} \frac{1}{t^{kn}}$$

$$= \sum_{k=1}^{\infty} \left(\frac{1}{t^k}\right)^n - \sum_{k=1}^{\infty} \left(\frac{1}{t^k} \left(\frac{z}{t}\right)^1\right)^n.$$

In this example have for $\tau > 1$

$$\widetilde{R}(\tau) = \tau \cdot \min_{k \geq 1} \left(\frac{\tau}{1}\right)^k = \tau^2.$$

Applying Theorem 3.3 we get the Walsh equiconvergence theorem.

EXAMPLE 7.2. Let $L_{n-1}(f; z)$ denote the Lagrange interpolant to f on the zeros of $z^n - 1$. Write

$$p_{n-1,j}(f; z) := \sum_{k=0}^{n-1} a_{k+jn} z^k, \quad j = 0, 1, 2, \ldots \tag{7.2}$$

By Chapter 1,(3.5) and (3.4) we have

$$L_{n-1}(f; z) - \sum_{j=0}^{\ell-1} p_{n-1,j}(f; z) = \frac{1}{2\pi i} \int_{\Gamma_R} \frac{f(t)}{t - z} K_n(t, z)\, dt. \tag{7.3}$$

where Γ_R is the circle $|t| = R$ and

$$K_n(t, z) := \frac{t^n - z^n}{(t^n - 1)t^{\ell n}} = \frac{t^n \left(1 - \left(\frac{z}{t}\right)^n\right)}{t^{(\ell+1)n} \left(1 - \frac{1}{t^n}\right)}$$

$$= \sum_{k=\ell}^{\infty} \left(\frac{1}{t^k}\right)^n - \sum_{k=\ell}^{\infty} \left(\frac{1}{t^k} \left(\frac{z}{t}\right)^1\right)^n. \tag{7.4}$$

In this example have for $\tau > 1$

$$\widetilde{R}(\tau) = \tau \cdot \min_{k \geq \ell} \left(\frac{\tau}{1}\right)^k = \tau^{\ell+1}.$$

Applying now Theorem 3.3 we obtain Theorem 7 of Chapter 1.

EXAMPLE 7.3. Let $f \in A_\rho$ ($\rho > 1$). As in Chapter 2, Sec. 2.1, (1.1), let $h_{r,rn-1}(f;z)$ denote the Hermite interpolant to f in the zeros of $(z^n - 1)^r$. Then $h_{r,rn-1}(f;z) \in \pi_{rn-1}$ and it satisfies the conditions

$$h_{rn-1}^{(j)}(f;w^k) = f^{(j)}(w^k), \quad j = 0,1,\ldots,r-1; \quad k = 0,1,\ldots,n-1$$

where w is a primitive n^{th} root of unity. Write

$$p_{rn-1,0}(f;z) := \sum_{k=0}^{rn-1} a_k z^k = \frac{1}{2\pi i} \int_{\Gamma_R} \frac{f(t)}{t-z} \cdot \frac{t^{rn} - z^{rn}}{t^{rn}} \, dt \qquad (7.5)$$

By Chapter 2, (1.4)), we have

$$\Delta_{rn-1,1}(f;z) := h_{r,rn-1}(f;z) - p_{rn-1,0}(f;z) = \frac{1}{2\pi i} \int_{\Gamma_R} \frac{f(t)}{t-z} K_n(t,z) \, dt \quad (7.6)$$

where

$$K_n(t,z) := \frac{z^{rn}}{t^{rn}} - \frac{(z^n - 1)^r}{(t^n - 1)^r} . \qquad (7.7)$$

By Chapter 2, Lemma 1, we have for all t with $|t| > 1$,

$$\frac{z^r}{t^r} - \frac{(z-1)^r}{(t-1)^r} = \frac{t-z}{t^r} \sum_{s=1}^{\infty} \frac{\beta_{s,r}(z)}{t^s} \qquad (7.8)$$

where $\beta_{s,r}(z)$ is a polynomial in z of degree $r-1$ given by

$$\beta_{s,r}(z) := \sum_{k=0}^{r-1} \binom{r+s-1}{k} (z-1)^k, \quad s = 1,2,\ldots . \qquad (7.9)$$

$$= \sum_{k=0}^{r-1} \binom{r+s-1}{k} \sum_{\sigma=0}^{k} (-1)^{k-\sigma} \binom{k}{\sigma} z^\sigma$$

$$= \sum_{\sigma=0}^{r-1} \binom{r+s-1}{\sigma} z^\sigma \sum_{k=\sigma}^{r-1} (-1)^{k-\sigma} \binom{r+s-\sigma-1}{k-\sigma}$$

$$= \sum_{\sigma=0}^{r-1} (-1)^{r-\sigma-1} \binom{r+s-\sigma-2}{s-1} \binom{s+r-1}{\sigma} z^\sigma$$

$$= \sum_{\sigma=0}^{r-1} (-1)^{r-\sigma-1} \frac{1}{s+r-\sigma-1} \binom{s+r-1}{r-1} \binom{r-1}{\sigma} z^\sigma .$$

Hence we have

$$K_n(t,z) = \frac{z^{rn}}{t^{rn}} - \frac{(z^n-1)^r}{(t^n-1)^r} = \frac{t^n - z^n}{t^{rn}} \sum_{s=1}^{\infty} \frac{\beta_{s,r}(z^n)}{t^{sn}}$$

$$= \frac{t^n - z^n}{t^{rn}} \sum_{s=1}^{\infty} \sum_{\sigma=0}^{r-1} (-1)^{r-\sigma-1} \binom{r+s-\sigma-2}{s-1} \binom{s+r-1}{\sigma} \left(\frac{z^\sigma}{t^s}\right)^n$$

$$= \sum_{s=1}^{\infty} \sum_{\sigma=0}^{r-1} (-1)^{r-\sigma-1} \binom{r+s-\sigma-2}{s-1} \binom{s+r-1}{\sigma} \left(\frac{z^{\sigma}}{t^{s+r-1}}\right)^{n}$$

$$- \sum_{s=1}^{\infty} \sum_{\sigma=0}^{r-1} (-1)^{r-\sigma-1} \binom{r+s-\sigma-2}{s-1} \binom{s+r-1}{\sigma} \left(\frac{z^{\sigma+1}}{t^{s+r}}\right)^{n}$$

$$= \sum_{s=1}^{\infty} \sum_{\sigma=0}^{r-1} (-1)^{r-\sigma-1} \binom{r+s-\sigma-2}{s-1} \binom{s+r-1}{\sigma} \left(\frac{1}{t^{s+r-\sigma-1}}\right)^{n} \left(\left(\frac{z}{t}\right)^{\sigma}\right)^{n}$$

$$- \sum_{s=1}^{\infty} \sum_{\sigma=0}^{r-1} (-1)^{r-\sigma-1} \binom{r+s-\sigma-2}{s-1} \binom{s+r-1}{\sigma} \left(\frac{1}{t^{s+r-\sigma-1}}\right)^{n} \times$$

$$\times \left(\left(\frac{z}{t}\right)^{\sigma+1}\right)^{n}$$

$$= \sum_{s=1}^{\infty} (-1)^{r-1} \binom{r+s-2}{s-1} \left(\frac{1}{t^{s+r-1}}\right)^{n}$$

$$+ \sum_{s=1}^{\infty} \sum_{\sigma=1}^{r-1} (-1)^{r-\sigma-1} \binom{r+s-\sigma-2}{s-1} \binom{s+r-1}{\sigma} \left(\frac{1}{t^{s+r-\sigma-1}}\right)^{n} \times$$

$$\times \left(\left(\frac{z}{t}\right)^{\sigma}\right)^{n}$$

$$- \sum_{s=1}^{\infty} \sum_{\sigma=1}^{r} (-1)^{r-\sigma} \binom{r+s-\sigma-1}{s-1} \binom{s+r-1}{\sigma-1} \left(\frac{1}{t^{s+r-\sigma}}\right)^{n} \left(\left(\frac{z}{t}\right)^{\sigma}\right)^{n}$$

$$= \sum_{k=1}^{\infty} (-1)^{r-1} \binom{k-1}{r-1} \left(\frac{1}{t^{k}}\right)^{n}$$

$$+ \sum_{\nu=1}^{r-1} \sum_{k=r-\nu}^{\infty} (-1)^{r-\nu-1} \binom{k-1}{\nu-1} \binom{k+\nu}{k} \left(\frac{1}{t^{k}} \left(\frac{z}{t}\right)^{\nu}\right)^{n}$$

$$- \sum_{\nu=1}^{r} \sum_{k=r-\nu+1}^{\infty} (-1)^{r-\nu} \binom{k-1}{r-\nu} \binom{k+\nu-1}{k} \left(\frac{1}{t^{k}} \left(\frac{z}{t}\right)^{\nu}\right)^{n}$$

In this example we get for $\tau > 1$

$$\widetilde{R}(\tau) = \tau \cdot \min \left(\min_{1 \leq \nu \leq r-1} \min_{k \geq r-\nu} \left(\frac{\tau}{1}\right)^{k/\nu}, \min_{1 \leq \nu \leq r} \min_{k \geq r-\nu+1} \left(\frac{\tau}{1}\right)^{k/\nu} \right)$$
$$= \tau^{\frac{1}{r}+1}.$$

EXAMPLE 7.4. For any positive integer $\ell \geq 1$, set, as in Chapter 2,(1.14),

$$\Delta_{rn-1,\ell}(f;z) := h_{r,rn-1}(f;z) - h_{r,rn-1,0}(f;z) - \sum_{j=1}^{\ell-1} h_{r,rn-1,j}(f;z) \quad (7.10)$$

where

$$h_{r,rn-1,0}(f;z) := p_{rn-1,0}(f;z) = \sum_{k=0}^{rn-1} a_k z^k$$

By Chapter 2, (1.15) and (1.16) we have (where $\Gamma_R := \{t : |t| = R\}$)

$$\Delta_{rn-1,\ell}(f; z) = \frac{1}{2\pi i} \int_{\Gamma_R} \frac{f(t)K(t, z)}{t - z} \, dt, \tag{7.11}$$

where

$$K(t, z) := \frac{z^{rn}}{t^{rn}} - \frac{(z^n - 1)^r}{(t^n - 1)^r} - \frac{t^n - z^n}{t^{rn}} \sum_{j=1}^{\ell-1} \frac{\beta_{j,r}(z^n)}{t^{jn}}. \tag{7.12}$$

$$= \frac{t^n - z^n}{t^{rn}} \left(\sum_{s=1}^{\infty} \frac{\beta_{s,r}(z^n)}{t^{sn}} - \sum_{s=1}^{\ell-1} \frac{\beta_{s,r}(z^n)}{t^{sn}} \right)$$

$$= \frac{t^n - z^n}{t^{rn}} \sum_{s=\ell}^{\infty} \frac{\beta_{s,r}(z^n)}{t^{sn}}$$

Applying now the argument used in the previous example we get

$$K_n(t, z) = \sum_{s=\ell}^{\infty} \sum_{\sigma=0}^{r-1} (-1)^{r-s-1} \binom{r+s-\sigma-2}{s-1} \binom{s+r-1}{\sigma} \left(\frac{1}{t^{s+r-\sigma-1}} \right)^n \times$$

$$\times \left(\left(\frac{z}{t} \right)^{\sigma} \right)^n$$

$$- \sum_{s=\ell}^{\infty} \sum_{\sigma=0}^{r-1} (-1)^{r-s-1} \binom{r+s-\sigma-2}{s-1} \binom{s+r-1}{\sigma} \left(\frac{1}{t^{s+r-\sigma-1}} \right)^n \times$$

$$\times \left(\left(\frac{z}{t} \right)^{\sigma+1} \right)^n$$

$$= \sum_{s=\ell}^{\infty} (-1)^{r-s-1} \binom{r+s-2}{s-1} \left(\frac{1}{t^{s+r-1}} \right)^n$$

$$+ \sum_{s=\ell}^{\infty} \sum_{\sigma=1}^{r-1} (-1)^{r-s-1} \binom{r+s-\sigma-2}{s-1} \binom{s+r-1}{\sigma} \left(\frac{1}{t^{s+r-\sigma-1}} \right)^n \times$$

$$\times \left(\left(\frac{z}{t} \right)^{\sigma} \right)^n$$

$$- \sum_{s=\ell}^{\infty} \sum_{\sigma=1}^{r} (-1)^{r-s-1} \binom{r+s-\sigma-1}{s-1} \binom{s+r-1}{\sigma-1} \left(\frac{1}{t^{s+r-\sigma}} \right)^n \left(\left(\frac{z}{t} \right)^{\sigma} \right)^n$$

$$= \sum_{k=\ell+r-1}^{\infty} (-1)^{2r-k} \binom{k-1}{r-1} \left(\frac{1}{t^k} \right)^n$$

$$+ \sum_{\nu=1}^{r-1} \sum_{k=\ell+r-\nu-1}^{\infty} (-1)^{2r-k-\nu-2} \binom{k-1}{r-\nu-1} \binom{k+\nu}{k} \left(\frac{1}{t^k} \left(\frac{z}{t} \right)^{\nu} \right)^n$$

$$- \sum_{\nu=1}^{r} \sum_{k=\ell+r-\nu}^{\infty} (-1)^{2r-k-\nu-1} \binom{k-1}{r-\nu} \binom{k+\nu-1}{k} \left(\frac{1}{t^k} \left(\frac{z}{t} \right)^{\nu} \right)^n$$

In this example we get for $\tau > 1$

$$\widetilde{R}(\tau) = \tau \cdot \min\left(\min_{1 \leq \nu \leq r-1} \min_{k \geq \ell+r-\nu-1} \left(\frac{\tau}{1}\right)^{k/\nu}, \min_{1 \leq \nu \leq r} \min_{k \geq \ell+r-\nu} \left(\frac{\tau}{1}\right)^{k/\nu}\right)$$
$$= \tau^{\frac{\ell}{r}+1}.$$

By Theorem 3.3 we obtain another proof of Theorem 1 of Chapter 2.

8. Historical Remarks

R. Brück [17] was the first who replaced the convergence in equiconvergence theorems by A-summability. R. Brück also determined the equisummability domains for several pairs of approximation operators. The ideas of Brück were continued by A. Jakimovski and A. Sharma in [54]. Some of the results of this chapter are given in [54].

The case of the $(0, 2)$ lacunary interpolation is dealt with in detail and its Walsh radius and domain of equisummability are obtained in [54]. The kernel in this case is similar but does not fall into the category of the kernels discussed here. It seems probable that other variations of the kernels considered here can also be dealt with.

References

[1]   Achieser, N.I., *Theory of Approximation*, Ungar, New York, 1956.

[2]   Akhlagi, M.R., Jakimovski, A. and Sharma, A., *Equiconvergens of some complex interpolatory polynomials*, Numer. Math. **57** (1990), 635-649.

[3]   Al'per, S.J., *Asymptotic values of best approximation of analytic functions in a complex domain,*, Uspehi Mat. Nauk **14** (1959), 131-134.

[4]   Baishanski, B.M., *Equiconvergence of interpolating processes*, Rocky Mountain J. Math. **11** (1981), 483-490.

[5]   Birkhoff, G.D., *General mean value and remainder theorems with applications to mechanical differentiation and quadrature*, Trans. Amer. Math. Soc. **7** (1906), 107-136.

[6]   Bokhari, M.A., *Equiconvergence of some interpolatory and best approximation processes*, Ph.D. Thesis, University of Alberta, 1985.

[7]   Bokhari, M.A., *Perturbation of nodes and poles in certain rational interpolants*, Bull. Austral. Math. Soc. **36** (1987), 177-185.

[8]   Bokhari, M.A., *Equiconvergence of some sequences of complex interpolating rational functions*, J. Approx. Theory **55** (1988), 205-219.

[9]   Bokhari, M.A., *On certain sequences of least squares approximants*, Bull. Austral. Math. Soc. **38** (1988), 415-422.

[10]   Bokhari, M.A., *Converse results in the theory of equiconvergence of interpolating rational functions*, Rocky Mountain J. Math. **19** (1989), 73-82.

[11]   Bokhari, M.A., *Extensions of certain results in Walsh-type equiconvergence*, Bull. Austral. Math. Soc. **44** (1991), 451-460.

[12]   Bokhari, M.A., *A note on certain next-to-interpolatory rational functions*, Acta Math. Hungar. **62** (1993), 49-55.

[13]   Bokhari, M.A., *Interpolation mixed with ℓ_2-approximation*, J. Approx. Theory **83** (1995), 255-265.

[14]   Bokhari, M.A., *On constrained L^2-approximation of complex functions*, Canad. Math. Bull. **38** (1995), 149-155.

[15]   Bokhari, M.A. and Sharma, A., *Equiconvergence of certain rational interpolants (Hermite case)*, Methods of Functional Analysis in Approx. Theory. (C.A. Micchelli, D.V. Pai and B.V. Limaye, eds.), 1986, pp. 281-292.

[16]   Bokhari, M.A. and Sharma, A., *Equiconvergence of some sequences of rational functions*, Internat. J. Math. and Math. Sci. **15** (1992), 221-228.

[17]   Brück, R., *Generalization of Walsh's equiconvergence theory by the application of summability methods*, Ph.D. Thesis Giessen Mitt. Math. Sem. **195** (1990), 1-84.

[18]   Brück, R., *Summability of sequences of polynomial interpolants in the roots of unity*, Analysis **11** (1991), 27-42.

[19]   Brück, R., *On the failure of Walsh's equiconvergence theory for Jordan domains*, Analysis **13** (1993), 229-234.

[20]   Brück, R., *On Walsh's equiconvergence theorem and applications of summability methods*, Arch. Math. **62** (1994), 321-330.

291

[21]   Brück, R., *Lagrange interpolation in non-uniformly distributed nodes on the unit circle*, Analysis **16** (1996), 273-282.

[22]   Brück, R., Sharma, A. and Varga, R.S., *An extension of a result of Rivlin on Walsh equiconvergence*, Advances in Computational Mathematics, New Delhi, India, 1993, pp. 225-234, World Scient. Pub. Co., Singapore, 1994 (H.P. Dikshit and C.A. Micchelli, eds.).

[23]   Brück, R., Sharma, A. and Varga, R.S., *An extension of a result of Rivlin on Walsh equiconvergence (Faber nodes)*, Birkhäuser Verlag, Basel, ISNM **119** (1994), 41-66.

[24]   Cavaretta, A.S., Jr., Dikshit, H.P. and Sharma, A, *An extension of a theorem of Walsh*, Resultate Math. **7** (1984), 154-163.

[25]   Cavaretta, A.S., Dikshit, H.P. and Sharma, A., *Convergence of certain polynomial interpolants to a function defined on the unit circle*, Acta Math. Hungar. **53** (1989), 143-147.

[26]   Cavaretta, A.S., Jr., Micchelli, C.A. and Sharma, A., *A multivariate extension of Walsh equiconvergence*, Approximation Theory VI (C.K. Chui, L.L. Schumaker and J.D. Ward, eds.), Academic Press, 1989, pp. 1-4.

[27]   Cavaretta, A.S., Jr., Micchelli, C.A. and Sharma, A., *Walsh equiconvergence theorem and optimal recovery*, Analysis **12** (1992), 271-302.

[28]   Cavaretta, A.S., Jr., Sharma, A. and Varga, R.S., *Hermite-Birkhoff interpolation in the n-th roots of unity*, Trans. Amer. Math. Soc. **259** (1980), 621-628; MR 81c: 30064.

[29]   Cavaretta, A.S., Jr., Sharma, A. and Varga, R.S., *Lacunary trigonometric interpolation on equidistant nodes*, Qualitative Approximation (R.A. DeVore and K. Scherer, eds.), Academic Press, Inc., New York, 1980, pp. 63-80, MR 81k: 42004.

[30]   Cavaretta, A.S., Jr., Sharma, A. and Varga, R.S., *Interpolation in the roots of unity: An extension of a theorem of J.L. Walsh*, Resultate Math. **3** (1981), 155-191.

[31]   Cavaretta, A.S., Jr., Sharma, A. and Varga R.S., *A theorem of J.L. Walsh revisited*, Pac. J. Math. **118** (1985), 313-322.

[32]   Cavaretta, A.S., Jr., Sharma, A. and Varga R.S., *Converse results in the Walsh theory of equiconvergence*, Mathematical Modeling and Numerical Analysis **19** (1985), 601-609.

[33]   Curtiss J.H., *Convergence of complex Lagrange interpolation polynomials on the locus of the interpolation points*, Duke Math. J., **32** (1965), 187-204.

[34]   Curtiss J.H., *Faber polynomials and the Faber series*, Amer. Math. Monthly, **78** (1971), 577-596.

[35]   de Bruin, M.G. and Sharma, A., *Equiconvergence of some simultaneous Hermite-Padé interpolants*, Math. Modelling and Numerical Analysis **29** (4) (1995), 477-503.

[36]   de Bruin, M.G. and Sharma, A., *Equiconvergence of some simultaneous Hermite-Padé interpolants: multiple nodes*, Approx. Theory VIII (Ch. K. Chui and L. Schumaker, eds.), 1995, pp. 103-110.

[37]   de Bruin, M.G. and Sharma, A., *Overconvergence of some simultaneous Hermite-Padé interpolants*, Annals of Numerical Mathematics **4** (1997), 239-259.

[38]   de Bruin, M.G., Sharma, A. and Szabados, J., *Birkhoff type interpolation on perturbed roots of unity*, Approximation Theory (N.K. Govil et al., eds.), M. Dekker, (1998), pp. 167-179.

[39]   de Montessus de Ballore, R., *Sur les fractions continues algebriques*, Bull. Soc. Math. France **30** (1902), 28-36.

[40]   Dikshit, H.P., Sharma, Singh, V. and Stenger, F., *Rivlin's theorem on Walsh equiconvergence*, J. Approx. Theory **52** (1988), no. (3), 339-349.

[41]   Duren, P.,, *Theory of H^p spaces*, Academic Press, New York, 1970.

[42]   Gaier, D., *Lectures on Complex Approximation*, Birkhauser, Basel, 1985.

[43]   Goodman, T.N.T., Ivanov, K.G. and Sharma, A., *Hermite interpolation in the roots of unity*, J. Approx. Theory **84** (1996), 41-60.

[44]   Goodman, T.N.T. and Sharma, A., *A property of Hermite interpolants in the roots of unity*, Ganita **43** (1992), 171-180.

[45]   Hardy, G.H., *Divergent series*, The Clarendon Press, London, 1949.

[46]   Hermann, T., *Some remarks on a theorem of Saff and Varga*, J. Approx. Theory **39** (1983), 241-246.

[47]   Hille, Einar., *Analytic Function Theory Vol.II* (1962), Grim & Co, New-York.

[48]   Ivanov, K.G. and Saff, E.B., *Behavior of the Lagrange interpolants in the roots of unity*, Computational Methods and Function Theory, Proceedings, Valparaiso (S. Ruscheweyh et al, Eds.), Springer Verlag, Berlin (1990), 81-87.

[49]   Ivanov, K.G. and Sharma, A, *Some new results on Walsh theory of equiconvergence*, Proc. of Haar Memorial Conference, Budapest, Aug. 11-18, 1985.

[50]   Ivanov, K.G. and Sharma, A., *Quantitative results on Walsh equiconverge II. Hermite interpolation and ℓ_2-approximation*, Approx. Theory and Its Appl. **2** (1986), 47-84.

[51]   Ivanov, K.G. and Sharma, A., *More quantitative results on Walsh equiconvergence I. Lagrange case*, Constructive Approx. **3** (1987), 265-280.

[52]   Ivanov, K.G. and Sharma, A., *Converse results on equiconvergence of interpolating polynomials*, Analysis Mathematica **14** (1988), 44-50.

[53]   Ivanov, K.G. and Sharma, A., *Walsh equiconvergence and (ℓ, p) distinguished points*, Constructive Approximation **7** (1991), 315-327.

[54]   Jakimovski, A. and Sharma, A., *Walsh equiconvergence and equisummability*, Journal d'Analysis, Madras (India) (1993), 1-52.

[55]   Jakimovski, A. and Sharma, A., *Asymptotic properties of differences of generalized Hermite-Padé interpolants*, Journal of Orissa Math. Soc. **12-15** (1993-96), 341-386.

[56]   Jakimovski, A. and Sharma, A., *Hermite interpolations on Chebyshev nodes and Walsh equiconvergence*, Approximation Theory (N.K. Govil et al., eds.), M. Dekker, (1998), pp. 293-333.

[57]   Jakimovski, A. and Sharma, A., *Hermite interpolation on Chebyshev nodes and Walsh equiconvergence II*, J. Analysis **7** (1991), 89-102.

[58]   Jakimovski, A. and Sharma, A., *Asymptotic properties of differences of Hermite-Padé interpolants,*, J. of Orissa Mathematical Soc. **12-15** (1993-1996), 339-386.

[59]   Jakimovski, A. and Sharma, A., *Quantitative results on equiconveregence of certain sequences of rational interpolants*, Matematica Balkanika **16** (2002), 1-36.

[60]   Juneja, D.P. and Pratibha Dua, *On Walsh equiconvergence by Chebyshev polynomials* (preprint).

[61]   Juneja, O.P. and Pratibha, Dua, *A note on Walsh equiconvergence* (preprint).

[62]   Kalmár, L., *On interpolation*, Matematikai és Physikai Lapok (1926), 129-149 (in Hungarian).

[63]   Kövari T. and Pommerenke Ch., *On Faber polynomials and Faber expansions*, Math. Z. **99** (1967), 193-206.

[64]   Lagomasino, G.L., *Sobre un teorema de sobreconvergencia de J.L. Walsh*, Revista Ciencias Matematicas **IV(3)** (1993), 67-78.

[65]   Landau, E., *Abschätzung der Koeffizientensumme einer Potenzreihe I, II*, Archiv der Math. und Physik **(3)21** (1913), 42-50, 250-255.

[66]   Lorentz, G.G., *Approximation of Functions* (1986), Holt, Rinehart and Winston, New York.

[67]   Lorentz, G.G., Jetter, K. and Riemenschneider, S.D., *Birkhoff Interpolation*, Addison-Wesley, 1983, pp. (Chapter 11).

[68]   Lou, Yuanren, *Extensions of a theorem of J.L. Walsh on the overconvergence*, Approx. Theory. and Its Appl. **2** (1986), 19-32.

[69]   Lou, Yuanren, *Some problems on Walsh equiconvergence*, Approx. Theory and Its Appl. **2** (1986), 21-37.

[70]   Lou, Yuanren, *On a theorem of Saff and Varga*, Approx. Theory and Its Appl. **4** (1988), 91-102.

[71]   Lou, Yuanren, *On Walsh theory of overconvergence of interpolating polynomials in the complex domain*, Advances in Math. **20** (1991), 257-276 (in Chinese).

[72]   Lou, Yuanren, *The research on Walsh overconvergence in China and its influence upon foreign countries*, Acta Sci. Naturalium Universitatis Pekinensis **30** (1994), 6-22 (in Chinese).

[73]   Lou, Yuanren, *Equiconvergence of Lagrange interpolating processes inside lemniscate*, Approx. Theory and its Appl. **11** (1995), 51-57.

[74]   Lou, Yuanren, *On the overconvergence of Hermite interpolations polynomials*, Advances in Mathematics **21** (1992), 79-92.

[75]   Lou, Yuanren, *Some remarks on overconvergence of Hermite interpolating polynomials*, Approx. Theory and its Appl. **10** (3) (1994), 79-92.

[76]   Lou,Yuanren, *The relations between the Hermite interpolants and Jacobi expansion on equiconvergence*, Acta Sci. Naturalium, Universitasis Pekinensis **35(5)** (1999), 602-608.

[77]   Lou, Yuanren, *Extension of a theorem of J.L. Walsh on overconvergence*, Approx. Theory and its Applications **2** (1986), no. (3), 19-32.

[78]   Marden, M., *Geometry of Polynomials*, Math. Surveys No. 3, Amer. Math. Soc., Providence, 1966.

[79]   Muir, T., *A Treatise on the Theory of Determinants*, Dover Publications, New York, 1960.

[80]   Nehari, Z., *Conformal mapping*, McGRAW-Hill, New York Toronto London, 1952.

[81]   Okada, Y., *On interpolation by polynomials*, Tohoku Math. J. **48** (1941), 69-70.

[82]   Pál, L.G., *A new modification of the Hermite-Fejér interpolation by polynomials*, Analysis Math. 1(1975) **1** (1975), 197-205.

[83]   Palagallo, J.A. and Price, T.E., *Generalization of results of Walsh equiconvergence*, (preprint) (1992), 1-10.

[84]   Pommerenke Ch., *Über die Verteilung der Fekete-Punkte*, Math. Ann. **188** (1987), 111-127.

[85]   Price, T.E., Jr., *Extensions of a theorem of J.L. Walsh*, J. Approx. Theory **43** (1985), 140-150.

[86]   Riemenschneider, S.D. and Sharma, A., *Birkhoff interpolation at the nth roots of unity: convergence*, Can. J. Math. **33** (1981), 362-371.

[87]   Rivlin, T.J., *Some explicit polynomial approximations in the complex domain*, Bull. Amer. Math. Soc. **73** (1967), 467-469.

[88]   Rivlin, T.J., *On Walsh equiconvergence*, J. Approx. Theory **36** (1982), 334-345.

[89]   Rivlin, T.J., *Chebyshev Polynomials, 2nd edn.*,, Wiley & Sons,Inc., Toronto, 1990.

[90]   Saff, E.B., *An extension of Montessus de Ballore's theorem on the convergence of interpolating rational functions*, J. Approx. Theory **6** (1972), 63-67.

[91]   Saff, E.B. and Sharma, A., *On equiconvergence of certain sequences of rational interpolants*, Proc. Rational Approx. and Interpolation (Graves Morris, Saff and Varga, eds.), (1983), pp. 256-271.

[92]   Saff, E.B., Sharma, A. and Varga, R.S., *An extension to rational functions of a theorem of J.L. Walsh on differences of interpolating polynomials*, Mathematical Modeling and Numerical Analysis R.A.I.R.O. **15** (1981), 371-390.

[93]   Saff, E.B. and Varga, R.S., *A note on the sharpness of J.L. Walsh theorem and its extension for interpolation*, Acta Math. Hungar. **42** (1983), no. (3-4), 371-377.

[94]   Saxena, R.B., Sharma, A. and Ziegler, Z., *Hermite-Birkhoff interpolation on the roots of unity and Walsh equiconvergence*, Linear Algebra and Applications **52/53** (1983), 603-615.

[95]   Sharma, A., *Some recent results on Walsh theory of equiconvergence*, Approximation Theory V (C.K. Chui, L.L. Schumaker and J.D. Ward, eds.), Academic Press, Inc., New York, 1986, pp. 173-190.

[96]   Sharma, A. and Szabados, J., *Quantitative results in some problems on Walsh equiconvergence*, East J. Approx. **4** (1998), 291-309.

[97]   Sharma, A. and Szabados, J., *A multivariate extension of Walsh's overconvergence theorem*, Rendiconti del Circolo Matematico di Palermo **68** (2002), 797-803.

[98]   Sharma, A. and Ziegler, Z., *Walsh equiconvergence for best ℓ_2-approximates*, Studia Math. **LXXVII** (1983), 523-528.

[99]   Sharma, A. and Ziegler, Z., *Hermite interpolation on some perturbed roots of unity*, Analysis **19** (1999), 1-12.

[100]   Smirnov, V. I. and Lebedev, N. A., *Functions of a Complex Variable. Constructive Theory* **24** (1990), London Iliffe Books Ltd., 51-84.

[101]   Stojanova, M.P., *Equiconvergence in rational approximation of meromorphic functions*, Constructive Approx. **4** (1988), 435-445.

[102]   Stojanova, M.P., *Overconvergence in rational approximation of meromorphic functions*, Mathematica Balkanica, New Series **3** (1989), 12-281.

[103]   Stojanova, M.P., *Overconvergence of some complex interpolants*, Mathematica Balka-nica, New Series **3** (1989), 149-171.

[104]   Stojanova, M.P., *Some remarks on Walsh equiconvergence*, Approx. Theory and its Appl. **6** (1990), 78-87.

[105]   Stojanova, M.P., *Ph.D. thesis*, (unpublished).

[106]   Szabados, J., *Converse results in the theory of overconvergence of complex interpo-lating polynomials*, Analysis **2** (1982), 267-280.

[107]   Szabados, J. and Varga, R.S., *On the overconvergence of complex interpolating poly-nomials*, J. Approx. Theory **36** (1982), 346-363.

[108]   Szabados, J. and Varga, R.S., *On the overconvergence of complex interpolating poly-nomials: II. Domain of geometric convergence to zero*, Acta. Sci. Math. (Szeged) **45** (1983), 377-380.

[109]   Szász, O., *Ungleichungen für die Koeffizienten einer Potenzreihe*, Math. Z. **1** (1918), 163-183.

[110]   Totik, V., *Solutions to three problems concerning the overconvergence of complex interpolating polynomials*, Acta Sci. Math. (Szeged) **45** (1983), 4115-418.

[111]   Totik, V., *Quantitative results in the theory of overconvergence of complex interpo-lating polynomials*, J. Approx. Theory **47** (1986), 173-183.

[112]   Varga, R.S., *Topics in polynomial and rational interpolation and approximations*, Seminaire de Math Superieures Montreal, 1982, (Chapter IV), pp. 69-93.

[113]   Walsh, J.L., *On polynomial interpolations to analytic functions with singularities*, Bull. Amer. Math. Soc. **38** (1932), 289-294.

[114]   Walsh, J.L., *On interpolation and approximation by rational functions with preas-signed poles*, Trans. A.M.S. **34** (1932), 22-74.

[115]   Walsh, J.L., *Interpolation and Approximation by Rational Functions in the Complex Domain*, A.M.S., Colloq. Publications X, Providence, R.I., 5th ed., 1969.

[116]   Xin, Li, *Rational interpolation to functions on the unit circle*, Method & Applications of Analysis **6** (1991), 81-96.